D0926714

Watershed Management

Watershed Management

Practice, Policies, and Coordination

Robert J. Reimold
Editor

McGraw-Hill
New York San Francisco Washington, D.C. Auckland Bogotá
Caracas Lisbon London Madrid Mexico City Milan
Montreal New Delhi San Juan Singapore
Sydney Tokyo Toronto

Library of Congress Cataloging-in-Publication Data

Watershed Management / Robert J. Reimold, editor.
p. cm.
Includes bibliographical references (p.).
ISBN 0-07-052299-5
1. Watershed Management. I. Reimold, Robert J.
Construction series (McGraw Hill, Inc.)
TC409. W3698 1998
333.73' 15—dc21 98-25965
CIP

McGraw-Hill

A Division of The ***McGraw-Hill*** *Companies*

1 2 3 4 5 6 7 8 9 0 DOC/DOC 9 0 3 2 1 0 9 8

ISBN 0-07-052299-5

The sponsoring editor for this book was Larry Hager, the editing supervisor was Scott Amerman, and the production supervisor was Sherri Souffrance. It was set in Palatino by Dina E. John of McGraw-Hill's Professional Book Group composition unit.

Printed and bound by R. R. Donnelley & Sons Company.

This book is printed on recycled, acid-free paper containing a minimum of 50% recycled de-inked fiber.

Contents

Foreword

Ecosystem has been widely accepted as an important concept in theory but, until recently, not in practice. One reason is that the quick-fix or piecemeal approach so often works well in the short run of political and economic worlds. Thus, timber management has increased the yield of wood, and deer management has increased the number of deer, often to the point where the deer graze down all the tree seedlings! So it is evident we must now move up the scale to more holistic levels of management in order to avoid what we might call the tyranny of small technologies. Since the ecosystem is the lowest complete unit in the ecological levels-of-organization hierarchy (it has all the components, biological and physical, necessary for survival!), it is a logical level around which to organize both theory and practice.

Too often the ecosystem is depicted as a closed geographic unit with discrete boundaries; rather, it is a functionally open unit far from thermodynamic equilibrium with inputs and outputs. An ecosystem is not just an ecoregion. The best graphic mode, the one I use in all my textbooks, consists of a box, which represents an area of interest that can have either natural or arbitrary boundaries, and two large funnels, which represent the input and output environments. Management at the ecosystem level considers the inflow and outflow "funnels," i.e., what (energy, materials, and organisms) goes in and comes out of the system as well as what goes on within the designated area. That is how ecosystem management differs from population or ecoregion management, which focuses mostly on internal affairs, as it were.

Where humans are numerous or very active, the environment

becomes so fragmented and cut up into patches and strips that management has to move up to the next level in the hierarchy, i.e., the landscape. The watershed as a landscape-level unit becomes an ideal management focus because it is both a geographic unit, usually with discrete boundaries, and a functional system unit with inputs and outputs.

Eugene P. Odum
Institute of Ecology
University of Georgia
Athens, GA

Preface

Watersheds are composed of ecosystems. Ecosystems are made up of organisms and their homes (including the homes of humans). To understand a watershed it is crucial that you understand what is meant by *ecosystem*. Often engineers, planners, politicians, teachers, bankers, and other stakeholders of renewable natural resources do not use either *ecosystem* or *watershed* as household words. This book is designed to bridge that gap.

Among the top singular important features of the landscape that significantly impact the future of the human population, watersheds and our stewardship for the resources that watersheds support (water, wetlands, wildlife, and habitat for plants, animals, and humans) will become more important everyday concepts for managing and protecting the diverse assemblages of life, water, soil, minerals, and their interactions.

All readers know their mailing and home addresses. The mission of this book is to engage them in one or more facets of watershed management (practice, policies, and coordination) so that they will all know the names of their own watersheds and become proactive participants in managing those watersheds for future generations.

Robert J. Reimold

Acknowledgments

World pioneers of ecology, such as my ecology mentor and friend, Professor Eugene P. Odum, Institute of Ecology, The University of Georgia, have taught many people and inspired me about the value of ecosystems, wetlands, and watersheds. Business leaders like my "motivating, self-reliant" business mentor, Dr. Donald A. Deieso, President, EA Engineering, Science, and Technology Inc., have helped me recognize the importance of innovative, team-led, forward-looking thinking in solving problematic yet practical environmental issues. Still others, including my loving wife, Mardie, and my "heirs," Elizabeth, Katherine, and Raymond, have taught me the "gift of today" each day I have spent with them, while my family matron and mother, Frances Rickard Reimold, has led by example regarding the importance of trying to be helpful to others.

It is to these specific family persons (Gene, Don, Mardie, Liz, Kath, Ray, and Frances), and what they have taught me about watersheds, innovation, gifts of life, and support, that I dedicate this book about watersheds.

Long before the current explosion of watershed-based thinking, McGraw-Hill's Larry Hager encouraged me to consider the development of a watershed-based book that would provide a window into how watershed management was done around the world. To both Larry and Scott Amerman at McGraw-Hill, who facilitated the motivation, development, and delivery of this book, and to my highly respected professional peers who authored and contributed chapters of this book, I also convey my most sincere gratitude for your support, contributions, and professional associations.

Watershed Management

1

Watersheds—An Introduction

Robert J. Reimold
EA Engineering, Science, and Technology, Inc.
Hingham, MA

Watersheds as a Unit of Measurement

Watersheds are areas delineated by natural hydrological boundaries and are used to manage water quality and develop solutions to environmental problems. These areas include natural assemblages of natural resources that rely on the type and quantity of water present within the watershed.

Watersheds (originating from the German word for "water parting") are regions or areas with natural hydrological boundaries draining to a particular watercourse or water body. Watersheds are catchment basins or drainage basins from which the waters of a stream or stream system are drawn. A watershed includes the surface and groundwater, soils, vegetation, and animals in the drainage basin, as well as humans and their anthropogenic impacts.

Why is there so much interest in watersheds? Over the past several decades, humans concerned with the environment have embraced the notion that one of the most important indicators of the health of our natural resources is the quality of the water. It follows that when the quality of rivers, lakes, streams, ponds, and wetlands is improved and protected, we will have more healthy lands, wildlife, air, and an overall environment. The integration, coordination, and management of human activities within the natural boundaries of a watershed to protect or improve water quality include a number of activities, such as river basin planning; water quality monitoring and assessment; water withdrawal; hydropower production; planning, or permitting, or management; wastewater discharge permitting; stormwater and other nonpoint pollution source management; critical-area protection; wetlands restoration and protection; and other related initiatives.

Concepts of Watershed Management

Effective management of watershed depends on a comprehensive human understanding of the components of watersheds and their interactions. The application of ecological principles to watershed planning has recently become one of the most important topics of natural resource management discussions. Traditionally, interest in balanced natural resources (land or water) management has come only after humans have first severely damaged a landscape. To paraphrase the world famous naturalist Aldo Leopold: Humans do not seem to be able to understand a system that they did not build; instead, they seemingly must partially destroy and rebuild the system before its use limitations are understood and appreciated.

For example, the first successful land-use planning in the United States followed the widespread destruction of much of the nation's agricultural soils, which occurred as a result of the unplanned one-crop system of agriculture. From that time of the dust bowl and the ensuing human misery created by soil erosion came a conservation effort to manage soils (the soil conservation movement). This conservation effort eventually became an outstanding example of an effective conservation program which involved cooperation of local people, their state educational system (schools and universities), and state and federal governments. The soil conservation initiative, which was created by congressional actions, mostly established conservation actions along natural boundaries (such as large watersheds). The long-term success of the soil conservation programs has been due to holistic approaches, such as watershed approaches implemented by the U.S. Department of Agriculture's Natural Resource Conservation Service and the U.S. Forest Service. In implementing the program, land-use maps were prepared; classifications were based on ecological features (such as soil, slope, and natural biotic communities). The emphasis of the soil conservation program became "extremely focused" on creating more farmland and urban development land by channeling streams, draining wetlands, etc. Some of these actions occurred at great expense to other valued natural resource components.

Simply stated, contemporary watershed-based decision making involves five simple steps:

- Information collection
- Problem assessment
- Prevention, control, restoration, and management alternatives
- Goal and priority setting and funding
- Implementation—defining roles and responsibilities of separate stakeholders

Consequently the watershed initiative has evolved beyond the nearly exclusive farm, range, forest, and development focus which characterized its early history. The watershed approach includes the whole urban-rural landscape complex. In many areas of the world, the deteriorating quality of urban and suburban areas threatens the entire economic and social system. Implementing effective management is infinitely more difficult in urban areas than rural areas because of the interconnected social and human aspects of urban areas, and because of the significant differences in monetary value placed on different types of land use (including natural systems). Differences in agricultural versus urban land use results in significant impacts on the quantity and quality of waters of the watershed as well as the stability and functions of the lands within the watershed.

The relative stability and function of a watershed are determined by the rate of water inflow and outflow, materials (substrate), and activity patterns of organisms living within the watershed. In other words, fields, forests, towns, and waters linked together by a stream or river flow interact and consequently are appropriately considered as one management unit. For example, past work in the Chesapeake Bay and ongoing environmental restoration of the Florida Everglades are both good examples of decisions being made and actions being taken on a "whole drainage basin" basis. In the Everglades, the interactions of agriculture, urbanization, development, and preservation are all being focused by the South Florida Water Management District to ecosystem restoration on a watershed basis.

Importance of Watershed Protection

While public attention has recently been focused on watersheds as units of management, consideration of watersheds as management entities is not new. All efforts to effectively manage watersheds involve an approach in which all stakeholders can share resources, expertise, and authority. In the United States, Virginia's Governor William Gage in 1610 issued the following proclamation:

> There shall be no man or woman dare to wash any unclean line, wash clothes,...nor rinse or make clean any kettle, pot, or pan, or any suchlike vessel within twenty feet of the old well or new pump. Nor shall anyone aforesaid, within less than a quarter mile of the fort, dare to do the necessities of nature, since by these unmanly, slothful, and loathsome immodesties, the whole fort may be choked and poisoned.

In the world of public health, it has become common practice to vaccinate. Clearly a vaccination to prevent illness is well worth the time, expense, and discomfort. For example, the U.S. Center for Disease Control estimates that in the United States, waterborne diseases transmitted through drinking water infect 940,000 people each year and are responsible for 900 deaths. Common chemical contaminants which cause acute effects (death) and chronic effects (such as birth defects, cancer, and organ, nervous system, and blood damage) include metals, volatile organic carbons, synthetic organic chemicals, and pesticides. It is true that an ounce of prevention is worth a pound of cure.

A similar approach to achieving healthy watersheds by managing both natural resources (air, water, fish, wildlife, and forests) and anthropogenic influences (impacts of development and expansion of the human population on these resources) merits the "prevention" approach. The vehicle for this approach in the United States has been the promulgation of restrictions and regulations.

In the United States, there has been a strong tradition against land-use planning at the national level because such planning is seen as abrogating the rights of individual ownership of land. National efforts have focused on protection and enhancement of public resources (air, water, or federally owned land). The U.S. Clean Water Act is a good example of regulations created to attempt to maintain and restore the physical, chemical, and biological integrity of the nation's waters (which include wetlands). Land management practices in uplands, which can also significantly affect downslope water quality, have not traditionally been regulated.

In order to address growing public concern about risks to human health or ecosystems, international attention has recently focused on watersheds, especially in relation to sustainable development. To develop concepts to address these concerns requires consideration of the following questions:

What is a watershed?

What are the current approaches in watershed management?

What has been accomplished?

Why has watershed management been successful, or why has it failed? (Was it due to lack of resources, lack of local support, absence of available management practices, or some other factor?)

What are the best watershed management options, and how can they be implemented?

What is the best mix of regulatory and voluntary approaches?

To understand the answers to these questions in a global perspective, it is essential to have a clear understanding of how contaminants and uses of the land and waters within a watershed can impact humans and other natural resources.

Common pollutants and water quality issues addressed in watershed planning include acid mine drainage; biological oxygen demand; cadmium; copper; dissolved oxygen; hydrogen sulfide; mercury; pesticides (including aldrin, dieldrin, chlordane, endrin, heptachlor, heptachlor epoxide, hexacholorocyclohexane plus lindane, endosulfan, and toxaphene); molybdenum; lead; polychlorinated biphenyls; selenium; sedimentation; temperature; total dissolved solids; unknown toxicants; and zinc. Common sources of watershed disruption include agriculture; dairies; dam construction and operation for hydropower, flood control, and/or recreation; grazing; hydro modifications for irrigation or fisheries enhancement; industrial point source discharges; land disposal; resource extraction; mine tailings; municipal point source discharges; on-site disposal systems; road construction; urban runoff; storm sewer overflow discharges; combined sewer overflow discharges; sanitary sewer overflow discharges; and silviculture.

Approaches to Watershed Protection

Common benefits of prudent watershed management include the safeguarding of natural resources for future generations (environmental stewardship); maintenance of stakeholder confidence in the political infrastructure responsible for stewardship of these resources: provision of a source of safe drinking water for the present and future populations; lower flood insurance costs; improved public safety; reduced damage to property from catastrophic precipitation events; and effective management of limited financial resources appropriated for environmental protection.

Watersheds have traditionally been protected by dispersing development and implementing best management practices. Low-density zoning results in large lots for dwelling units. For example, some communities may minimize potential watershed impacts by decreasing allowable density from, say, two dwelling units per acre to one unit per acre, or even one dwelling unit per 5 acres. Larger lots are expected to dissipate development impacts because they theoretically produce less stormwater and pollutant runoff.

Such an approach is often ineffective. While large-lot zoning reduces rooftop impervious cover, large lots often increase the total area of impervious surfaces created for each dwelling unit due to extended road networks. One way to minimize the creation of extensive impervious cover (on a regional scale) is to concentrate development in a high-density cluster and thus eliminate the establishment of impervious surfaces in many of the subwatersheds. Large-lot, low-density zoning is also more expensive to construct because of the added infrastructure costs and reduced reliability of some infrastructure. For example, on-site septic systems are often the only cost-effective method for wastewater treatment in large-lot developments, yet such systems often have the greatest probability of failure and environmental contamination.

Another strategy for watershed protection is implementation of best management practices (BMPs). Permit efficacy monitoring studies have shown that stormwater BMPs can achieve significant pollutant removal rates, yet their long-term performance success is often impaired by poor design, inadequate construction, or lack of proper maintenance. Even with BMPs, it is challenging to replicate predevelopment hydrology, and thus downstream channel erosion is often the result.

Watersheds are often defined as all the land areas contributing runoff or draining to a singular watercourse. Consequently, it is possible to define a nearly infinite number of subwatershed boundaries. Schueler (1995) divided watersheds into five different basic units: catchment, subwatershed, watershed, subbasin, and basin (Table 1.1).

Such approaches to artificially subdivide the watershed based on the size of the catchment area often result in arbitrary and shortsighted considerations. In most of these urban planning focused studies, the models for watershed classifications have been based on stream sensitivity and degraded conditions, with the majority of the emphasis on impervious cover. In such instances, impervious cover rather than population density is seen as the favored measure of growth and impact.

Contemporary recommendations to zone watersheds include a variety of similar approaches. Typical tasks include these:

Table 1.1 Watershed Management Unit Characteristics (Schueler, 1995)

Watershed management unit	Typical area, mi^2	Influence of impervious cover	Primary planning authority	Management focus
Catchment	0.05–0.50	Very strong	Property owner (local)	BMP and site design
Subwatershed	1–10	Strong	Local government	Stream classification and management
Watershed	10–100	Moderate	Local or multi-local government	Watershed-based zoning
Subbasin	100–1,000	Weak	Local, regional, or state	Basin planning
Basin	1,000–10,000	Very weak	State, multi-state, or federal	Basin planning

- Conduct comprehensive inventory.
- Determine impervious cover–stream quality relationships.
- Map existing and future impervious cover.
- Define subwatersheds based on growth patterns and attainable stream quality.
- Develop land-use master plan.
- Refine management priorities based on larger watershed planning.
- Adopt specific protection strategies for each subwatershed.
- Implement efficacy monitoring and management feedback.

Key components of effective management include land-use controls to protect sensitive streams. These land-use controls are usually formulated based on a number of basic watershed variables including hydrology, channel morphology, water quality, habitat, and ecology, flood frequency, and rainfall (frequency, duration, and volume).

In some instances, cluster development is favored as a means of affording cost-effective watershed protection. *Cluster development* refers to a compact form of development allowing a higher density of dwelling units on one portion of the site while reserving larger parts of the site for preservation. This is not a new concept. Clustering is accomplished by reducing the dimensions and geometry of individual lots and shortening road networks. Stream protection clusters are designed to reduce the amount of impervious surface (compared to traditional development) and to retain a significant area of the site as a permanently protected green space. Such development reduces site imperviousness by 10 to 50 percent, reduces stormwater and pollutant runoff loads, and reduces soil erosion potential. As a result, capital development and public infrastructure costs are reduced, and runoff and pollutants are concentrated in locations where management and treatment can be implemented. Such development supports other community planning goals including affordable housing, diversity of architecture, and farmland preservation.

Buffers are another effective means of dealing with potential watershed impacts from development. Shoreline buffers, stream buffers, and wetland buffers all are effective in protecting natural areas. Buffers can include sensitive habitats, steep slopes, floodplains, and other special resource areas. Terms used in dealing with buffers includes *stream order, drainage density,* and types and frequency of waterways flow. A stream with no tributaries is a first-order stream. One which is formed as a result of two first-order streams combining is a second-order stream. Drainage density relates to the length of stream channel per unit area. Areas with dense vegetative cover, permeable soils, and gentle topography have significantly less stream mileage than areas with more relief and less permeable soils. The distance between a watershed divide and the first observable stream is the overland flow path, which is the area across which water flows as sheet flow. Once the uniform, very shallow flow of water concentrates to form channels, intermittent channels are formed. At some point farther downstream, groundwater supplies contribute to water running in the channel on a year-round basis, resulting in the formulation of perennial streams.

Since the concept of a watershed involves ecosystem thinking, pollutant inputs, biological, physical, and chemical features, and assimilative capacities, watershed-based management involves a wealth of adaptive management approaches which includes cooperation and partnering among all stakeholders. Watershed-based regulatory permitting has the potential to be more equitable because the most easily controlled sources are more easily targeted for management.

The U.S. Environmental Protection Agency (EPA) has endorsed a watershed approach that results in integrated environmental management (EPA, 1996). This initiative involves local community-based consensus building to make environmental management decisions as well as a watershed-based effluent trading strategy which includes environmental, economic, and social benefits of such trading. The watershed-based planning focuses not on specific chemical constituents of water, but on the general suitability of a watershed for recreation, aquatic life, and other uses. As a result, a number of risk-based management frameworks have evolved (Chen et al., 1996). These and other world perspectives are highlighted in subsequent chapters of this book.

References

Chen, C. W., J. Herr, L. Gomez, and R. A. Goldstein, Watershed risk analysis tool, *Proceedings of the First International Conference on Water Resources Engineering,* American Society of Civil Engineers, Reston, VA, 1996, pp. 1347–1354.

Environmental Protection Agency, *Watershed Protection: A Statewide Approach,* EPA 841-R-95-004, Office of Water, Washington, 1995.

Environmental Protection Agency, *Watershed Approach Framework—1996,* EPA 800-R-96-001, Office of Water, Washington, 1996 (available on the Internet at http://www.epa.gov/OWOW).

Schueler, T., *Site Planning for Urban Stream Protection,* Metropolitan Washington Council of Governments, Washington, 1995.

Young, T., and C. Congdon, *Plowing New Ground: Using Economic Incentives to Control Water Pollution from Agriculture,* New York, NY, Environmental Defense Fund, 1994, pp. 126–127.

2

Regional Assessment for Watershed Management in the Mid-Atlantic States

Carolyn T. Hunsaker,
Barbara L. Jackson,
and Adam Simcock

Ecologists and many natural resource managers recognize that often the environmental hazards or stresses that we currently have to address impact large geographic areas (e.g., acid deposition, nonpoint source pollution, introduction of exotic species, increased global CO_2, alteration of hydrologic regimes), yet traditional concepts and methods in ecology and assessment are relevant mainly to single sites or small geographic areas. Also, monitoring programs and assessment needs often have not been integrated, and rigorous quantitative assessments have been the exception rather than the norm for large geographic areas. Effective watershed management requires approaches and tools for the small watershed or catchment, the intermediate watershed, and the large river basins and regions. We have the most information and tools for the small and intermediate-size watersheds; however, cumulative effects are manifested at the intermediate and large watershed scales (Hunsaker, 1993). Therefore, this chapter introduces fundamental issues for watershed and regional assessment and presents ongoing research to establish relationships between land use and water quality for the mid-Atlantic states.

Regional Assessment

When one moves from monitoring and analyzing data at the site-specific or local scale to the watershed or regional scale, it is not obvious what is the appropriate resolution for data or how to bound the activity with regard to spatial and temporal scales. Regional management requires organizing paradigms or conceptual models, and an assessment framework can serve this purpose; for quantitative assessments we consider the risk assessment framework to be effective (EPA, 1992; Suter et al., 1993). Spatial scale can also serve as a very strong organizing paradigm for regional management and is compatible with a risk assessment framework (Hunsaker and Graham, 1991). We briefly discuss the risk assessment framework, spatial constructs for aquatic assessments—ecoregions and watersheds—and ecological paradigms including landscape ecology.

The factors that are similar and different between local (site-specific) and regional risk assessment have been identified (Hunsaker et al., 1990) and illustrated (Graham et al., 1991). Regional risk assessment has some attributes in common with local risk assessment but has others that are unique. The general theoretical framework for doing the two types of environmental risk assessment is the same. Both have two phases: first, the *hazard definition,* in which the *endpoint, source terms,* and *reference environments* are defined and described; and second, the *problem solution,* in which the exposure and effect on the endpoint are assessed by using models and the risk and its associated uncertainty are determined (Table 2.1). Regional risk assessment differs in (1) the extent of interaction between the source terms, endpoints, and reference environment and (2) the degree to which boundary definition and spatial heterogeneity are significant in determining uncertainty. Although local risk assessments involve the development of databases and the use of models, these steps may be more significant in regional risk assessments. Few regional-level databases of biological variables exist; furthermore, unique problems arise in aggregating or integrating dissimilar local data into regional databases. Regional models of ecological

Table 2.1 Comparison of Kuchler's Potential Vegetation to Current Land Cover by Ecoregions of the United States

The percentage of land cover given is from Kuchler's map (1964). Change is the subtraction of the estimated amount of current land cover from potential land cover; a negative value indicates a loss of that land cover.

Omernik ecoregion	Rangeland		Forest		Wetland		Barren		Agriculture*	Urban*	Water	
	Percent	Change	Percent	Change	Percent	Change	Percent	Change	Change	Change	Percent	Change
Ia	0.4	−0.3	98.6	−28.6	0.0	0.0	0.0	0.0	25.1	0.3	1.0	3.5
Ib	13.0	−12.8	86.8	−48.6	0.0	0.0	0.0	0.0	55.6	3.5	0.2	2.3
Ic	66.0	−62.8	33.9	−28.1	0.0	0.0	0.0	88.7	1.4	1.4	0.1	0.7
II	3.2	−3.1	96.7	−4.5	0.0	0.0	0.0	0.0	6.0	0.9	0.2	0.7
III	3.3	−1.4	96.2	−28.5	0.0	0.0	0.0	0.0	27.4	1.8	0.4	0.7
IVa	95.0	−42.1	5.0	−4.4	0.0	0.0	0.0	0.0	46.1	0.0	0.0	0.4
IVb	92.9	−65.9	6.8	−4.3	0.0	0.0	0.0	0.0	69.0	1.0	0.4	0.2
Va	87.8	−16.1	11.9	−4.8	0.0	0.0	0.0	0.0	19.8	0.8	0.1	0.2
Vb	78.4	−5.0	18.4	−12.3	0.2	−0.2	1.4	11.1	6.0	0.2	0.5	0.2
VI	12.2	1.7	84.0	−10.4	0.0	0.0	3.6	−1.0	8.8	0.2	0.2	0.0
VII	12.4	−7.0	78.9	−48.6	7.6	−1.8	0.0	0.4	54.7	1.9	1.1	0.7
VIIa	75.6	−44.7	10.6	−5.0	13.8	−13.8	0.0	0.4	60.4	2.2	0.0	0.5
VIIb	0.0	−3.1	100.0	63.2	0.0	0.0	0.0	0.0	26.6	6.6	0.0	0.4

*Since agriculture and urban land uses are created by humans, they are not present in Kuchler's map of potential natural vegetation.

processes are much less common and can be difficult to validate. Hunsaker and Graham (1991) illustrate the influence of boundary definition and spatial heterogeneity on modeling acidic deposition effects on Adirondack lakes.

As the issues being addressed by ecological research and resource management have become more complex and integrative, an interesting question has surfaced with regard to the spatial construct we use to characterize regions. Geographers have struggled for a long time with spatial interrelationships and questions of geographic characterization. As Hunsaker et al. (1996) discuss, a regional characterization scheme is especially difficult for aquatic ecosystems as two accepted approaches exist—watersheds and ecoregions—and both can be hierarchically constructed. An *ecoregion* is an area (region) of relative homogeneity based on one or more attributes in ecological systems. A *watershed* is an area of land draining to a specific point on a stream or to a lake or wetland; watersheds are based on topography and the fact that water flows downhill because of gravity. Watersheds are a logical analysis unit if chemical pollutants from point sources are the primary concern because the distribution and concentrations of these pollutants are primarily governed by the hydrologic network and dilution. However, if nonpoint source pollution (which is governed by land-use pattern, geology, and soils) and ecosystem condition as measured by aquatic organisms are of primary interest, then ecoregions (Gallant et al., 1989) may serve as a more useful analysis unit (Fig. 2.1). Ecoregions are defined by characteristics such as physiography, climate, and land cover, and they better characterize ecosystem homogeneities (e.g., fish species found in headwater streams across watersheds in a similar physiographic region are similar to species in the large rivers of the downstream portions of the same watersheds, e.g., the river continuum concept, Vannote et al., 1980).

Watersheds, on the other hand, have been used by ecologists for a long time as an organizing paradigm for spatial scale. Biogeochemical cycling research is usually carried out on small experimental watersheds (Johnson and Van Hook, 1989), and many of the National Research Foundation's long-term ecological research sites are based around watersheds such as Hubbard Brook (Likens and Bormann, 1974; Bormann and Likens, 1981), Coweeta (Swank and Crossley, 1987), and other long-term study sites (Johnson and Van Hook, 1989; Correll, 1986). Naiman (1992) presents current knowledge and future trends in watershed management and research with a focus on watershed and regional models, indicators of environmental change, and new techniques for integrated resource management. Ecoregions have received more attention recently (Bailey, 1976; Wiken, 1986). Omernik (1987) developed a set

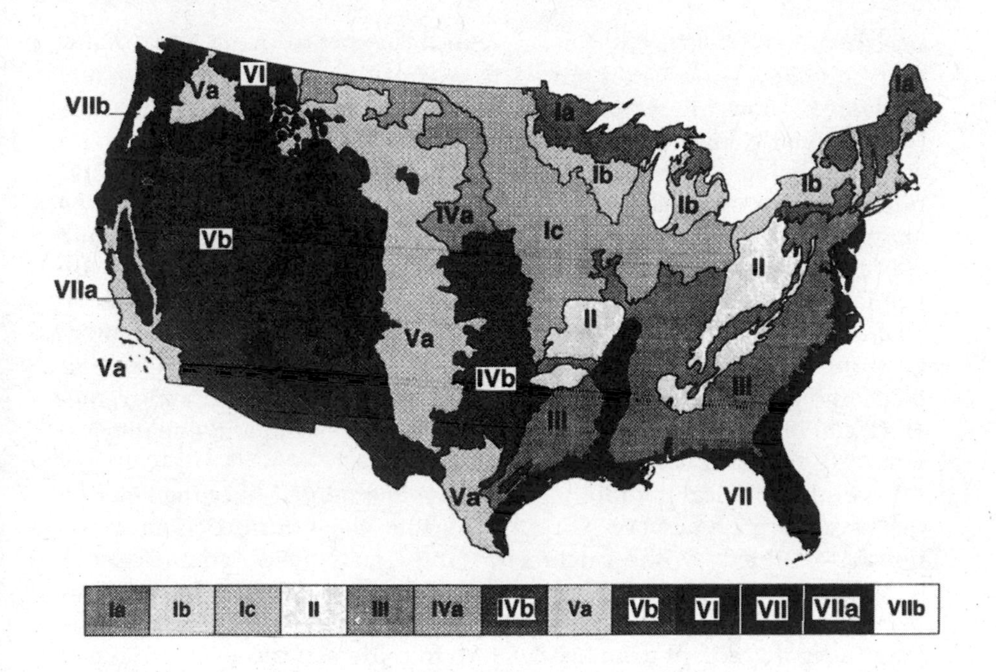

Ia	Northern glaciated, nonagricultural
Ib	Northern glaciated, mixed use
Ic	Northern glaciated, agricultural
II	Central and eastern forested hills and mountains
III	South central and humid mixed use
IVa	Subhumid agricultural plains, northern
IVb	Subhumid agricultural plains, southern
Va	Western xeric, semiarid
Vb	Western xeric, arid
VI	Western forested mountains
VII	Unique alluvial and coastal plains
VIIa	Unique alluvial and coastal plains, central California valley
VIIb	Unique alluvial and coastal plains, Willamette valley

Figure 2.1 Omernik aggregated ecoregions.

of ecoregions (1:7,500,000 scale map) for the conterminous United States that he believes are better suited to regional assessments for aquatic ecosystems than are watersheds such as the U.S. Geological Survey's (USGS's) hydrologic units (Gallant et al., 1989; Omernik, 1987). Several states are using ecoregions for establishing variable water-quality criteria (e.g., Ohio, Arkansas, and Minnesota), including biocriteria (Davis and Simon, 1995). Ongoing ecoregion work to refine ecoregions, define subregions at 1:250,000 scale maps, and locate sets of reference sites are being done in all of Iowa, Florida, and Massachusetts and parts of Alabama, Mississippi, Virginia, West Virginia, Maryland, Pennsylvania, Oregon, and Washington. The correspondence between fish assemblages and water chemistry data suggests that an ecoregional framework can be a useful context for integrating terrestrial and aquatic systems (Hughes and Larsen, 1988).

Several ecological paradigms about rivers emerged over the last two decades since Hynes (1975) proposed that the nature of streams is tightly coupled with catchment and watershed characteristics (Cummins et al., 1984). The river continuum concept provided a template for examining how biotic attributes of rivers change within the longitudinal gradient from headwaters to outlets at the sea (Vannote et al., 1980; Minshall et al., 1985). The serial discontinuity concept (Ward and Stanford, 1983) provided a construct for the likelihood of rivers to predictably reset biophysical attributes in relation to distance downstream from on-channel impoundments. Comparison of organic matter budgets in streams in different biomes provided the basis for the riparian control concept and demonstrated the importance of wood and leaves in lotic systems (Harmon et al., 1986; Ward et al., 1990; Gregory et al., 1991). The nutrient spiraling concept (Webster and Patten, 1979; Newbold et al., 1983) led to an understanding of how plant nutrients are transformed from dissolved to particulate states during movement from upstream to downstream reaches. The ecotone concept (Naiman and Decamps, 1990; Holland et al., 1991) elevated interest in the importance of transformations and fluxes of materials that occur within boundaries between functionally interconnected patches that form the riverine landscape. Stanford and Ward (1992) believe that the ecotone concept integrates the other paradigms by emphasizing the functional connectivity inherent in all ecosystems (Hunsaker et al., 1996).

Landscape ecologists seek to better understand the relationships between landscape structure and ecosystem processes at various spatial scales (Turner, 1989). The word *structure* refers to the spatial relationships of ecosystem characteristics such as vegetation, animal distributions, and soil types. *Processes* or *function* refers to the interactions—i.e., the flow of energy, materials, and organisms—between the spatial elements. Because landscapes are spatially heterogeneous areas, or environmental mosaics, the structure and function of landscapes are themselves scale-dependent. Hunsaker and Levine (1995) developed methods to characterize landscape attributes that influence water quality at various spatial scales. Understanding how scale, both data resolution and geographic extent, influences landscape characterization and how terrestrial processes affect water quality is critically important for model development and translation of research results from experimental watersheds to management of large drainage basins (Hunsaker et al., 1996).

Land-Use Change and Water Quality

Streams and rivers serve as integrators of terrestrial landscape characteristics and as recipients of pollutants from both the atmosphere and the land; thus, large rivers are especially good indicators of cumulative impacts (Cada and Hunsaker, 1990). Cumulative impacts were defined in the implementing regulations of the National Environmental Policy Act, and Dickert and Tuttle (1985) provide a somewhat more detailed definition:

> ...cumulative impacts are those that result from the interactions of many incremental activities, each of which may have an insignificant effect when viewed alone, but which become cumulatively significant when seen in the aggregate. Cumulative effects may interact in an additive or a synergistic way, may occur onsite or offsite, may have short-term or long-term effects, and may appear soon after disturbance or be delayed.

Land-use change is a good example of a cumulative effect in that what can seem like small changes taking place in a local area can actually result in major changes at the regional scale. Land-use change may be the single greatest factor affecting ecological resources. Allan and Flecker (1993), who have identified six major factors threatening the destruction of river ecosystems, state that various transformations of the landscape—hydrologic changes to streams and rivers resulting from changes in land use, habitat alteration, and nonpoint source pollution—are probably the most widespread and potent threats to the well-being of lotic ecosystems. Measures of landscape structure are useful to monitor change and assess the risks it poses to ecological resources. Many studies have shown that the proportion of different land uses within a watershed can account for some of the variability in surface water quality (DelRegno and Atkinson, 1988; Omernik, 1977; Reckhow et al., 1980; Sivertun et al., 1988). For example, the water quality in a watershed with 50 percent agricultural land use and an intact forest riparian zone may be expected to be better (e.g., lower turbidity and nutrients) than that in a similar watershed without any riparian zone.

It is difficult to know to what degree and in what spatial pattern land cover has changed across the United States or what water quality might be like without the changes humans have created. We performed an exercise to provide a general idea of how different land cover is now from what it might be without human intervention. Kuchler (1964) developed a map for the United States of "potential natural vegetation," which is the vegetation that would exist today if humans were removed from the scene and if the resulting plant succession were telescoped into a single moment. For current land cover data we used the Loveland et al. (1991) classification developed from advanced very high-resolution radiometer (AVHRR) satellite imagery (1 km^2 resolution or grid size) augmented with an urban layer taken from the Digital Chart of the World. Kuchler's and Loveland's land cover classes (117 and 167, respectively) were aggregated to an Anderson level I set of seven classes: rangeland, forestland, wetland, barren land, water, urban land, and agricultural land. Using a geographic information system (GIS), the proportion of each land cover was determined for Omernik's 13 aggregated ecoregions (Fig. 2.1). While the spatial accuracy of the Kuchler map is not expected to be especially good, we believe that comparing it to current data at 1-km resolution for large regions is reasonable. (The difference in the proportion of water is usually less than 1 percent and never larger than 3 percent, thus substantiating the validity of our exercise.) In general, agricultural, and to a lesser extent, urban land use has replaced forest in the eastern United States and rangeland in the western United States. Large areas of wetland have virtually disappeared at this coarse data resolution.

The mid-Atlantic region is primarily represented by ecoregion II which contains the Appalachian Mountains (central and eastern forested hills and mountains, ecoregion III (south central and humid mixed use) which is basically the Piedmont, and ecoregion VII (unique alluvial and coastal plains), which is the eastern coastal plain. Ecoregion II has not experienced large land-use changes, but in ecoregion VII agricultural land has increased to 55 percent and urban land has increased to 2 percent while forestland and rangeland have decreased. Wetlands have decreased by 2 percent in this ecoregion. Ecoregion III has also lost a significant amount of forest (28 percent) mostly to agricultural production and an increase of 2 percent in urban land.

Regional Landscape Characterization for Water Quality

From the literature, we know that nonpoint source pollution (sediment and nutrients) is a problem (USGS estimates 29 percent of all water bodies are affected) and that proportion of land use is highly correlated with water quality. So what assessment tools and approaches can help us better manage for desired water quality? And how important are spatial pattern, boundary definition, data resolution and aggregation, and other issues that contribute to uncertainty in an assessment?

In addition to overall proportion of land use in a watershed, several researchers have addressed the issue of whether land use close to streams is a better predictor of water quality than land use over the entire watershed (Omernik et al., 1981; Osborne and Wiley, 1988; Wilkin and Jackson, 1983). Research on nutrient and sediment movement within small watersheds with forest or grass buffer areas between streams and disturbed uplands generally supports such statements (Cooper et al., 1987; Lowrance et al., 1984; Peterjohn and Correll, 1984; Schlosser and Karr, 1981). However, conclusions by Omernik et al. (1981) for larger watersheds from a wide variety of hydrologic settings suggest that upland land uses are as important as near-stream land uses.

Research such as that by Hunsaker and Levine (1995) aids in the understanding of how existing data, tools such as geographic information systems (GISs) and models, and the synthesis of research results facilitate the management of nonpoint source pollution for large geographic areas. Using nested watersheds as the hierarchical regional characterization scheme, they addressed three questions relevant to characterizing landscape attributes important to water quality:

- Are both the proportions of land uses and the spatial pattern of land uses important for characterizing and modeling river water quality in watersheds of different sizes?
- Can land use near the stream better account for the variability in water quality than land use for the entire watershed?
- Does the size of the watershed influence statistical relationships between landscape characteristics and water quality or model performance?

Hunsaker and Levine (1995) compared two different studies to evaluate the use of land-use data (both proportions and spatial pattern) for modeling water quality and to explore how scale and data resolution influence the type of spatial analyses performed. They concluded that both proportion of land uses and the spatial pattern of land uses are important for characterizing and modeling water quality; however, proportion consistently accounted for the greatest variance (40 to 86 percent) across a range of watershed sizes (1000 to 1.35 million ha). The Illinois study and work by Omernik (1977) indicated that proximity to streams is not a critical factor in modeling water quality. However, the success of the distributed model in the Texas study indicated that the location of various types of land use in the watershed is critical to modeling. Hunsaker and Levine (1995) concluded that differences in data resolution and extent of analysis units were contributing to these seemingly contradictory conclusions and that further work should be done with intermediate data resolutions to determine a breakpoint for the effectiveness of using hydrologically active areas.

The lumped modeling approach employed in the Illinois study by Hunsaker and Levine (1995) did not show any bias between watershed size and model performance, while the Texas study began to show a trend in decreasing model performance with the largest watersheds. The Texas study demonstrated that it is useful to calibrate models within similar physiographic regions (ecoregions). This finding also suggests that when the lumped modeling approach is employed in an area spanning a number of ecoregions, different models should be developed for each ecoregion.

Mid-Atlantic Case Study

The Environmental Protection Agency (EPA) is performing a mid-Atlantic integrated assessment activity (Kepner et al., 1995) for EPA region III (Fig. 2.2). One aspect of this assessment is landscape characterization for watershed management; this work builds from many years of research on landscape characterization for water quality assessment and management (Hunsaker et al., 1996; O'Neill et al., 1996; Hunsaker and Levine, 1995; Riitters, et al., 1995; Hunsaker, 1993; Levine et al., 1993; Hunsaker et al., 1992; and McDaniel et al., 1987). As Hunsaker and Levine (1995) suggested, we are taking a multistage approach. They advised using a lumped modeling approach with coarse resolution as a screening method to identify watersheds making the most significant pollutant contributions. Then a high-resolution distributed modeling technique could be used for those smaller watersheds identified as critical for specific management actions.

Our initial work focuses on establishing relationships between commonly measured water quality parameters and landscape (land cover

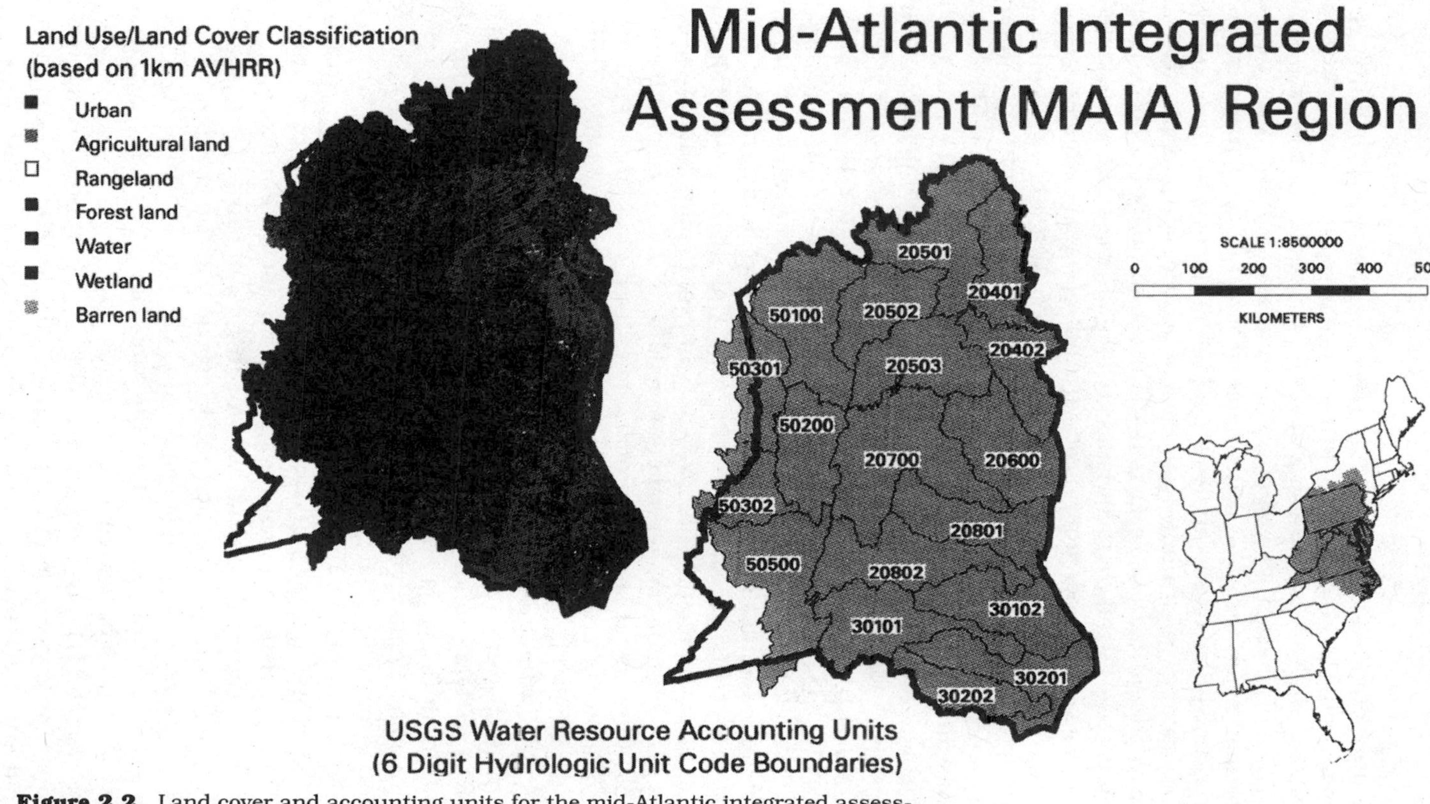

Figure 2.2 Land cover and accounting units for the mid-Atlantic integrated assessment region.

and use) proportion and spatial pattern. This research employs a lumped, empirical modeling approach and is reported here. The same approach will be repeated using several hundred stations and a set of hierarchically nested watersheds. A similar statistical analysis may be performed using the same water chemistry stations but with Omernik (1987) ecoregions as the analysis units. The next research phase will include landscape attributes other than land cover and use that are known to be important because of their use in nonpoint source water quality models—soil characteristics such as soil mean particle diameter and permeability and ground slope. Anthropogenic stresses such as permitted point source discharges, distance to roads, or number of times roads cross streams may also be included in the statistical analysis. Later research phases will compare our results using a limited number of water quality parameters with results from a designed survey for the region (Paulsen et al., 1991) which has more water quality parameters and includes biological monitoring data.

The mid-Atlantic study region contains 18 of the U.S. Geological Survey (USGS) hydrologic units at the accounting unit level (12,550 to 38,890 km^2 for the region). USGS developed a four-level, hierarchical watershed classification scheme for the United States (Seaber et al., 1984); there are 352 accounting units in the United States and 2149 cataloging units. Cataloging units range from 930 to 6000 km^2, at least an order of magnitude smaller than the accounting units. For simplicity, we discuss the general landscape characteristics by the larger accounting units (Tables 2.2 and 2.3); however, the multivariate statistical analysis uses the smaller cataloging units as the analysis unit.

Regional water quality was characterized using commonly measured parameters; water chemistry data were retrieved from the EPA's STORET (STOrage and RETrieval) database for the years 1990 through 1994. Monitoring data are from the states of Maryland, Virginia, West Virginia, and Pennsylvania and from USGS. Stations were considered adequate for multivariate analysis if they had measurements for at least three-quarters of each of at least 3 years. This screening procedure resulted in 594 stations that consistently monitored three of the following four chemistry parameters: specific conductance, nitrogen (either total nitrate nitrogen, total nitrite plus nitrate, or nitrite plus dissolved nitrate), total phosphorus, and total residue. Land-use and land cover data (Loveland et al., 1991) came from the AVHRR satellite imagery with a resolution of 1 km^2 and are the same data we used to get an estimate of overall land-use change by ecoregion.

A watershed is typically characterized by the proportion of the watershed covered by each land use of interest (Table 2.2); however, the spatial pattern of that land use is thought to be equally important for some ecological processes (Hunsaker and Levine, 1995). Landscape ecologists

Table 2.2 Percent of Watershed in Each Land-Use Class for Mid-Atlantic Region

Watersheds are the USGS accounting units. If no value is given, the percent is less than 1.

Accounting unit	Urban land	Agricultural land	Rangeland	Forestland	Water	Maximum patch dominance*
20401	2.0	22.2		75.6		69
20402	11.1	52.0		25.3	10.9	13
20501	1.5	19.3		79.0		74
20502		20.5		78.9		75
20503	1.6	37.9		59.6		41
20600	3.8	42.9	1.1	25.4	26.6	34
20700	3.3	12.0		81.1	3.6	76
20801	1.2	12.8		59.6	25.1	55
20802	2.7	7.4		87.9	1.9	80
30101	1.5	18.1		79.8		74
30102		39.2		47.2	12.9	25
30201		33.2	1.0	26.5	38.0	37
30202	2.2	59.0		36.0	2.7	46
50100	1.6	10.5		87.6		87
50200	2.9	4.9		92.0		92
50301	5.4	21.9		71.6	1.1	68
50302		3.6		95.9		96
50500		1.0		98.4		98

*Area of largest patch divided by area of hydrologic unit.

have proposed many metrics of spatial pattern that may be useful for monitoring ecological conditions (Riitters et al., 1995, Hunsaker et al., 1994). Landscape pattern in the mid-Atlantic region was characterized by proportion of the seven land-use types and several integrative metrics. Dominance measures the extent to which one or a few land uses dominate the landscape, and contagion measures the extent to which the landscape is fragmented. These metrics can range from 0 to 1. Shape complexity is a perimeter-to-area ratio for each patch of a land use; natural land covers such as forest are expected to have higher shape complexity than agriculture patches, which usually have more uniform, linear edges. Shape complexity can range from 0 to 2, but we seldom see

Table 2.3 Integrative Pattern Metrics by USGS Accounting Units in Mid-Atlantic Region

Accounting unit	Number of land uses	Dominance	Contagion	Shape complexity	Number of patches	Percent of potential edges
20401	5	0.6	0.62	1.45	468	90
20402	5	0.25	0.62	1.46	604	90
20501	5	0.64	0.65	1.5	946	80
20502	4	0.61	0.57	1.49	387	83
20503	5	0.51	0.6	1.49	761	80
20600	7	0.36	0.65	1.49	810	71
20700	5	0.59	0.61	1.45	764	100
20801	7	0.47	0.59	1.45	588	86
20802	6	0.73	0.59	1.39	323	93
30101	5	0.64	0.53	1.5	560	90
30102	7	0.47	0.68	1.51	765	71
30201	7	0.38	0.62	1.44	707	81
30202	6	0.52	0.63	1.49	470	93
50100	4	0.69	0.46	1.48	706	100
50200	5	0.79	0.61	1.35	362	80
50301	5	0.52	0.52	1.52	724	90
50302	4	0.86	0.49	1.34	140	100
50500	4	0.94	0.61	1.26	127	100

values less than 1.0 or larger than 1.8 with real landscapes. These metrics are calculated after identification of all patches of the same land use; a patch is an area or polygon of the same, contiguous land-use class. The percentage of potential edges tells us how many of the edge types (i.e., forest and wetland edge or agriculture and urban edge) that could exist, given the number of land uses, actually do exist; one can think of this as a measure of edge heterogeneity.

Table 2.3 lists some of these landscape metrics for the mid-Atlantic region. Disturbed land covers such as agriculture, barren, and rangeland have positive associations with water quality parameters; that is, as the proportion of agriculture increases, so does the amount of nitrogen or sediments. Contagion and proportion of forest were found to be negatively correlated with water quality parameters (Hunsaker and Levine, 1995). Thus, an area that has contiguous land covers (is not fragmented) or that is dominated by forests tends to have better water quality.

The mid-Atlantic region is heavily dominated by forests when characterized by AVHRR data and seven land-use classes. In general, agriculture is the second largest land use, although water makes up a large proportion of some of the hydrologic units for cataloging (HUCs) that contain the Chesapeake Bay. We focus on describing a few of the hydrologic units to highlight their similarities and differences, but all the data are presented in Tables 2.2 and 2.3. The upper Ohio–Little Kanawha (50302) and the Kanawha (50500) are almost totally dominated by forests, the largest patch accounts for more than 95 percent of the watershed (Table 2.2), and there are a small number of total patches compared to the other watersheds (Table 2.3). Thus it is not surprising that these hydrologic units have a high dominance value (Table 2.3). They contain only four of the seven land uses and have all their potential edge types. One difference between the two watersheds is that the patches in 50500 are significantly more contiguous with a contagion value of 0.61, compared to 0.49 for 50302. The shape complexity values are low, 1.26 and 1.34, considering that forest patches are very dominant. The lower Delaware watershed (20402) is extremely different from the upper Ohio and Kanawha watersheds. It has a lot of patches (604) that are very contiguous (0.62), but it is not dominated by a single land use with a dominance value of 0.25 for five land uses. Its largest patch only makes up 13 percent of the watershed. The lower Delaware has the highest proportion of disturbed land use with 11 percent urban and 52 percent agriculture; we expect that it will have poor water quality compared to those watersheds that are dominated by contiguous forest patches. The Albemarle-Chowan watershed (30102) contains all seven land uses, has similar amounts of forest and agriculture, and thus has a moderate dominance value (0.47). It has a large number of patches with the largest patch accounting for 25 percent of the watershed. It has high contagion and shape complexity and has low edge heterogeneity.

For our multivariate statistical analysis using 1-km land cover data, we chose to work with the cataloging units; this provided enough cells per watershed to maintain landscape pattern metric stability. The GIS allowed us to select the most downstream station to represent a cataloging unit. At the cataloging unit level, 34 stations have nitrogen, phosphorus, and conductivity data, and 26 stations have these three chemistry parameters plus residue (Fig. 2.3). The input value to the statistical analysis is the mean of all observations for a chemistry parameter for a station during our period of interest, 1990 to 1994. A quality insurance inspection of the chemistry data led us to eliminate three of the stations from further use. This inspection included evaluating the variability of the chemistry data within a station, residuals from the individual regressions of landscape metrics with each chemistry parameter, and bivariate plots of the major canonical variables to identify possible out-

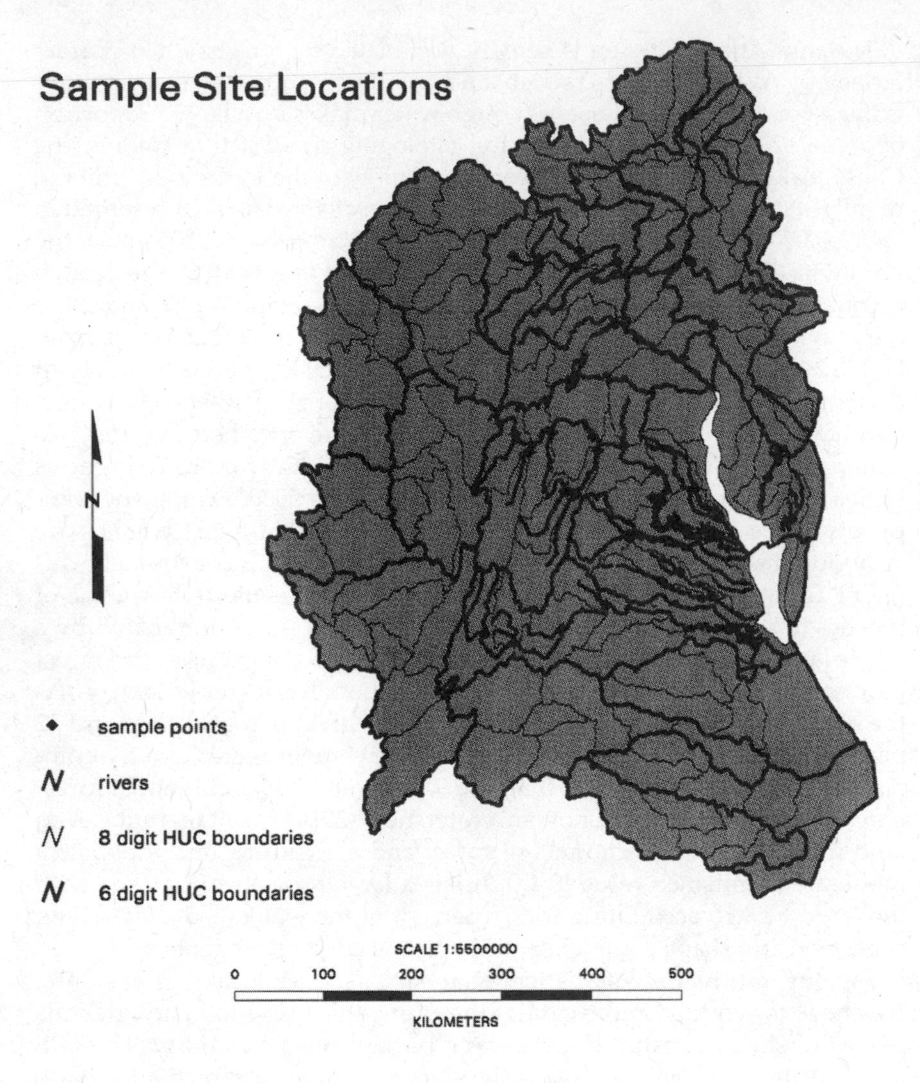

Figure 2.3 Water quality stations used in the statistical analysis of watershed characterization and water quality for the mid-Atlantic integrated assessment.

lier stations. Because of the small sample size, we limited the number of landscape metrics to no more than five at any one time for analyses.

We chose canonical correlation analysis (CCA) (Johnson and Wichern, 1992) for our multivariate approach because it provides a powerful and integrated look at the relationship of landscape pattern to water quality. In CCA, we examine linear relationships between the *X* (landscape pattern metrics) and *Y* (water chemistry parameters) variables. We are trying to determine the linear combination of the *X* variables that best correlates with the single *Y* variable. This is the same context as that of multiple linear regression, so CCA is an extension of multiple linear regression. The first pair of canonical variables identifies the linear combination of the *Y* variables and the linear combination of the *X* variables that maximizes the correlation between these new variables (the linear combinations). Several pairs of canonical variables can be identified, but not all will be significant. In our analyses only the first canonical variable was ever significant, thus simplifying interpretation.

When we evaluated the amount of disturbed edge, the proportion of forest, and the proportion of urban land in a watershed, the canonical correlation was 0.63 for conductivity, nitrogen, and phosphorus, and 0.62 for conductivity, nitrogen, phosphorus, and residue. In these analyses, the first pair of canonical variables accounts for between 42 and 46 percent of the variation in the chemistry parameters and between 55 and 56 percent of the variation in the landscape parameters. The canonical correlations were somewhat higher when we used some of the integrative pattern metrics such as contagion and shape complexity. The canonical correlation for dominance, contagion, shape complexity, amount of disturbed edge, and proportion of forest in a watershed was 0.67 for conductivity, nitrogen, and phosphorus. The same watershed characteristics gave a canonical correlation of 0.77 for conductivity, nitrogen, phosphorus, and residue. In these analyses, the first pair of canonical variables accounts for between 47 and 59 percent of the variation in the chemistry parameters and between 21 and 46 percent of the variation in the landscape parameters. Usually nitrogen and residue are highly correlated with the chemistry canonical variable, and proportion of forest and amount of disturbed edge are highly correlated with the landscape canonical variable. As expected, the proportion of forest is negatively correlated with water chemistry, meaning that the more forest in a watershed, the better the water quality.

Summary and Conclusions

The literature provides effective assessment frameworks, spatial constructs, and ecological paradigms to aid in the assessment and management at regional scales for desired water quality and aquatic habitat. It is still open to debate as to when watersheds or ecoregions are the most effective spatial units for managing aquatic resources; it is likely that only more experience and some designed studies will provide the answer. GIS, the availability of spatial data for the entire nation, and continued use of satellite imagery are making regional analyses more tractable, but they require close attention to uncertainty from model characteristics, boundary definition and data resolution, and aggregation, as we have tried to illustrate. The concept of scale is important to ensure that analyses are highlighting issues rather than obscuring them. We have also tried to show how ecological theory can play an important part in the development of applied techniques. The river continuum and landscape ecology concepts are examples. Focusing on the heterogeneity and pattern of the landscape in a watershed or region rather than forcing it into a homogeneous paradigm may be more difficult but should ultimately be rewarding for management.

Both the literature and our mid-Atlantic study show that both the proportion of land cover and use and the spatial pattern are good predictors of nonpoint source pollution as represented by sediment, nitrogen, and phosphorus parameters. Land use and its effect on water quality provide an easy-to-understand example of the interactive processes between humans, other organisms, and their environment, i.e., the holistic ecosystem approach. The hierarchical structure of cumulative effects is also well illustrated by land-use change. Regional assessments are critical to understanding cumulative effects, and ideally one should look at the spatial scales above and below the scale of interest (Hunsaker and Graham, 1991). Since watersheds are hierarchically organized, they lend themselves nicely to a multiscale assessment.

Although some citizens do not seem to want to acknowledge that everyone's actions have to be considered for effective management of water quality, studies such as our mid-Atlantic assessment can be powerful illustrations for those trying to manage watersheds. Illustrations like Table 2.1, where large geographic areas have shifted from being dominated by natural land cover to be dominated by human land cover (urban and agriculture), can also be effective management tools.

Working with existing data, as we have for the mid-Atlantic, also identifies the need for physical, chemical, and biological monitoring and research to be designed and integrated to support regional assessments. The fact that our initial analysis phase only had a sample size of 31 stations with three or four chemistry parameters is fairly limiting. Canonical correlation did prove to be an effective statistical method, given the data limitations, as it provides an integrative look at water quality. Correlations between 0.62 and 0.77 are reasonable for environmental data, and the amount of variance explained is acceptable, given that there are multiple stressors (not just land use) that contribute to the overall water quality condition.

Acknowledgments

Statistical advice from John Beauchamp and mentoring by Robert O'Neill have been especially helpful over the years. Our ability to present these findings is the result of working over the years with many students for various institutions, and we wish to thank them. We also appreciate Sumner Cosby, U.S. Environmental Protection Agency, Philadelphia, for providing the water chemistry data for the mid-Atlantic study. Research was sponsored jointly by the U.S. Environmental Protection Agency, Office of Research and Development, under Interagency Agreement DW89937287-01-2 and the U.S. Department of Energy, under contract number DE-AC05-96OR22464, with Lockheed Martin Energy Research Corp.

References

Allan, J. D., and A. S. Flecker. 1993. Biodiversity conservation in running waters. *BioScience* **43**:32–43.

Bailey, R. G. 1976. *Ecoregions of the United States* (map 1:7,500,000), U.S. Department of Agriculture, Forest Service, Intermountain Region, Ogden, UT.

Bormann, F. H., and G. E. Likens. 1981. *Pattern and Process in a Forested Ecosystem,* Springer-Verlag, New York.

Cada, G. F., and C. T. Hunsaker. 1990. Cumulative impacts of hydropower development: Reaching a watershed in impact assessment. *The Environmental Professional* **12**(1): 2–8.

Cooper, J. R., J. W. Gilliam, R. B. Daniels, and W. P. Robarge. 1987. Riparian areas as filters for agricultural sediment. *Soil Science Society of America Journal* **51**:416–420.

Correll, D. L. (ed.) 1986. *Watershed Research Perspectives,* Smithsonian Environmental Research Center, Smithsonian Institution Press, Washington.

Cummins, K. W., G. W. Minshall, J. R. Sedell, C. E. Cushing, and R. C. Petersen. 1984. Stream ecosystem theory, Internationale Vereinigung fur theoretische und angewandte Limnologie. *Verhandlungen,* **22**:1818–1827.

Davis, W. S., and T. P. Simon (eds.). 1995. *Biological Assessment and Criteria Tools for Water Resource Planning and Decision Making,* Lewis Publishers, Boca Raton, FL.

DelRegno, K. J., and S. F. Atkinson. 1988. Nonpoint pollution and watershed management: A remote sensing and geographic information system (GIS) approach. *Lake Reservoir Management* **4**:17–25.

Dickert, T. G., and A. E. Tuttle. 1985. Cumulative impact assessment in environmental planning: A coastal wetland watershed example. *Environmental Impact Assessment Review* **5**:37–64.

Environmental Protection Agency. 1992. *Framework for Ecological Risk Assessment,* EPA 630/R-92/001. Risk Assessment Forum, Washington.

Gallant, A. L., T. R. Whittier, D. P. Larsen, J. M. Omernik, and R. M. Hughes. 1989. *Regionalization as a Tool for Managing Environmental Resources,* EPA/600/3-89/060. Environmental Protection Agency, Environmental Research Laboratory, Corvallis, OR.

Graham, R. L., C. T. Hunsaker, R. V. O'Neill, and B. L. Jackson. 1991. Ecological risk assessment at the regional scale. *Ecological Applications* **1**:196–206.

Gregory, S. V., F. J. Swanson, W. A. McKee, and K. W. Cummins. 1991. An ecosystem perspective of riparian zones. *BioScience* **41**:540–551.

Harmon, M. E., F. J. Franklin, F. J. Swanson, P. Sollins, S. V. Gregory, J. D. Lattin, N. H. Anderson, S. P. Cline, N. G. Aumen, J. R. Sedell, G. W. Lienkaemper, K. Cromack, Jr., and K. W. Cummins. 1986. Ecology of coarse woody debris in temperate ecosystems. *Advances in Ecological Research* **15**:133–302.

Holland, M. M., P. G. Risser, and R. J. Naiman (eds.). 1991. *The Role of Landscape Boundaries in the Management and Restoration of Changing Environments.* Chapman and Hall, New York.

Hughes, R. M., and D. P. Larsen. 1988. Ecoregions: An approach to surface water protection. *Journal of Water Pollution Control Federation* **60**:486–493.

Hunsaker, C. T. 1993. Ecosystem assessment methods for cumulative effects at regional and global scales. In S. G. Hildebrand, and J. B. Cannon (eds.), *Environmental Analysis: The NEPA Experience.* Lewis Publishers, Boca Raton, FL, pp. 480–493.

Hunsaker, C. T., and R. L. Graham. 1991. Regional ecological assessment for air pollution. In S. K. Majumdar, E. W. Miller, and J. Cahir (eds.), *Air Pollution: Environmental Issues and Health Effects.* Pennsylvania Academy of Science, Easton, pp. 312–334.

Hunsaker, C. T., R. L. Graham, G. W. Suter, II, R. V. O'Neill, L. W. Barnthouse, and R. H. Gardner. 1990. Assessing ecological risk on a regional scale. *Environmental Management* **14**:325–332.

Hunsaker, C. T., R. V. O'Neill, B. L. Jackson, S. P. Timmins, D. A. Levine, and D. J. Norton. 1994. Sampling to characterize landscape pattern. *Landscape Ecology* **9**:207–226.

Hunsaker, C. T., and D. A. Levine. 1995. Hierarchical approaches to the study of water quality in rivers. *BioScience* **45**:193–203.

Hunsaker, C. T., D. A. Levine, S. P. Timmins, B. L. Jackson, and R. V. O'Neill. 1992. Landscape characterization for assessing regional water quality. In D. McKenzie, E. Hyatt, and J. McDonald (eds.), *Ecological Indicators.* Elsevier, New York, pp. 997–1006.

Hunsaker, C. T., K. Dickson, W. Waller, and E. Morgan. 1996. Watershed/regional assessment and in-stream monitoring. In *Biomonitoring in the Water Environment.* Water Environment Federation, Alexandria, VA. 1997.

Hynes, H. B. N. 1975. The stream and its valley. Internationale Vereinigung fur theoretische und angewandte Limnologie. *Verhandlungen* **19**:1–15.

Johnson, D. W., and R. I. Van Hook (eds.). 1989. *Analysis of Biogeochemical Cycling Processes in Walker Branch Watershed.* Springer-Verlag, New York.

Johnson, R. A., and D. W. Wichern. 1992. *Applied Multivariate Statistical Analysis,* 3d ed., Prentice-Hall, Englewood Cliffs, NJ.

Kepner, W. G., K. B. Jones, D. J. Chaloud, J. D. Wickham, K. H. Riitters, and R. V. O'Neill. 1995. *Mid-Atlantic Landscape Indicators Project Plan, Environmental Monitoring and Assessment Program,* EPA 620/R-95/003. National Exposure Research Laboratory, Environmental Protection Agency, Research Triangle Park, NC.

Kuchler, A. W. 1964. *Manual to Accompany the Map: Potential Natural Vegetation of the Conterminous United States.* American Geographical Society, Special Publication 36, New York.

Levine, D. A., C. T. Hunsaker, S. P. Timmins, and J. J. Beauchamp. 1993. A geographic information system approach to modeling nutrient and sediment transport. ORNL Report 6736. Oak Ridge National Laboratory, Oak Ridge, TN.

Likens, G. E., and F. H. Bormann. 1974. Linkages between terrestrial and aquatic ecosystems. *BioScience* **24**:447–456.

Loveland, T. R., J. W. Merchant, D. O. Ohlen, and J. F. Brown. 1991. Development of a land-cover characteristics database for the conterminous U.S. *Photogrammetric Engineering and Remote Sensing* **57**:1453–1463.

Lowrance, R., R. Todd, J. Fair, Jr., O. Hendrickson, Jr., R. Leonard, and L. Asmussen. 1984. Riparian forests as nutrient filters in agricultural watersheds. *BioScience* **34**:274–377.

McDaniel, T. W., C. T. Hunsaker, and J. J. Beauchamp. 1987. Determining regional water quality patterns and their ecological relationships. *Environmental Management* **11**(4):507–518.

Minshall, G. W., K. W. Cummins, R. C. Peterson, C. E. Cushing, D. A. Burns, J. R. Sedell, and R. L. Vannote. 1985. Developments in stream ecosystem theory. *Canadian Journal of Fisheries and Aquatic Sciences* **42**:1045–1055.

Naiman, R. J. (ed.). 1992. *Watershed Management: Balancing Sustainability and Environmental Change,* Springer-Verlag, New York.

Naiman, R. J., and H. Decamps (eds.). 1990. *The Ecology and Management of Aquatic-Terrestrial Ecotones.* UNESCO, Paris, and Parthenon Publishing Group, Carnforth, United Kingdom.

Newbold, J. D., J. W. Elwood, R. V. O'Neill, and A. L. Sheldon. 1983. Phosphorus dynamics in a woodland stream ecosystem: A study of nutrient spiraling. *Ecology* **64**:1249–1265.

Omernik, J. M. 1977. *Nonpoint Source-Stream Nutrient Level Relationships: A Nationwide Study,* EPA-600/3-77-105. Environmental Protection Agency, Corvallis, OR.

Omernik, J. M. 1987. Ecoregions of the conterminous United States. *Annals of the Association of American Geographers* **77**:118–125.

Omernik, J. M., A. R. Abernathy, and L. M. Male. 1981. Stream nutrient levels and proximity of agricultural and forest land to streams: Some relationships. *Journal Soil Water Conservation* **36**:227–231.

O'Neill, R. V., C. T. Hunsaker, S. P. Timmins, B. L. Jackson, K. B. Jones, K. H.

Riitters, and J. D. Wickham. 1996. Scale problems in reporting landscape patterns at the regional scale. *Landscape Ecology* **11**(3):169–180.

Osborne, L. L., and M. J. Wiley. 1988. Empirical relationship between land use/cover and stream water quality in an agricultural watershed. *Journal Environmental Management* **26**:9–27.

Paulsen, S. G., D. P. Larsen, P. R. Kaufmann, T. R. Whittier, J. R. Baker, D. V. Peck, J. McGue, R. M. Hughes, D. McMullen, D. Stevens, J. L. Stoddard, J. Lazorchak, W. Kinney, A. R. Selle, and R. Hjort. 1991. *Environmental Monitoring and Assessment Program, Surface Waters Monitoring and Research Strategy Fiscal Year 1991*, EPA/600/3-91/022, EPA, Corvallis, OR.

Peterjohn, W. T., and D. L. Correll. 1984. Nutrient dynamics in an agricultural watershed: Observations on the role of a riparian forest. *Ecology* **65**:1466–1475.

Reckhow, K. H., M. N. Beaulac, and J. T. Simpson. 1980. *Modeling Phosphorus Loading and Lake Response under Uncertainty: A Manual and Compilation of Export Coefficients,* EPA 440/5-80-011. Environmental Protection Agency, Washington.

Riitters, K. H., R. V. O'Neill, C. T. Hunsaker, J. D. Wickham, D. H. Yankee, S. P. Timmins, K. B. Jones, and B. L. Jackson. 1995. A factor analysis of landscape pattern and structure metrics. *Landscape Ecology* **10**(1):23–39.

Schlosser, I. J., and J. R. Karr. 1981. Water quality in agricultural watersheds: Impact of riparian vegetation during base flow. *Water Resources Bulletin* **17**:233–240.

Seaber, P. R., F. P. Kapinos, and G. L. Knapp. 1984. *State Hydrologic Units Maps.* U.S. Geological Survey Open-File Report 84-708, Department of the Interior, Reston, VA.

Sivertun, A., L. E. Reinelt, and R. Castensson. 1988. A GIS method to aid in non-point source critical area analysis. *International Journal of Geographic Information Systems* **2**:365–378.

Stanford, J. A., and J. V. Ward, 1992. Aquatic resources in large catchments: Recognizing interactions between ecosystem connectivity and environmental disturbance. In R. J. Naiman (ed.), *Watershed Management: Balancing Sustainability and Environmental Change.* Springer-Verlag, New York, pp. 91–124.

Suter, G. W., II, L. W. Barnthouse, S. M. Bartell, T. Mill, D. Mackay, and S. Paterson. 1993. *Ecological Risk Assessment*, Lewis Publishers, Boca Raton, FL.

Swank, W. T., and D. A. Crossley, Jr. 1987. *Forest Hydrology and Ecology at Coweeta,* Springer-Verlag, New York.

Turner, M. G. 1989. Landscape ecology: The effect of pattern on process. *Annual Review of Ecological Systems* **20**:171–197.

Vannote, R. L., G. W. Minshall, K. W. Cummins, J. R. Sedell, and C. E. Cushing. 1980. The river continuum concept. *Canadian Journal of Fisheries and Aquatic Science* **37**:130–137.

Ward, G. M., A. K. Ward, C. N. Dahm, and N. G. Aumen. 1990. Origin and formation of organic and inorganic particles in aquatic systems. In R. S. Wotton (ed.), *The Biology of Particles in Aquatic Systems.* CRC Press, Boca Raton, FL, pp. 27–56.

Ward, J. V., and J. A. Stanford. 1983. The intermediate disturbance hypothesis: An explanation for biotic diversity patterns in lotic ecosystems. In T. D. Fontaine, III, and S. M. Bartell (eds.), *Dynamics of Lotic Ecosystems.* Ann Arbor Science, Ann Arbor, MI, pp. 347–356.

Webster, J. R., and B. C. Patten. 1979. Effects of watershed perturbation on stream potassium and calcium dynamics. *Ecological Monographs,* **49:**51–72.

Wiken, E. 1986. Terrestrial ecozones of Canada. *Ecological Land Classification Series Number 19,* Environment Canada, Ottawa, Ontario.

Wilkin, D. C., and R. W. Jackson. 1983. Nonpoint water quality contributions from land use. *Journal of Environmental Systems* **13:**127–136.

3

Watershed Tool Kit

Robert J. Reimold
EA Engineering, Science, and Technology, Inc.
Hingham, Massachusetts

Robert Singer
Metcalf & Eddy, Inc.
Queensbury, New York

To effectively protect and manage aquatic ecosystems and human health, the watershed management approach is a contemporary method that is likely to be effective for coming centuries. To restore, maintain, and enhance the physical, biological, and chemical quality of surface waters of planet earth, the premise of watershed protection focuses on solving water quality and ecosystem problems rather than individual discharges or individual water bodies. This technique takes into account all threats to the environment and human health instead of regulating specific pollutants and pollutant sources. It includes identifying existing priority and potential future problems which afford a measurable risk to human health, ecological resources, desired uses of the waters, or a combination of these. Section 101 of the Clean Water Act establishes the physical, chemical, and biological integrity of the nation's waters as the primary goal of the national water quality program.

According to the 1992 edition of the National Water Quality Inventory: Report to Congress (EPA, 1994), the major impairments of streams and rivers are due to siltation, excessive nutrients, and other pollutants from nonpoint sources. Nonpoint source pollution is generated from a diversity of diffuse sources—runoff from farm fields (carrying nutrients and pesticides), runoff from city streets (carrying metals, hydrocarbons, and pathogens), and sediment-laden runoff from logging and construction activities. The impacts of such stressors may be acute (cause immediate death) or chronic (exhibit long-term debilitation of the organism, such as reproductive impairment and weight loss) to both humans and aquatic life within the watersheds. The impacts also may cause significant degradation to the habitat. As a result, the watershed approach to environmental problem solving started with a focus on nonpoint source controls. Through various federal funding programs such as the Nonpoint Source Program, Comprehensive State Ground Water Protection Programs, National Estuary Program, Wellhead Protection Program, Clean Lake Program, Advanced Identification and Special Area Management Plans for Wetlands, and related cooperation by the U.S. Environmental Protection Agency (EPA), U.S. Forest Service, U.S. Department of Agriculture (Natural Resource Conservation Service), U.S. Department of Interior (including U.S. Geological Survey, Bureau of Reclamation, Bureau of Land Management, and U.S. Fish and Wildlife Service) and Native American tribal governments, watershed protection approaches evolved to complement other environmental and natural resource management activities.

More recent applications of watershed initiatives include the reorientation of traditional federal regulatory programs to a watershed basis. For example, in many states, the National Pollution Discharge Elimination System (NPDES) permit program is now becoming watershed-oriented. Instead of randomly issuing 5-year NPDES permits for different discharges, permits for a given watershed are all being evaluated and issued at one time (on a watershed basis). Some of the individual states have also offered financial incentives toward watershed management, thus fostering the formation of private, nonprofit watershed management associations.

The contemporary approach involves interactions between control of industrial and municipal pollution (through the NPDES program) and the management of persistent issues related to nonpoint sources, sewer overflows, and habitat degradation. There are environmental, community, and economic benefits to such an approach. Environmental benefits are maximized because there is a focus on the resources. Enhanced public attention focused on the actual resources results in the achievement of tangible ecological results instead of administrative requirements. Stakeholder cooperation can be the controlling factor in debates over the control of a watershed (Stave, 1996). Community building occurs through cooperation and collaboration because stakeholders gain a sense of common purpose to develop lasting solutions. Streamlined regulatory requirements result in cost savings. Watershed-based permitting, monitoring, and reporting save valuable time and money. The comprehensive, long-term nature of watershed plans provides predictability for the regulated community and a better understanding of needed environmental policies and how they can be achieved.

Watershed management plans are formulated based on a number of key terms, including these:

Management units—large hydrologic units, major river basins or aquifers, delineated by the state as a basin, each containing multiple watersheds (see subsequent discussion in this chapter regarding delineation of watersheds).

Management cycles—a state's grouping of basins in a sequence so that the entire state is studied and management plans are developed and updated (typically at 5-year intervals). Management cycles typically have two features: a specific time period and a sequence for addressing watersheds. Selection of management cycles depends on availability of data, workload requirements, degree of water quality impairment, environmental risk, and stakeholder support.

Stakeholder involvement—agencies, organizations (government and nongovernment), businesses, and individuals interested in water quality, ecosystem health, and management strategies included in watershed management activities. Typical stakeholders include the applicable state water quality agency; state public health, agriculture, forestry, fish, and wildlife agencies; municipal and industrial dischargers, city and county governments, trade associations, environmental and conservation groups, Chambers of Commerce, local offices of federal agencies, EPA regions, and individual landowners within a watershed.

Strategic monitoring—water quality and ecological health monitored to measure the nature and extent of problems and stressors, typically done on a rotating basis (such as two summers of sampling every 5 years, for a given basin).

Assessment—analyses of data through statistical procedures and best professional judgment to identify problems, sources, and stressors. Water quality standards (at the state level) and criteria (at the federal level) are integral to assessments because they reflect criteria for restoring and maintaining the physical, chemical, and biological integrity of the water.

Prioritization and targeting—a ranking of watersheds according to resource value, degree of use impairment, or other factors; rankings are used for prioritization for special management attention. Targeting is the process of deciding how resources should be allocated to address priority concerns. Targeting typically includes ranking water bodies according to such factors as

- severity of risk to human health and aquatic community
- impairment to water body (documented or potential)
- resource value of the water body to the public

Targeting (EPA, 1993) involves selection of specific watersheds for special management attention based on ranking from the above prioritization step, availability of staff and financial resources, overall planning goal(s), and willingness of stakeholders to proceed.

Development of management strategies—realistic goals set for specific watersheds; management strategies are then developed prior to allocation of scarce resources.

Watershed plans—documents that summarize the assessment results, establish goals, and identify management strategies for a given watershed. These plans are often issued in conjunction with NPDES permits and are revised periodically (typically every 5 years). The plans are also used to educate the public on watershed-specific issues. The plans must specify how goals will be achieved, who is responsible for implementation, and the schedule for implementation. Criteria for measuring the effectiveness of the plan must also be developed. Experience dictates that formal written commitments for all stakeholders are essential prior to plan implementation.

Implementation—activation of management strategies based on the plan.

Through development and implementation of a plan, and with sufficient commitment by stakeholders, all priority problems can be solved. There are multiple benefits to watershed management. Benefits include

- More direct focus by stakeholders on achieving ecological goals and water quality standards rather than on measurement of program activities such as numbers of permits or samples
- Improved basis for management decisions through consideration of traditional stressors (e.g., toxics, biochemical oxygen demand, nutrients, from point sources) and nonchemical stressors (e.g., habitat loss, temperature change, sedimentation, or flow alteration)
- Enhanced program efficacy due to the fact that permits and required monitoring are focused on a limited number of watersheds at a given time

- Improved coordination among federal, state, and local agencies and other organizations (including increased data sharing and pooling of resources)
- Enhanced public involvement, including better relations with permittee due to increased involvement and greater consistency and equitability in permit conditions
- Innovative solutions such as ecological restoration, wetlands mitigation banking, and market-based solutions (e.g., pollutant trading or restoration in lieu of advanced wastewater treatment)

Defining a Watershed

Succinctly defined, watersheds are nature's boundaries, i.e., wetlands, streams, lakes, rivers, estuaries, surrounding landscape, and groundwater recharge areas which drain to a surface water body. The EPA defined *watershed* as "a geographic area in which water, sediments, and dissolved materials drain into a common outlet" such as a stream, lake, estuary, or ocean. Based on the observation that water flows downhill, watersheds are defined by topography. It is common to think of watersheds within watersheds.

For example, the North American Continental Divide serves as a boundary between waters flowing to the Atlantic and Pacific Oceans. Within such large areas, there are numerous subdivisions. In the Mississippi River watershed (which drains to the Atlantic Ocean via the Gulf of Mexico) there are many additional subwatersheds. The waters of any large tributary to the Mississippi (such as the Ohio River) can also be divided into subwatersheds. Within the Ohio River watershed, there are numerous rivers and lakes each with its own watershed. Whether naturally created (such as Conneaut Lake, in western Pennsylvania) or anthropogenically created (such as Pymatuning Lake, in western Pennsylvania) both bodies of water eventually drain to the Ohio River watershed.

Since watersheds are defined based on topography, a topographic map and some simple instructions are all that is needed to delineate a watershed. For example, U.S. Geological Survey topographic maps (available by telephone at 1-800-USA-MAPS) or many state conservation agencies, bookstores, etc., are a good source of topographic maps in the United States. Once you have a topographic map that covers the general area of the watershed you desire to delineate, there are several easy steps:

1. Determine and mark the downstream outlet or *bottom* of the watershed you wish to delineate. This could be the outlet of a lake or pond,

the mouth of a river or stream, or a geographic landmark (such as a bridge) representing the farthest downstream point in which you are interested.

2. Use a colored highlighter to locate all the streams, wetlands, lakes, etc., that eventually flow to this downstream outlet or bottom point. In addition to following the continuous blue lines of streams and the broken blue lines of intermittent streams, comparisons should be made of the land elevation features surrounding each watercourse or water body. Use of an arrow on the highlighted watercourses is also helpful to indicate direction of flow.
3. From the identified watercourses, find and mark the highest nearby hills, ridges, etc. Connect the points (following ridge lines, crossing slopes at right angles to contour lines) until the perimeter of the watershed has been outlined.

Depending on the level of detail needed, field checking of the boundary is not only useful but also helpful to gain experience in judgment regarding location of the highest points. Field confirmation is also useful to facilitate identification of recent anthropogenic alterations (ditches, dikes, roads, etc.) not visible based on the topographic map, which could significantly alter the direction of water flow and thus impact the watershed boundary. Detailed approaches (Ammann and Stone, 1991, and Williams, 1992) are available to provide further refinement and guidance.

The U.S. Geological Survey (USGS) has designated and mapped a natural system of hydrologic units for cataloging, called HUCs, that provide a common national framework for delineating watersheds and their boundaries at various geographic scales. In this hierarchical system, the largest units are called *water resource regions* and are designated by a two-digit code. Each regional unit is subdivided into a four-digit subregion, and then further subdivided into six-digit and eight-digit units representing smaller and smaller watersheds. The eight-digit units, typically fairly large watersheds averaging thousands of square miles, are the most detailed delineations currently available in geographic information system (GIS) format. Some states have carried this hierarchical system further to the 11-digit and 14-digit level to delineate watersheds averaging approximately 100 and 30 mi^2, respectively. These hydrologic units are an important data set within the National Spatial Data Infrastructure data set; watershed programs delineating smaller-scale watersheds can collaborate with this existing national framework.

The development of fully compatible watershed boundaries involves close coordination between USGS, Natural Resource Conservation

Service (NRCS), state water quality, coastal management, GIS agencies, and others. Using such an approach, North Carolina (Division of Environmental Management) has divided the state into 18 basins. Using the *nested* hierarchy of watershed, North Carolina has delineated over 1600 fourteen-digit watersheds, each averaging 30 mi^2. Access to the latest watershed maps using this approach can be found on the Internet at *http://www.nrcs.usda.gov/*.

Watershed Tools via the Internet

The Internet contains vast resources for watershed approaches. The Water Environment Federation (*http://www.wef.org*) has a Web site devoted to technical assistance for watersheds. In addition to listing publications, public information materials, and conferences related to watersheds, this site has technical discussion groups to track watershed management initiatives, public discussion groups (to facilitate public involvement), products and supplier search engine, and links to related government and private sites. For example, Purdue University hosts the Conservation Technology Information Center's "Know Your Watershed." This site (*http://www.ctic.purdue.edu.watershed/watershedoptions.html*) tracks watershed projects that have effectively attained broad stakeholder involvement. In addition, the Purdue site is home for the National Watershed Network Mentor Search. The site contains a "National Watershed Network Mentor Search" engine which facilitates quick acquisition of watershed-related data by

Waterbody type: aquifers, coastal waters, estuaries, ponds, lakes, rivers, reservoirs, streams, and wetlands.

Water use: drinking water, flood retention, fisheries, hydroelectric, irrigation, navigation, recreation, shellfishing, finfishing, wetlands habitat, and wildlife habitat

Scope and consideration: air quality, endangered species, fisheries, local economy, pristine resources, recreation, soil quality, water quality, and wildlife habitat

Stressor and pollutant: atmospheric deposition, bacteria, biochemical oxygen demand, dissolved oxygen, exotic species, flooding, heavy metals, in-stream flows, land use and development, nitrogen, noxious weeds, odor, open space, pathogens, pesticides, phosphorus, salinity, sediment, temperature, toxic substances, turbidity, and wildlife habitat

Focuses and sources: abandoned mines, abandoned wells, acid rain, animal manure, buried tanks, channelization, cropland, dams, dredging, grazing, irrigation, landfill, logging, septic tanks and systems, riparian areas, shorelines, stream banks, and urban runoff

Partners, stakeholders, and audiences: absent landowners, agricultural interests, conservations, developers, builders, environmental activists, forestry interests, government, mining, municipalities, recreation, rural residents, schools, suburban residents, tribes, and urban residents

Information available: annual reports, assessments, inventories, reports, newsletters, water data, and other Internet sites

From each of these various subdivisions, the Internet user can select connections to sources of libraries, videos, workshops, national networks, calendars of events, partners, and suggestions for improvements.

As an example, we select the Putuxent River watershed in Maryland. This Internet site provides information on USGS topographic map numbers, county location, size of watershed, local coordinator (name, affiliation, address, phone, and fax) of watershed, percentage of area of watershed used for cropland, forestry, grazing, mining, tribal uses, passive recreation, federal ownership, state/county ownership, etc. The site also provides information about that specific watershed for each of the above-named categories including water uses, stressors, etc. For partners associated with the designated watershed, information is provided about the names of the partners, their goals, their meeting frequencies, their activities, and locations for additional information. In addition, a summary is provided regarding the financial, in-kind support provided by local, state, and federal government, industries, foundations, and individuals.

The Purdue University Internet site also contains the National Watershed Library. This site can be accessed by keyword. Library contents are categorized by:

Audiences: including children, construction, farmers, forestry residents industry, miners, natural resource professionals, outdoor recreation residents, ranchers, residents, rural small businesses, suburban watershed residents, urban residents, and several others

Description of library holdings: including annual reports, booklets, brochures, curricula, fact sheets, newsletters, posters, proceedings, slide shows, videos, workshop materials, and others

Their database is closely linked with the Environmental Protection Agency's Internet access site "Surf Your Watershed" (http://www.epa.gov/OWOW/WhatsNew.html) which allows users to access numerous federal, state, and other natural resource databases. This site contains the National Watershed Assessment Data Profiles and National Watershed Assessment Data Fact Sheets (in WordPerfect format).

At the Internet site www.epa.gov/surf/ you can download watershed data on-line. In addition, at the above-referenced OW/OWOW Internet address, by entering a Zip code, hydrologic unit code, or a place name, users can "point and click" their way to key watershed information including water quality, toxics release inventories, municipal and industrial point source discharges, demographic data, habitat data, and nonpoint source information. Also through the OWOW Internet address, you can read the most recent watershed case studies from selected sites throughout the United States.

Because of the efficacy of watershed management in formulating, implementing, and monitoring cost-effective nonpoint source controls, the OWOW Internet address http://www.epa.gov/OWOW/watershed/index.html is also linked to federal efforts that track cost-effectiveness associated with watershed initiatives. The same Internet address facilitates tracking of the EPA's "National Watershed Assessment Project" in which EPA is working with states, tribes, and others to characterize the health of every U.S. watershed and identify those watersheds that are at greatest risk. Using water quality, public health, and ecosystem data, this EPA project (highlighted on the Internet) summarizes key watershed factors such as fish consumption advisories, aquatic species at risk, contaminated sediments, drinking water quality, discharges of toxics and pollutants, impacts due to hydro modification, wetland losses, etc.

The EPA also maintains a summary of "Training Tools" which consists of a catalog of watershed training opportunities. This is a collection of current educational courses available on watershed topics. The purpose of the on-line catalog is to direct readers to training and educational opportunities on watershed protection by providing summaries and contact information for watershed-based training. The catalog also includes other educational products such as videos, guidance documents, and other information that is easily sorted by topics and keywords. Each catalog entry includes a list of "tool users" (such as businesses, federal agencies), "keywords" (such as estuaries, habitat, bioassessments), and "sources for additional information." Representative catalog entries summarize training opportunities related to evaluation of dredged materials, contaminant source inventories for drinking water, ecological restoration, wetland restoration, stream bank restoration, total maximum daily load development, water quality standards, groundwater supply contingency planning, lake restoration, land stewardship and watershed planning, international conferences on water-related topics, and watershed protection techniques.

The EPA has categorized a wide variety of Watershed Tools on the Internet at the address http://www.epa.gov/OWOW/watershed/tools/index.html. There are updated, extensive lists of data collection, measurement, and assessment tools including use of aerial photography, analytical methods for determining pollutants in wastewater, and procedures to assess uncertainty and variability of wildlife and aquatic organism toxicity. Information is provided on establishing fish advisories due to chemical contaminants; statistical analyses of water quality; determination of pathological indicators of fresh, estuarine, and marine waters; and conduct of sanitary surveys. The site also summarizes a wide variety of database tools including environmental monitoring methods index, national directories of volunteer monitoring programs, grants reporting and tracking systems, guides to technical literature on biological monitoring criteria and environmental monitoring methods, and economic cost-effectiveness and cost-benefit analytical approaches. Environmental goal-setting tools and financial assistance tools are summarized, and current copies of pertinent literature can be easily ordered. In addition, this Internet site lists current mathematical modeling tools with links to sources of the models and on-line help. Outreach / Education tools include links to other sites (such as Clean Water Online), and sources of water quality criteria and standards program videotapes. Policy and Planning tools include the California Ocean Plan, Guidance Management Measures for Nonpoint Pollution in Coastal Waters, and manuals on volunteer watershed monitoring. Reference Reports and Studies include numerous development documents and guidelines for standards, ecosystem inventory tools, sediment classification methods, wildlife exposure factors handbooks, and methods for xeriscape landscaping. This site, updated by EPA's Office of Water, provides a wealth of watershed tools for the novice as well as the advance watershed planner or manager.

Conventional Routes to Watershed Tools

A number of valuable watershed publications are available from the EPA. A summary of watershed developments and protection approaches is available in the bulletin *Watershed Events,* available from Environmental Protection Agency, 4501 F, 401 M Street, SW, Washington, DC 20460. In addition, available from the same address are an annual summary document entitled *Watershed Protection: Catalog of Federal Programs* and an annual summary of EPA watershed-related activities entitled *The Watershed Protection Approach: Activity Approach.*

In addition to federal government publications, many conservation groups have developed a variety of manuals and bulletins. Representative significant examples include

Hands On Save Our Streams: The Save Our Streams Teachers Manual and *Save Our Streams Volunteer Trainers Handbook,* available for purchase from the Izaak Walton League, 707 Conservation Lane, Gaithersburg, MD 20878.

How to Save a River: A Handbook for Citizen Action, written by David Bolling, and *River Voices* (a quarterly newsletter devoted to watershed protection), available for purchase from the River Network, P.O. Box 8787, Portland, OR 97207-8787.

Understanding Lake Data (1994), a "how to" manual to interpret results of physical and chemical water testing, and *Life on the Edge...Owning Waterfront Property,* both available for purchase from the University of Wisconsin Extension—Lake Management, CNR-UWSP, Stevens Point, WI 54481.

Environmental Resources Guide: Nonpoint Source Pollution Prevention, a supplemental curriculum for grades 9 to 12 detailing what nonpoint source pollution is, how it results from land-use practices, and how to make decisions to manage it; available for purchase from Air and Waste Management Association, Publications Order Department, One Gateway Center, third floor, Pittsburgh, PA 15222.

Proceedings of Watershed '96: A National Conference on Watershed Management, Water Environment Federation, Baltimore, MD (703-684-2400).

The EPA also provides assistance to the "Internet-impaired community" through regional watershed coordinators. EPA watershed assistance is available from

Headquarters: 202/260-9108 and 202/260-7040

Region 1 (CT, ME, MA, NH, RI, VT): 617/573-5700

Region 2 (NJ, NY, PR, VI): 212/637-3724

Region 3 (DE, DC, MD, PA, VA, WV): 215/597-9911

Region 4 (AL, FL, GA, KY, MS, NC, SC, TN): 404/347-4450

Region 5 (IL, IN, MI, MN, OH, WI): 312/353-2147

Region 6 (AR, LA, NM, OK, TX): 214/665-7101

Region 7 (IA, KS, MO, NE): 913/551-7030

Region 8 (CO, MT, NE, SD, UT, WY): 303/293-1603

Region 9 (AZ, CA, HI, NV, AS, GU): 415/744-2125

Region 10 (AK, ID, OR, WA): 206/553-1793

or through the conventional mail at National Center for Environmental Publications, P.O. Box 42419, Cincinnati, OH 45242.

To define the watershed in terms of potential water quality problems, an inventory and assessment of baseline conditions are essential. Typically there is a wealth of data about a particular watershed, but the information is widely disseminated. Consequently, a checklist (Table 3.1) can be extremely useful in identifying and assembling information.

Sources for these data include state and local GIS databases, state surface and groundwater databases and reports, and aerial photography. In addition, county-level National Resources Conservation Service Field

Table 3.1 Checklist of Background Watershed Information to Assemble

1. Sizes, locations, and designated uses of all water bodies
2. Water bodies having impaired use support
3. Causes of impairment (pollutants, habitat limits, etc.)
4. Physical water attributes
5. Biological water attributes
6. Chemical water quality
7. Locations, sources, and loadings of point source discharges
8. Categories of nonpoint sources and estimates of loadings
9. Groundwater quality
10. Sources impacting groundwater
11. Fish and wildlife surveys
12. Topographic and hydrologic maps
13. Land use and cover maps
14. Detailed soil surveys
15. Demographic data and growth projections
16. Economic conditions—income, employment
17. Threatened and endangered species and habitat
18. Wetland maps
19. Riparian land maps
20. List of relevant local stakeholders (agencies, nongovernment organizations, businesses, individuals)

Office Technical Guides (available from the local county NRCS office) are valuable sources of information on soils, water, plants, animals, non-point source best management practices, and other topics.

Many examples exist regarding how to plan and implement a watershed-wide volunteer monitoring program. The River Watch Network (153 State Street, Montpelier, VT 05602) is an EPA award-winning example of a model program to implement volunteer watershed monitoring. This two-tier system included watershed-wide support system and study design, and individual monitoring programs in subwatersheds or river reaches. The monitoring consisted of a core set of studies to answer watershed-wide questions such as whether water meets state water quality standards. Optional study packages were developed to answer specific questions about a tributary or river reach, such as assessing the impacts of a certain wastewater treatment plant. This model involves a broad spectrum of the watershed community—citizens, businesses, and government and nongovernment organizations.

Habitat Monitoring

One typical watershed management tool is habitat monitoring. Ecological concepts regarding the "lebensraum" or living space are similar to interior decorators' beliefs about the space in which we live being as important to life as what we eat and breathe.

Based on the dictionary definition of *habitat* as "the place or type of site where a plant or animal naturally or normally occurs," ecologists have traditionally defined habitat as the "home address." For example, Odum (1997) defined habitat as the place where a certain species can be found, i.e., the address at which the organism lives. There is worldwide consensus that habitat degradation is a primary factor limiting the beneficial uses of the planet's surface waters.

Consequently because of the human drive to enjoy quality of life (and thus the focus on protection of public health and safety as the driving force behind most environmental laws), and because beneficial uses come with good water quality, habitat surveys have become the standard measure for assessing the watershed's ability to support life.

Anthropogenic influences on habitats of plants and animals (including humans) have been both physical and chemical. Humans have diked, drained, and "straightened" wetlands and watercourses. Land has been logged, cleared, plowed, and paved. Humans have used pesticides, and herbicides, to "manage" nuisance organisms. All these "impacts of progress" have had the potential for significant adverse impact on habitat. Habitat monitoring has evolved as one of the key tools of watershed management because by measuring the physical, chemical, and biological attributes of water bodies, one can get a picture of the water's ability to support life.

Many of the habitat assessment tools are developed based on habitat functions and community structure. Habitats are places for plant and animal growth and reproduction. Animal habitats also provide places for feeding and resting. Habitat functions can be measured in quantitative terms involving assessments of change over time (biological, chemical, and physical transformations). Guidance for developing biological criteria should be consulted and modified for use by each state and regional watershed management plan (EPA, 1996).

The communities assessed include macroinvertebrates and/or fish habitats in rivers and streams (Plafkin et al., 1989). Other specialists often focus on habitat for birds, amphibians, and other wildlife. Typically, any imbalance in the biological community is indicative of habitat limitations (such as excessive sediment deposition) or water quality problems (such as low dissolved oxygen).

Methods for habitat assessment and monitoring commonly are based on ecosystem perspectives. Direct measures of habitat include site morphology or species (of plants or animals), numbers of individuals of each species, and related measurements of diversity. Natural and anthropogenically created changes in watershed use result in changes in habitat functions. Such concepts have been viewed from the perspective of landscape ecology (Odum and Turner, 1990) and in terms of physical changes (Kentula et al., 1992). Habitat measurements typically include the species and density (Hook et al., 1988), the areal extent (Darnell, 1976), and diversity of fauna (Brooks et al., 1991).

Watershed hydrology is typically assessed by measurements of water quality, depth, flow rate, flow patterns, and potential for groundwater recharge. Watershed morphometric measurements include drainage area, surrounding land use, area, slope of wetland, slope of surrounding land, channel morphology, and perimeter to slope area. These physical attributes are often used to monitor changes in watershed habitats.

Approaches to watershed habitat evaluation systems evolved from systems based on cover types, i.e., the vegetative cover of the area such as deciduous forest, coniferous forest, and residential woodland (U.S. Fish and Wildlife Service, 1977, 1980), to community-based models (Roberts et al., 1987; and Schroeder, 1987), to the habitat evaluation procedure (HEP), which compares relative habitat values at different times (e.g., present and future). Using hypotheses of species-habitat interactions, HEP employs habitat suitability indices (HSI) based on models of species distribution, life history, and specific requirements for food, water, cover, etc., to compare habitats. There are also habitat gradient models (Short, 1982) based on assumptions of potential natural vegetation type, i.e., what vegetative community would become established over a finite time, given satisfactory growing conditions. An alternative risk-based approach to watershed habitat monitoring (Johnson and Wichren, 1992) involves multiple linear regressions, discriminant analysis, and visual inspection of graphical data. Use of this approach, however, requires extensive data sets in order to develop reliable estimates of the risk functions and sound ecological judgment.

The preponderance of the habitat-monitoring approaches incorporate parameter measurements, classification schemes, and multivariate ecosystem models. The EPA has developed a rapid bioassessment protocol (Plafkin et al., 1989) and specifically developed an approach for watersheds. Although all these watershed habitat-monitoring approaches have many constraints (Schamberger and O'Neil, 1986), there are ways to increase efficiency and reduce effort associated with use of these procedures (Wakley and O'Neil, 1988).

Initiating Watershed Management—What to Do

Watershed management is a strategy for effectively preserving and restoring aquatic ecosystems and protecting human health. The strategy is based on the premise that many water quality and ecosystem problems are best solved at the watershed level rather than at an individual water body or discharger level. The watershed approach has four major features: targeting of priority problems, a high level of stakeholder involvement, integrated solutions that make use of the expertise and authority of multiple agencies, and success measured through monitoring and other data gathering.

The flexible framework of watershed management encompasses management and protection of ecosystems and human health at two different levels, i.e., the state and the watershed. Some issues are best addressed at the watershed level (such as controlling nutrients or restoring headwater riparian habitat quality). Other issues, such as phosphate detergent bans or wetland mitigation banking, may be better addressed on a state or regional level.

To make watershed management work, managers must "walk the walk." Switching from program-centered to watershed-centered management involves a functional change for most managers (although it need not involve an organizational change in structure). Important steps in implementation include budgeting sufficient time for key staff who will develop the approach, education of all parties on the principles of watershed management, and establishing an efficient means of communication among staff. Several states have used outside facilitators to bring staff from various program areas together to agree on common purposes and work out potential "turf" issues.

The lead agency must prepare a framework document defining the overall goals and objectives; the specific schedules, roles, and responsibilities for each organizational unit; procedures for developing plans; and guidelines for involvement of the public. The typical outline of a watershed management approach includes the following:

Executive summary
Watershed management approach vision

- Long-term management vision
- Relationship of specific watershed to vision

1. Introduction
   1.1 Objectives
   1.2 Rationale for approach
   1.3 Federal CWA mandate for approach
2. Coordination and integration of agency programs and functions
   2.1 NPDES permitting
   2.2 Monitoring
   2.3 Financial planning and grants
   2.4 Water resource planning
   2.5 Nonpoint source programs
   2.6 Coastal zone management
   2.7 Drinking water
   2.8 Groundwater
   2.9 Fish and wildlife
3. Transition issues and solutions
   3.1 EPA flexibility
   3.2 Organizational structures
   3.3 Coordination with local planning agencies
   3.4 Watershed scheduling process
   3.5 Other issues
4. Major components of a watershed management plan (see App. A in this book for further details)
5. Procedures for developing watershed management plans
6. Statewide monitoring plan
7. Data analysis, modeling, and presentation [including total maximum daily loads (TMDLs)]
8. State financial assistance
9. Roles and responsibilities in watershed planning
   9.1 Surface and groundwater

9.2 Soil and water resources
9.3 Other divisions
10. Implementation schedule
11. Data management
11.1 GIS
11.2 Existing data management structures
11.3 Recommended data management structures

Watershed management recognizes that water quality management must embrace human and ecosystem health and that managing for one without considering the other is usually detrimental to both. Watershed management allows managing a range of inputs for specific outputs. It equally emphasizes all aspects of water quality including chemical water quality (toxic and conventional pollutants), physical water quality (temperature, flow, circulation, groundwater and surface water interaction), habitat quality (channel morphology, substrate composition, and riparian zone characteristics), biological health and biodiversity (species abundance, diversity, and range), and subsurface biogeochemistry.

There are four major features of watershed management: targeting of priority pollutants, stakeholder involvement, integrated solutions, and measuring of success. Effective development of watershed plans involves a strong monitoring and evaluation component. Using these data, stakeholders identify stressors that may pose health and ecological risks in the watershed and any related aquifers, and they prioritize the stressors. Monitoring is also essential to determine the effectiveness of management options selected by the stakeholders to address high-priority stressors.

Watershed protection activities require long-term commitments from stakeholders. Consequently, stakeholders need to know whether their efforts are achieving real water quality improvements. One of the key goals of every watershed management plan must be to implement the goals of the Clean Water Act (i.e., to restore, protect, and maintain the physical, chemical, and biological integrity of the nation's water) and the Safe Drinking Water Act (i.e., to protect human health through source water protection). Subsequent chapters of this book illustrate these "how to do" and "how it has been done" in various worldwide locations.

References

Ammann, A., and A. L. Stone. 1991. *Method for the Comparative Evaluation of Nontidal Wetlands in New Hampshire.* New Hampshire Department of Environmental Sciences. Concord, NH.

Brooks, R. P, E. D. Bellis, C. S. Keener, M. J. Croonquist, and D. E. Arnold. 1991. A methodology for biological monitoring of cumulative impacts on wetland, stream, and riparian components of watersheds. In *Proceedings of the International Symposium: Wetlands and River Corridor Management.* J. A. Kusler and S. Daly, eds. Association of State Wetland Managers. Berne, NY. Pp. 3872–3998.

Darnell, R. M. 1976. *Impacts of Construction Activities in Wetlands of the United States,* EPA-600/3-76-045. Office of Research and Development, Environmental Protection Agency. Ecological Research Series.

Environmental Protection Agency. 1993. *Geographic Targeting: Selecting State Examples,* EPA-841-B-93-001. Office of Wetlands, Oceans, and Watersheds. Washington.

Environmental Protection Agency. 1994. *National Water Quality Inventory: 1992 Report to Congress,* EPA 841-R-94-001. Office of Wetlands, Oceans, and Watersheds. Washington.

Environmental Protection Agency. 1996. *Biological Criteria: Technical Guidance for Streams and Small Rivers,* rev. ed., EPA 822-B-96-001. Washington.

Hook, D. D., W. H. McKee, Jr., H. K. Smith, J. Gregory, V. G. Burrell, Jr., M. R. DeVoe, R. E. Sojka, S. Gilbert, R. Banks, L. H. Stolzy, C. Grooks, T. D. Matthews, and T. H. Shear. 1988. *The Ecology and Management of Wetlands.* Timber Press. Portland, OR.

Johnson, R. A., and D. W. Wichren. 1992. *Applied Multivariate Statistical Analyses.* Prentice-Hall, Englewood Cliffs, NJ.

Kentula, M. E., R. P. Brooks, S. Gwin, C. Holland, A. D. Sherman, and J. Sifneos. 1992. *An Approach to Decision Making in Wetlands Restoration and Creation,* EPA/600/R-92/150. Environmental Research Laboratory. Environmental Protection Agency. Corvallis, OR.

Odum, E. P. 1997. *Ecology: A Budge Between Science and Society.* Sinauer Associates, Inc. Sunderland, MA.

Odum, H. T., and M. G. Turner. 1990. The Georgia landscape: A changing resource. In *Changing Landscapes: An Ecological Perspective.* I. S. Zonneveld and R. T. T. Forman, eds. Springer-Verlag. New York. Pp. 137–164.

Plafkin, J. L., M. T. Barbour, K. D. Porter, S. K. Gross, and R. M. Hughes. 1989. *Rapid Bioassessment Protocols for Use in Streams and Rivers: Benthic Macroinvertebrates and Fish,* EPA/444/4-89-001. Environmental Protection Agency, Washington.

Roberts, T. H., L. J. O'Neil, and W. E. Jabour. 1987. Status and source of habitat models and literature reviews. U.S. Army Engineer Waterways Experiment Station. Miscellaneous Paper EL-85-1. Vicksburg, MS.

Schamberger, M. L., and L. J. O'Neil. 1986. Concepts and constraints of habitat [model testing]. In *Wildlife 2000: Modeling Habitat Relationships of Terrestrial Vertebrates.* J. Verner, M. L. Morrison, and C. J. Falph, eds. University of Wisconsin Press. Madison, WI. Pp: 5–10.

Schroeder, R. L. 1987. Community models for wildlife impact assessment: A review of concepts and approaches. National Ecology Center. U.S. Fish and Wildlife Service. Washington. *Biological Report* **87**(2).

Short, H. L. 1982. Development and use of a habitat gradient model to evaluate wildlife habitat. In *Transactions of the Forty-Seventh North American Wildlife and Natural Resources Conference.* Wildlife Management Institute. Washington. Pp. 57–72.

Stave, K. A. 1996. Describing the elephant: Multiple perspectives in New York City's watershed protection conflict. In *Watershed '96: A National Conference on Watershed Management.* Water Environment Federation. Baltimore, MD.

U.S. Fish and Wildlife Service. 1977. *A Handbook for Habitat Evaluation Procedures.* U.S. Department of Interior. Washington. Resource Publication 132.

U.S. Fish and Wildlife Service. 1980. *Habitat Evaluation Procedures.* Division of Ecological Services. U.S. Department of Interior. Washington. ESM 102.

Wakley, J. S., and L. J. O'Neil. 1988. Alternatives to increase efficiency and reduce effort in application of the habitat evaluation procedures (HEP). U.S. Army Engineer. Waterways Experiment Station. Vicksburg, MS. Technical Report.

Williams, S. 1992. *A Citizens' Guide to Lake Watershed Surveys: How to Conduct a Nonpoint Source Phosphorus Survey.* Maine Department of Environmental Protection. Yarmouth, ME.

4

Developing a Watershed Management Plan with a Sanitary Survey

H. L. Selznick
David Evans and Associates, San Jose, California

S. Palmer
Metcalf & Eddy, Inc. Palo Alto, California

Watershed management is a fundamental concern of all water supply agencies. These agencies are now realizing that they are responsible for stewardship of the hydrological cycle, from the moment the rainfall hits the ground to when a faucet turns on in a home. During the time water travels through a watershed, quality can be affected in numerous ways. To document and better understand the mechanisms of water quality changes, water supply agencies have developed watershed management plans. One common goal of watershed management is to be certain that the quality of water supply is maintained and protected from needless degradation. The first step in developing such a plan is to perform a watershed sanitary survey. The specific objectives of this survey are to:

- Inventory potential sources of contamination and degradation and their pathways of movement.
- Develop procedures (best management practices or other regulations) to maintain water quality and prevent degradation or contamination.

One example of developing a watershed management plan based on a watershed sanitary survey is based on a case study of four watersheds in San Luis Obispo County, California (Fig. 4.1). The four watersheds

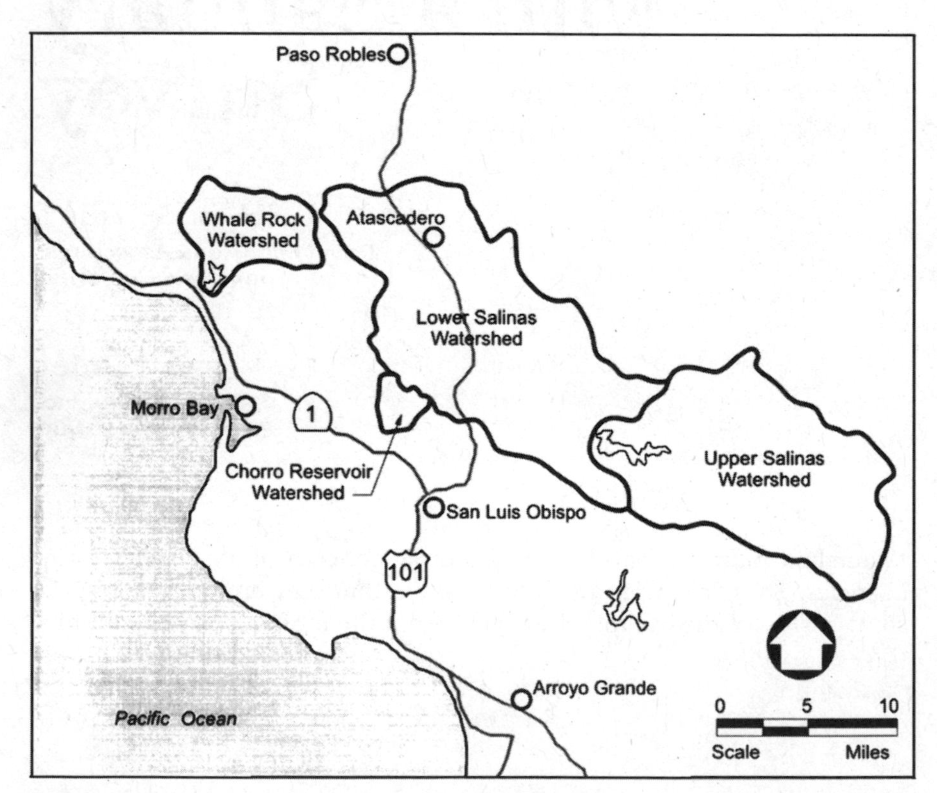

Figure 4.1 Location of watersheds, San Luis Obispo County, California.

Table 4.1 Watershed Characteristics

Characteristics	Watershed: Upper Salinas	Lower Salinas	Whale Rock	Chorro
Area, hectares	29,000	35,000	5260	930
Predominant land use(s)	Cattle grazing, recreation	Open space, cattle grazing, urban development	Cattle grazing, agriculture	Prison, cattle grazing
Agency using runoff from watershed for potable water use				
City of San Luis Obispo*	Yes	No	Yes	No
Atascadero Mutual Water Company (AMWC)	Yes	Yes	No	No
California Men's Colony	No	No	Yes	Yes
Whale Rock Commission	No	No	Yes	No
Water supply facilities	Salinas Dam and Reservoir†	AMWC wells‡	Whale Rock Dam and Reservoir	Chorro Reservoir

*City of San Luis Obispo has water treatment plant for water from upper Salinas and Whale Rock watersheds.

†Also known as Santa Margarita Reservoir.

‡Wells under the direct influence of surface water.

ranged from 930 to 35,000 ha, from rural and sparsely populated to urban and suburban. Selected characteristics are shown in Table 4.1.

The California Surface Water Treatment Regulation requires domestic water supplies using surface sources to conduct sanitary surveys of their watersheds every 5 years, beginning in 1996. These surveys are designed to be the basis of watershed management plans. However, the underlying purpose is to have local water supply agencies think in a broader and more holistic fashion about activities and practices that occur in watersheds that could affect quantity and quality of their water supplies. One result of the survey is the formulation of recommended best management practices (BMPs) that are the basis for watershed management plans.

Specific guidelines for conduct of these surveys have been prepared by the American Water Works Association, California-Nevada Section, Source Water Quality Committee (American Water Works Association, 1993). Much of the material in this chapter is drawn from the authors' practical experience in using these guidelines to prepare watershed surveys and management plans.

The watershed sanitary survey consists of the following five major steps:

1. Describe the existing watershed and water supply system.
2. Document potential contaminant sources in the watershed.
3. Describe the existing watershed control and management practices.
4. Evaluate existing water quality and ability to meet applicable regulations.
5. Recommend measures for watershed protection.

The first three steps involve an intensive effort of data collection. The fourth and fifth steps involve mainly data analyses. Each of these steps is discussed below.

Existing Watershed and Water Supply System

Watershed

To provide a baseline for subsequent evaluation, the existing watershed is described in terms of:

- Land uses
- Population
- Natural physical characteristics such as topography, soil types, underlying geology
- Biological characteristics such as vegetation, habitat, and wildlife
- Hydrological characteristics such as climate, precipitation, and stream flow

Of special interest are existing and proposed land uses and land ownership within the watershed, i.e., private or public. Another major concern is the areal extent (if any) of the land that the water supply agency owns (and therefore has some control over) in the watershed.

Much of this information was obtained from existing publications such as local general plans, environmental impact assessments, maps, and studies by other agencies. Such publications are usually available at local libraries or agencies and are public domain, so access is usually not an issue. Furthermore, in the San Luis Obispo case, the library at California Polytechnic State University is a depository for many documents (which would otherwise be unavailable since they are out of print or had limited distribution). Some field work was necessary to verify or update data.

While a great deal of time was spent collecting data, a few shortcuts were developed. In California, all local governments are required to prepare General Plans to include a land-use element; consequently, much of the descriptive material was available or referenced in these General Plans. If the watershed is urbanized, Environmental Impact Reports or Statements (EIR/EIS) for proposed projects or developments are also a worthwhile source of data. In less developed watersheds, there may be studies by environmental resource government agencies, e.g., the U.S. Forest Service (USFS) prepared as part of efforts to preserve the area as open space or recreational land. In one case (Chorro), a watershed was located above a state prison operated by the California Department of Corrections, but most of the land was owned by the California National Guard. These agencies had prepared inventories and studies of the watershed lands and were interested in keeping the land undeveloped for their own purposes.

Water supply system

The description of the existing water supply system should include sources, facilities, and emergency plans. The sources can be groundwater as well as surface water, since some groundwater may be considered under the direct influence of surface water as determined by the regulatory agency. Documenting the rights and entitlements to that water is also useful in determining the potential future use of the water, especially if the agency is not using its full entitlement. Water supply facilities include reservoirs, dams, intake structures, pumps, pipelines, recharge basins, chlorination stations, and treatment plants. Practices that may affect water quality should also be documented, such as backwash at existing potable water treatment plants. Emergency plans for accidents or disasters (such as earthquakes or floods), both natural and human-caused, should be described. Most agencies have contingency plans to deal with events that could result in contamination of water supplies. While the water agency should have these data available, supplementary information may need to be obtained from state agencies that handle water rights or supplies. In California, these agencies include the Department of Health Services (inspection reports),

Department of Water Resources (facilities data), and State Water Resources Control Board (water rights and entitlements).

Potential Contaminant Sources

The most critical part of the watershed sanitary survey is to determine the potential for future contamination or degradation of the water supply. The following sources of potential degradation are the most obvious and should be documented.

- *Regular wastewater discharges.* Major municipal and industrial discharges have permits issued through the National Pollutant Discharge Elimination System (NPDES); in California, these are also called waste discharge requirements (WDRs). Each WDR typically has specific quality standards that cannot be exceeded; often these standards are tied to water supply, i.e., a maximum increment over the quality in the water supply. Other sources of contamination, such as from septic tanks, leach fields, or seepage pits, may not have specific WDRs but must conform to specific construction standards so that wastewater is treated properly.
- *Nonregular discharges.* These include spills from such sources as leaky sewers; malfunctions at pumping stations or treatment plants; failed septic tanks, leach fields, or leach pits; leaking storage tanks; and traffic accidents. State or local government agencies typically have records of these incidents.
- *Nonpoint sources.* These include discharges from large areas that may contain contaminants, such as urban and industrial development, and agricultural land (crops and livestock) and logging. Such activities have the potential of delivering rainfall- or runoff-derived contaminants into rivers and lakes. These activities seldom have WDRs, except that some may be covered under a general NPDES stormwater permit. In the case of urban runoff, the types of land use should be described, especially the uses of impervious areas. Agricultural data should include types of crops grown; pesticides used; and number, type, and density of animals grazing or kept in intensive-use areas (such as feedlots or zoos). Such data may be obtained from state or local agricultural departments or agents. Runoff from irrigation using reclaimed or recycled water is also a potential nonpoint source of contamination.
- *Other.* Other potential sources of watershed contamination include inadvertent or nonauthorized activities such as drainage from aban-

doned mines, fire management practices, erosion, landslides and other geologic hazards, and uncontrolled dumping. Recreation is another potential source of contamination and includes such activities as picnicking, boating, horseback riding, and use of off-road vehicles. State or local regulatory agencies should have data on these activities. Recreational activities can be documented from locally available data and should include park areas, allowed and restricted uses, list of restroom facilities, and annual number of visitors. The U.S. Geological Survey, state geology agencies, Environmental Impact Reports/Studies, or geotechnical reports often have information on geological hazards.

Once the potential contaminant sources have been identified and enumerated, there should be an evaluation of their current and future significance in the watershed. If the watershed is entirely rural, sparsely populated, and likely to remain so, then sources such as urban runoff and watershed discharges are probably not important contaminant sources. If this watershed is mostly agricultural, pesticides and/or grazing and feedlot activities (resulting in nutrients or pathogens) may be of greater significance. If the area is subject to intensive growth and development, urban sources will have greater importance. The potential impacts of future population growth and economic development should be assessed based on review of general plans and developers' plans (EIR/EIS or other documents), and information developed by federal agency landowners, such as the USFS and Bureau of Land Management.

Watershed Control and Management Practices

Once the watershed has been described and the potential contamination and degradation sources determined, the next step is to compile the *existing* control and management practices for dealing with water quality maintenance. Such practices can be implemented by the water agency itself and by other agencies. The degree of control that a water agency has over a watershed, and ultimately water quality, depends on the extent of ownership or jurisdiction over watershed lands. The most critical case occurs when watershed land is owned by other agencies or private parties so that the water agency has no direct jurisdiction over the watershed land. In the case of the four San Luis Obispo watersheds, landowners and other agencies not directly involved with water supply and distribution seemed to perform their roles and handle their responsibilities with the understanding that runoff and stream flow in these watersheds eventually result in potable water supply. This is accomplished by land-use policies and standards, permits, regulations, or lease arrangements. Table 4.2 illustrates the range of agencies directly and indirectly involved with watershed management. Specific methods of watershed management are presented later in this chapter.

In some cases, a watershed management plan may already exist. If so, that plan should be referenced and summarized. A more critical (and more likely) case is that no such plan exists, but there may be policies and ordinances that have the same effect as elements of a watershed management plan. Some examples are:

- Local general plan policies precluding urban and industrial development that could result in potential water contamination. These include special land-use designations such as *sensitive resource area, geological study area, flood hazard area,* or *open space.*
- Local general plan standards and zoning and subdivision ordinances regulating land use and development, such as minimum developable parcel size, maximum animal density, and specific construction standards for waste disposal.

If federal land is involved, such as a national forest, then the federal agency will probably have some type of management plan. In San Luis Obispo County, three of the watersheds were partially in the Los Padres National Forest, for which the U.S. Forest Service prepared a land and resource management plan (U.S. Department of Agriculture, 1993). This plan had policies on recreation, mining, erosion control, and fire management which impacted existing water quality; and identified specific management practices for use and protection of forest resources, one of which was to maintain wilderness values and improve water quality.

In addition to land use–related policies, there may be permit procedures that implement policies, laws, or regulations which directly affect water quality. These include waste discharge permits issued under NPDES or other WDRs. Typically, these permits are developed to protect, preserve, or maintain water quality in a certain body of water, which may include a surface water or groundwater supply source. The permits may be for point or nonpoint sources, as previously discussed.

In the Whale Rock and Chorro watersheds, the land-owning agency leased land to ranchers. This represented a degree of control over how land is used in relation to watershed water quality. Thus, all leases should be reviewed for any stipulations relating to watershed management.

Table 4.3 is a summary of practices and typical methods affecting watershed water quality.

Table 4.2 Agencies Involved with Watershed Management

Watersheds	Agency/utility	Principal role/responsibility
Whale Rock	California Department of Water Resources	Owns land surrounding Whale Rock Reservoir; conducts annual inspections of dam
Whale Rock	Whale Rock Commission (WRC)*	Owns Whale Rock Dam and reservoir facilities
Whale Rock	City of San Luis Obispo	Operates Whale Rock Dam, Reservoir, and recreational facilities (as designated by the WRC)†
Whale Rock, upper Salinas	City of San Luis Obispo	Owns and operates water treatment plant
All	County of San Luis Obispo	Administers land-use planning, zoning, use permits, septic tank regulations
All	California Regional Water Quality Control Board (RWQCB)	Issues waste discharge requirements (NPDES permits) and guidelines for waste disposal
All	U.S. Forest Service	Administers land use in Los Padres National Forest, including management practices for recreation, erosion control, fire
Lower Salinas	City of Atascadero	Administers land-use planning, zoning, use permits
Chorro	California Department of Corrections (Men's Colony)	Owns and operates water treatment plant, patrols accessible portions of watershed
Chorro	California National Guard	Owns land, administers leases, conducts routine inspections
Upper Salinas	County of San Luis Obispo Parks and Open Space Division	Administers public recreation areas and cattle leases
Upper Salinas	U.S. Army Corps of Engineers	Owns land which is administered by San Luis Obispo County

*Joint powers agreement among the city of San Luis Obispo, California Men's Colony, and California State Polytechnic University.

†Includes water quality monitoring and access control.

Table 4.3 Summary of Watershed Control and Management Practices

Practice	Responsible agency or organization	Typical methods
Inspection/surveillance	Reservoir owner,* landowners*	Boat, vehicle trips
Land-use planning and review	Local government	General plans, zoning and subdivision ordinances
Land ownership and right-of-way	Local government	
Access control	Landowners*	Fence, gates, signs
Septic tank regulations	Local government, California Regional Water Quality Control Board (RWQCB)	Construction standards, policies, discharge permits
Stormwater runoff	RWQCB	Discharge permits
Grazing practices	Landowners*	Leases, regulations
Wildlife management	Landowners*	Policies, regulations
Pesticide and herbicide applications	State, federal governments	Policies, regulations
Domestic animal use and control	Local government	Ordinances, regulations
Mining and mine runoff	RWQCB	Discharge permit
Forest management and logging practices	California Department of Forestry (CDF), USFS	Permits, leases
Erosion control/soil management	Local government	Ordinances, regulations
Road maintenance and winter salt storage	Local government, landowners*	Policies, procedures
Off-road-vehicle use	Landowners*	Regulations, policies
Recreation	Landowners*	Policies, regulations
Reservoir use restrictions	Landowners*	Policies, regulations
Water quality monitoring	RWQCB, discharger	Monitoring program in discharge permit
Vegetation management	Landowners*	Policies, regulations
Riparian management	Landowners*	Policies, regulations
Wetland management	Landowners*	Policies, regulations
Public education and relations	Landowners*	Brochures, posters, advertising
Wastewater discharges	RWQCB	Discharge permit
Fire management	Local fire departments, CDF, USFS	Policies, procedures
Open-space policies	Local government	General plan, ordinances

*Could be a private party or a government agency such as USFS or San Luis Obispo County.

Since many agencies and water utilities can be involved in watershed management, there is always a need for coordination measures. The water supply agency or utility is in an excellent position to facilitate this coordination, since it has the ultimate responsibility of protecting watershed water quality for its customers. The water agency's or utility's role as a facilitator is especially important if the watershed lands lie outside its ownership or direct jurisdiction.

An interagency coordinating committee is sometimes suggested, but this may not be necessary since all agencies and parties are given opportunities to comment and judge developments or other proposals that may affect watershed water quality through public meetings or hearings for certifying EIR/EIS, adopting ordinances and regulations, or issuing of permits.

Existing Water Quality

The "bottom line" of any watershed management plan is the assurance of water quality. Consequently, the sanitary survey must contain a discussion of applicable water quality regulations and the ability of the water agency or utility to meet them. The first step is to determine the applicable regulations, either federal (Safe Drinking Water Act) or state and local [in California, Surface Water Treatment Rule (SWTR)]. In California, water suppliers must conform to the SWTR, which has the following key elements:

- Provide multibarrier treatment to reliably ensure 99.9 percent (3 log) removal and inactivation of *Giardia lamblia* cysts and 99.99 percent (4 log) removal and inactivation of enteric viruses, i.e., filtration and disinfection.
- Implement monitoring program to demonstrate compliance with filtration and disinfection requirements.
- Submit technical report on how new filtration and disinfection treatment facilities will be designed to comply with SWTR.
- Operate a potable water treatment plant with a state-approved operations plan and employ state-certified staff.
- Meet notification requirements for exceedances of turbidity and disinfectant residual concentrations and monthly reporting requirements.

Additionally, there are maximum contaminant levels (MCLs) and maximum contaminant level goals (MCLGs) in both state and federal regulations.

The second step is to compile the agency's raw water quality data and compare the data to the established MCLs and MCLGs. This is best done in a series of tables or charts to summarize temporal and seasonal variations in different parts of the watershed, e.g., upstream to downstream related to land use or specific contaminant sources such as discharges. If there are any gaps in data, a revised monitoring program may be in order.

The third step is to determine—from the first two steps—the specific chemical constituents of potential concern to the agency. These constituents are typically ones that historically may have been exceeded or ones that may not meet future MCLs because of anticipated future changes in the watershed. In some cases, there may be constituents of concern which have no health implications, such as hardness and taste or odor issues.

The fourth step is to evaluate the ability of the agency or utility to meet all relevant water quality regulations. This evaluation should be based on the following factors:

- Results of the analyses of treated water quality
- Design and operation of existing potable water treatment plants, recent improvements, and future anticipated improvements
- Controls (or lack thereof) over current and future sources of pathogens and other contamination in the watershed

If there are any potential problems in meeting these regulations, a program for compliance should be recommended. The program may consist of the following elements:

- New facilities
- New or revised monitoring programs
- Special monitoring studies for specific contaminant sources
- New or revised operations plans or emergency and contingency plans

Conclusions and Recommendations

Integration of the previously described components serves as the foundation of the conclusions and recommendations of the sanitary survey as well as the first step toward developing a complete watershed management plan. This section includes a compilation of findings on the existing watershed and water supply system, existing and potential contaminant sources in the watershed, existing control and management practices in the watershed, and existing and potential water quality control issues. For each specific problem or issue, there should be a recommended solution or plan of action. Some typical examples are shown in Table 4.4.

Additionally, in the resultant watershed management plan, the agency or utility could prepare and distribute educational materials for landowners and residents in the watershed. These materials typically include:

- Emergency telephone numbers for reporting spills, illegal dumping, or unauthorized use of pesticides or fertilizers

Table 4.4 Examples of Recommendations

Issue or problem	Constituent of concern	Solution or plan of action
Cattle having direct access to a lake or reservoir for drinking or bathing	*Giardia lamblia* and enteric viruses	Fencing or modification of grazing practices
Horse stables, corrals, or feedlots near a river or lake	Nitrates, phosphorus, pathogens	Drainage and runoff control measures; fencing
Eroding sediments from earthwork	Turbidity	Drainage and erosion control measures
Abandoned mine drainage	Metals, organics	Risk assessment and remediation and closure plans
Numerous agencies with authority over land use in the watershed		Interjurisdictional committee, comments in EIR/EIS or permit process
Significant urban and suburban growth and development in the watershed		Comments in EIR/EIS or permit process
Widespread septic tank and leach field failure (nitrates, pathogens)	Nitrates, pathogens	Inspection program; discharge requirements; sewers and treatment plant
Other contaminants at certain times in urban areas	Metals, pathogens	Street-sweeping program; increased monitoring and/or surveillance

- Instructions on best practices for lawn, gardening, and agricultural activities
- Locations of authorized landfills or dropoff points for hazardous materials
- Appropriate waste disposal methods, including human waste in recreational areas
- Runoff controls for horse paddocks and other land uses which could contaminate lakes and rivers
- Techniques of vehicle maintenance and leak prevention

References

American Water Works Association. 1993. California-Nevada Section, Source Water Quality Committee, *Watershed Sanitary Survey Guidance Manual.*

U.S. Department of Agriculture. 1993. U.S. Forest Service. Pacific Southwest Region. *Land and Resource Management Plan, Los Padres National Forest.*

5

Facilitating Natural Resource Dialogues as the Foundation for Watershed Approaches

Staci J. Pratt
Public Policy Research Coordinator
Texas Institute for Applied Environmental Research
Tarleton State University, Stephenville

Jan M. McNitt
Public Policy Analyst
Texas Institute for Applied Environmental Research
Tarleton State University, Stephenville

Historical Background

When Congress passed the federal Clean Water Act Amendments in 1972, it focused on the pollution stemming from *point sources*, as natural targets in a command and control style regulatory regime.[1] The amendments required that point sources achieve maximum *effluent limitations* as well as comply with acceptable water quality standards. These amendments also established the National Pollutant Discharge Elimination System (NPDES) permitting scheme as a means of enforcing effluent limitations. The NPDES approach targeted "the most visible and easily regulated causes of water pollution."[2] This command and control system has achieved significant progress in regulating point source pollution, such as heavy industry and other discrete pollution sources, for over 20 years.[3]

Nonetheless, recent water quality surveys still indicate areas of concern, particularly with respect to agricultural pollution and nonpoint sources. As reported in the 1994 *National Water Quality Inventory*, many water quality and habitat degradation issues still present a challenge. Forty percent of the nation's assessed waters are too polluted for basic uses, such as fishing and swimming. Agriculture is identified as the leading source of impairment to U.S. rivers and lakes, affecting 60 percent of impaired river miles and 50 percent of impaired lake acres.[4] Addressing these remaining sources demands that policymakers initiate new approaches for handling water pollution. Essentially, the problem stems from the fact that diffuse sources do not lend themselves to command-and-control style of oversight. Instead of attending to the effluent discharged by obvious sources with discrete pipes, regulators face endless miles of fields and cropland dispersing runoff during random rainfall events. As Dana Rasmussen, regional administrator for region 10 of the Environmental Protection Agency (EPA), observed,

> [t]he widespread problem of nonpoint source pollution—runoff and deposition of air pollution to land and water...underscores the limits of effective enforcement. Our society does not have the resources to police each citizen's behavior and lifestyle in order to prevent or punish our polluting habits.[5]

New Approaches

Water management and stakeholder participation

Watershed approaches may address many interrelated factors affecting water quality. As John Burt at the U.S. Department of Agriculture (USDA) emphasized,

> Watersheds usually encompass multiple land uses, soils, geology, and biota, and receive both point and nonpoint source pollution. The watershed approach to controlling agricultural pollution is an example of ecosystem management which coordinates the conservation of water, air, plants, and animals, as well as their interactions. The watershed approach holds the most promise for agriculture, solicit-

> ing local input in developing corrective actions. Farmers, ranchers, and rural communities would be encouraged to participate in the planning process. And their participation could help them understand the water quality concerns, sources of water pollution, and the corrective actions needed.[6]

From the scientific perspective, a watershed is defined as the drainage area of a stream.[7] The implications of this fact for human activities regarding water have, however, existed in seeming independence from various efforts to manage and protect water resources. The control, management, and study of water have, until recently, occurred within a number of distinct disciplines working in relative isolation from one another. So organized, these programs have facilitated basic scientific research, offered simple administration and management, and produced some good results. They have also, for the last 20 years, fit in well with the general direction of federal water quality laws. Despite these successes, it has become increasingly evident that we have addressed the more visible and easily remedied problems and now are left to face the radical, structural problems which generate continued poor water quality nationally. Those working directly with water quality have come to believe that completing the work already begun will require broader, more integrated strategies to achieve long-term, sustainable success. Growing recognition of the need for a more holistic perspective for water resources underlies increasing calls for adopting "watershed approaches."

As commonly used, the concept of a watershed approach seems to involve a loosely linked set of evolving perspectives and methodologies; in this sense there is no single approach, but rather there are a number of different programs based around the central principle that water quality improvement and protection can best be achieved if all actors use the geohydrological unit of watersheds at their basic orienting framework. Beyond using watersheds as a starting point, there is not yet consensus on what the essential components should be for a watershed approach. Identification of the essential components and their integration into overall programs are vital to developing viable new approaches for addressing the next generation of water quality issues.

The EPA has recognized the need for new frameworks capable of addressing land-use management issues over vast areas. To handle remaining pollution issues and build upon progress achieved thus far, the EPA has promoted a watershed approach for water quality management. Generally, EPA views the watershed approach as "a coordinating framework for environmental management that focuses public and private sector efforts to address the highest priority water-related problems within geographic areas, taking into consideration both surface and ground water flow."[8] EPA has articulated its evolving vision, in *Watershed Approach Framework—1996,* as "people working together to achieve clean and safe water and healthy aquatic ecosystems through comprehensive management approaches tailored to the needs within watersheds."[9] The involvement of people, then, constitutes an essential element for any watershed approach. As stated in *Framework*:

> Broad involvement is critical; the watershed approach relies on community-based environmental protection. In many cases, the solutions to water resource problems depend on voluntary actions on the part of many people. Besides improving coordination among their own agencies, the watershed approach calls upon states...to fully engage local government and other affected parties in the watershed management process to help them better understand the problems, identify goals, select priorities, and choose and implement solutions.[10]

For these reasons, EPA encourages the formation of watershed management teams and partnerships. Only through watershed teams may the next frontier of nonpoint sources receive adequate treatment.

The addition of local perspectives to watershed programs has important implications for program design and implementation. The foundation of the watershed approach articulated above, particularly regarding nonpoint source pollution, represents a bottom-up perspective grounded in local activities, oversight, and control. Successful integration of local perspectives with national and regional programs may present the most difficult and important challenge to those implementing watershed strategies.

The North Bosque River watershed: A case study in the application of the watershed approach

Growing recognition of the role agriculture plays in water pollution has increased attention on agricultural livestock operations. Reports of severe water pollution due to livestock operations coincide with a distinct national trend in all livestock production sectors toward expanding operations which concentrate large numbers of animals on limited acreage. These concentrated animal feeding operations, commonly known as CAFOs,[11] are now prevalent in all major livestock production sectors. For example, in the United States in 1980 only 1 percent of all cattle feedlots handled 32,000 head, yet by 1992 nearly one-third of all these feedlots contained at least 32,000 head.[12] While the number of hog producers nationally declined from 670,000 in 1980 to 256,000 by 1992, the total number of hogs increased in that period with nearly 80 percent of hogs being raised on operations that housed 1000 animals or more.[13] This trend also applies to the dairy industry; the number of dairies with over 500 head nearly doubled between 1974 and 1987.[14]

On a more local level, the strength of the dairy industry in Erath County, Texas (see Fig. 5.1), the number one milk-producing county in the state since 1989,[15] draws attention to agriculture's potential influence on water quality. From 1986 to 1992 the number of dairy cows housed in Erath County rose from 31,500 to 70,900, an increase of almost 125 percent.[16] In 1989, the Texas State Soil and Water Conservation Board (TSSWCB) set up a specific nonpoint source management project for the North Bosque River, located within Erath County, recognizing it as a "known water quality problem" where "the category of agricultural nonpoint source activity of concern is dairy operations."[17] By the following year, state of Texas water quality assessments, conducted by the Texas Water Commission,[18] and TSSWCB, identified the North Bosque River as a known problem watershed as the result of dairy waste.[19] The

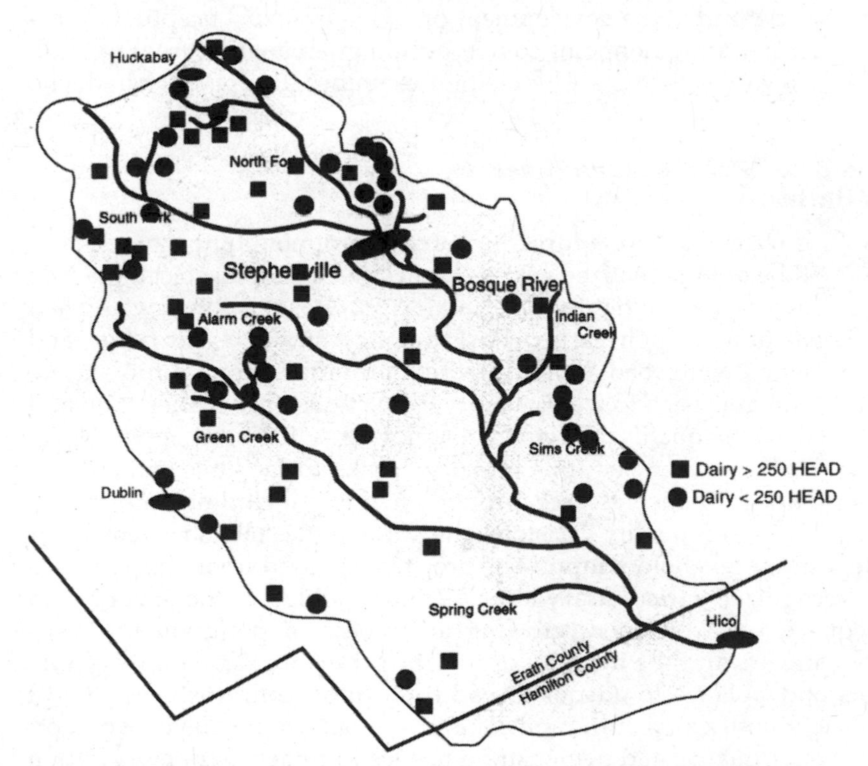

Figure 5.1 Dairy locations in the upper North Bosque River watershed.

majority of these dairy operations were located near the headwaters of the North Bosque River, within a drainage area referred to as the upper North Bosque River watershed, an area encompassing 230,000 acres (395 mi^2).[20] In June 1989, a subcommittee of the U.S. House of Representatives Committee on Agriculture held hearings in Erath County to investigate reported pollution attributed to local dairy operations.[21] Late in 1989, the Texas Water Commission levied fines on six dairy operators in the area, ranging from $23,000 to $75,000 for illegal waste dumping.[22] These developments highlighted the need for a comprehensive approach for effectively addressing the problem and to channel energy and concerns of local individuals impacted by changing conditions and events.

These events led to the participation of the Texas Institute for Applied Environmental Research (TIAER), and the EPA-sponsored project *Livestock and the Environment, a National Pilot Project* (NPP). The goal of the NPP has been focused upon determining technologies, management methods, policies, and institutional settings that can reduce the negative environmental impacts of livestock production while promoting the economic viability of the livestock industry. TIAER's work pursuant to the NPP has permitted the development of an integrated conceptual framework for handling nonpoint source pollution emanating from agriculture on a watershed basis. This chapter examines the approach in detail.

The Lake Waco/Bosque River initiative

TIAER's framework for addressing agricultural nonpoint source pollution will be operationalized as part of a USDA-funded project, the *Lake Waco/Bosque River Initiative.* The Lake Waco/Bosque River watershed initiative builds upon earlier TIAER policy strategies to refine and implement a watershed approach which provides a community-based, bottom-up approach to water quality protection that satisfies state and national water quality objectives. Operating within the upper North Bosque River watershed, this initiative seeks to facilitate stakeholder participation in the research process; and by including individuals directly in water quality decisions, the study will establish a vehicle for integrating stakeholder input. The program is based upon the existence of a recognized agricultural water pollution problem in the upper North Bosque River watershed, growing awareness of pollution problems downstream, and the inability of traditional institutional delivery systems and policies to adequately address these concerns. The project aims to identify a scientific and community-determined level of acceptable water quality and demonstrate reduced impacts in demonstration microwatersheds. This strategy will be refined through operationalizing TIAER's conceptual approach that integrates analysis of institutional, scientific, economic, and stakeholder factors. The project comprises of a number of closely coordinated activities that will be undertaken by state and federal agencies and entities. The conceptual framework is presented in the outline form below.

Water quality assessment. Water quality monitoring has been conducted within portions of the watershed by several agencies for some time. In support of the initiative, several new tasks will be undertaken. First, representatives from all entities involved with water quality–related studies within the area will form a technical work group to coordinate data collection and to evaluate long- and short-term data needs. Second, monitoring activity will be expanded, incorporating portions of some existing sampling programs and adding new monitoring sites to evaluate other potential pollution sources. Added sites will include Lake Waco, a flood control reservoir located at the base of the watershed which serves as the sole source of drinking water for the city of Waco, with a population of approximately 100,000 people.[23] Monitoring results will provide initial indications of actual pollution loading rates at different points throughout the watershed. Third, existing biophysical process models will be improved and adapted to allow comprehensive evaluation of the entire watershed. The goal of monitoring activities is to examine and evaluate the existence and role of nutrient loads, the pollutants of concern from agricultural runoff, extending work already completed in the upper watershed, and to create a comprehensive picture of nutrient runoff within the watershed as a whole.

Stakeholder input. The participation of watershed stakeholders is crucial for achieving and maintaining desired water quality in the watershed. Through the constituency committee and microwatershed council processes, individuals representing all true stakeholder groups from the entire watershed will provide input into project tasks. A constituency committee has been formed and has begun participating in the initiative.

Development of in-stream measures of success. Through earlier work, project personnel have developed the capacity to target areas producing excessive nonpoint source pollution loads. To gauge overall project success and to facilitate a watershed plan, efforts are required to develop information that can form the basis for nonregulatory, nutrient in-stream measures of success. This work will include analysis of existing biological water quality indicators in pertinent points of the watershed, input from stakeholders, and a tentative assimilative capacity evaluation for Lake Waco. Targets will evolve in a stepwise progression, proceeding from initial efforts to obtain simple reductions in relevant nutrient loadings, to percentage reductions, and ultimately to in-stream targets.

Microwatershed consortia. In microwatersheds targeted as producing excessive pollutant loads, local conservation districts will organize local landowners for collective discussion of pollution problems and solutions for reducing loads. The group will participate in prioritizing pollution abatement activities and informing development of a microwatershed plan to reduce loadings. Several critical activities surround microwatershed organization: (1) selection of two impaired microwatersheds, by local conservation districts, for project study; (2) definition of a list of pollution issues; (3) creation of a list of solutions to reduce pollutant loads; (4) discussion to achieve agreement regarding jurisdiction and oversight of manure application fields; (5) development of interim water quality goals for selected microwatersheds; and (6) development of a microwatershed plan to reduce loadings within target areas.

Development and implementation of management measures. Based on the microwatershed plan, individual landowners will develop and implement management measures to reduce runoff from identified priority areas. Facilitation of farm-level plans to implement reductions will be advanced by local conservation districts and the USDA Natural Resource Conservation Service (NRCS) through three phases. In phase 1, plans will be developed for the six highest-priority landowners. Under phase 2, initial BMPs will be implemented for the phase 1 landowners. During phase 3, plan development will continue for lower-priority landowners.

The USDA project will also be advanced through ongoing project management and coordination established at the outset of work. All watershed inhabitants will be offered an education program to explain nonpoint source pollution and steps to reduce pollutant loads. Supporting agencies and entities will provide economic analysis to assist in evaluating a variety of options for addressing loadings. They will also provide vital information to stakeholders and agencies on public sector costs and funding mechanisms. Water quality monitoring will continue to evaluate program success. At the final stage, project personnel will assemble materials for the application of program techniques, for dissemination to other areas and problems. As envisioned, this approach will be operationalized on a pilot basis and then evaluated for merit in improving water quality and potential value as a national model.

Scientific underpinnings for the

conceptual framework

Development of TIAER's conceptual framework has been supported and informed by significant scientific and economic research and information. This research was guided by the goal of obtaining a thorough understanding of the nature of the pollution problems associated with this area of the river basin. Through a number of studies, TIAER established a water quality monitoring system for examining the watershed. The study area contains a diversity of land uses with dairying as the major agricultural enterprise; 94 dairies with a combined milking herd size of approximately 34,000 cows reside within the watershed.

Due to the presence of numerous dairies, nutrient pollution of surface waters has provided the primary water quality monitoring focus.[24] Project personnel monitor water quality at approximately 19 stream sites, during storm events and at regular intervals, and monitor nutrient runoff from manure application fields to calibrate biophysical process models. Sampling sites reflecting diverse land uses were selected to monitor influences of agricultural practices and to distinguish these impacts from traditional urban and other rural sources of pollutant loadings. Land use was identified by a geographic information system, which includes detailed land-use and soil profile information. The concentrations of various water quality constituents measured through the monitoring system presented an image of pollution problems that could be correlated with land use in areas upstream of the sites.

Comparisons of TIAER monitoring data with land uses in the watershed indicate strong correlations between the percentage of land area used for waste applications fields in the drainage basin above individual monitoring sites and the concentration of in-stream phosphorus constituents. Similar relationships were found based on dairy cow density. These significant positive correlations indicate that as the percentage of land used for waste application (or dairy cow density) in a drainage basin increases, the concentration of phosphorus in stormwater runoff also increases. (See Figs. 5.2 and 5.3.)

Because few other watersheds have the resources to implement similarly intensive monitoring efforts, TIAER and its research partners—the Center for Agricultural and Rural Development (CARD), Iowa State University, and the Blacklands Research Center (BRC), Texas A&M University System—have entered into development of an integrated modeling framework for evaluating the impacts of environmental, economic, and policy decisions on water quality. The framework employs economic and biophysical process models, calibrated with monitoring data and validated through research. Economic models examine farm-level and regional-level economic impacts and evaluate different envi-

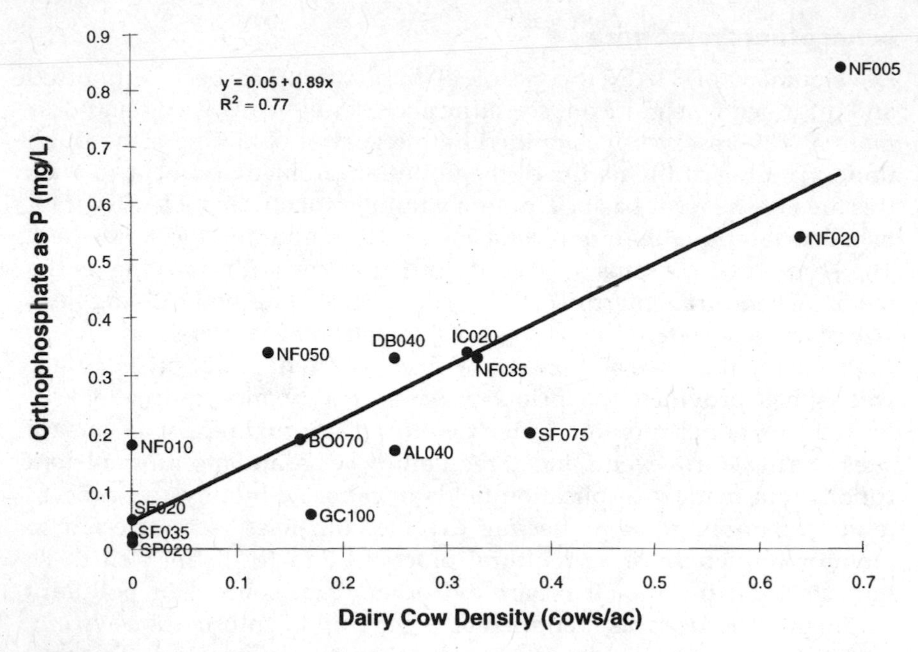

Figure 5.2 Comparison of cow density to water quality at stormwater monitoring sites, sample period November 1992–March 1994.

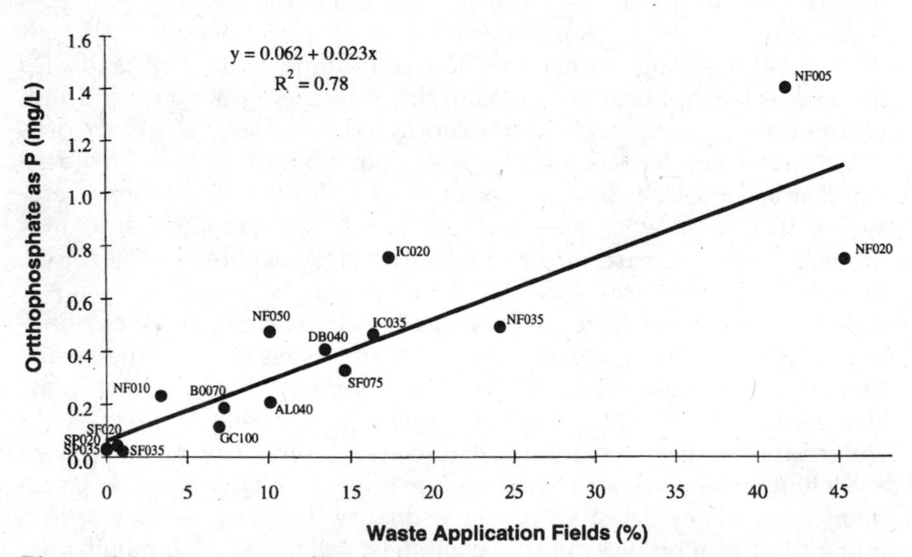

Figure 5.3 Comparison of land occupied by waste application fields to water quality, sample period November 1922–March 1994.

ronmental policy options to simulate economic impacts on producers. Biophysical process models, such as SWAT, are used to examine field- and watershed-level impacts. These models receive output from the farm-level model and predict the fate and transport of nutrients and sediment on the farm and throughout the microwatershed and greater watershed. Biophysical process models operating at the field, multifield, and watershed levels allow simulation of many subbasins and routing of surface runoff between areas with differing soils, slopes, crops, and management strategies. Results can be aggregated to evaluate water quality given all sources within a set area. Information from individual models is incorporated into the modeling framework to analyze alternative policies requiring implementation of specific management practices. A specific policy scenario, utilizing a particular best management practice, can be introduced into the modeling framework for evaluation. In this manner evaluations can be made of each alternative policy scenario and results obtained on the cost of proposed policies versus expected environmental benefits. The integrated modeling framework can assist in developing watershed plans by providing information to stakeholders and decision makers for evaluating alternative strategies, and when fully operational, they can be used in different locales as tools for evaluating the water quality impacts of different scenarios.

Microwatersheds as the foundation[25]

Before implementing a watershed approach, policymakers must identify the necessary building blocks for action and the means for grounding watershed approaches at the local level. Microwatersheds provide the natural focal point in this developmental process. They provide a convenient forum for local participation in setting priorities and determining treatment measures, as well as a discrete area for handling water-related issues. Essentially the term microwatershed refers to an area, within a watershed, that has identifiable hydrologic boundaries and is sufficiently small and discrete to allow for targeting of limited resources, manageable analysis, and natural resource problem amelioration. As used herein, the term is intentionally flexible to respond to differences in the size of watersheds, number of landowners, types of soils, number of involved water bodies, and number of other interested actors. Within the context of agricultural nonpoint source pollution, microwatersheds represent a natural stepping stone in the effort to aggregate individual farm plans and achieve larger policy objectives.

The microwatershed approach assumes that the best way to identify and control sources of polluted runoff is to subdivide watersheds into targeted microwatersheds. Because individual sources of nonpoint source pollution are difficult and costly to isolate and control, a watershed approach to water pollution prevention and abatement represents a promising methodology for addressing landscape issues. A watershed perspective views all nonpoint sources of pollution within the basin as part of a holistic hydrologic system, and broader ecosystem, producing overall water quality. Obtaining a handle on the source of pollution problems within a larger watershed, however, requires narrowing one's inquiry to smaller, more manageable units, which contain relatively homogeneous uses. In this way, one may discern the most likely actors in pollution loading. It will then become possible to develop an effective plan for action. Stream monitoring can identify water quality degradation and specific sections of the watershed can be targeted for amelioration. Large watersheds can be broken down into smaller subwatersheds and even smaller microwatersheds. Water quality at the confluence of watersheds, subwatersheds, and microwatersheds can be measured to identify the location of suspected pollution sources upstream. Within the upper North Bosque River watershed (see Fig. 5.4), tributaries representing the largest sources of pollutant loading were identified through water quality sampling and physical process modeling. Land uses within this area

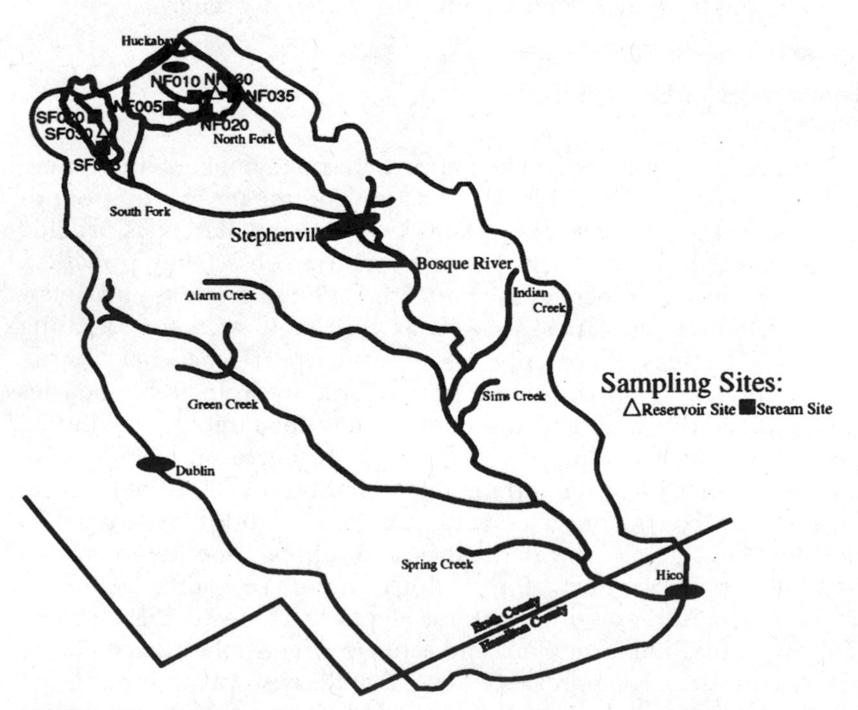

Figure 5.4 Selected microwatersheds in the North Bosque River watershed.

were determined[26] and correlated to water quality monitoring data to draw an image of pollution problems in areas upstream of monitoring sites. From examining stormwater runoff, the dairy industry emerged as the major contributor to nutrient loadings, and several microwatersheds exhibited disproportionate impacts.[27] Those microwatersheds hosting the largest number of waste application fields consistently showed high nutrient loadings. These sites become obvious candidates for pollution abatement efforts.

In choosing target microwatersheds, several principles emerge. First, target watersheds must be sufficiently small to allow for the interaction of local stakeholders, the identification of pollution sources, and the development of appropriate criteria for measuring successful pollution prevention and abatement. Second, the economic impact on both the regulated community and program implementation agencies must be reasonable in light of market forces. In this way, prioritizing pollution prevention and abatement efforts at the microwatershed level provides a method for solving water quality problems within existing government budgetary constraints. Public-sector resources are directed to those areas in the watershed that are the most severely impacted and that provide the greatest opportunity for environmental remediation. Less impacted areas can be addressed as water quality assessments indicate the need and as funding becomes available.

Applying stakeholder concepts—planned intervention and microwatershed approach

The Lake Waco/Bosque River initiative builds on the policy approach developed by TIAER, which harnesses evolving priorities to avoid the narrow confines of command-and-control regulation and to feed into the

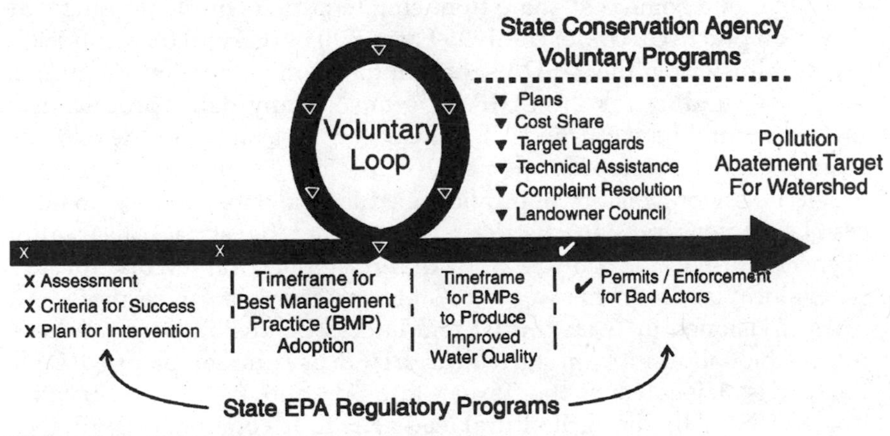

Figure 5.5 Planned intervention voluntary loop.

concept of watershed teams and partnerships. This approach, entitled the "planned intervention/microwatershed approach,"[28] has been developed within the context of addressing environmental issues that have evolved within the upper North Bosque River watershed (see Fig. 5.5). As developed, the planned intervention and microwatershed strategy is a community-based approach to environmental policy for nonpoint source water pollution, which responds to lessons from prior initiatives by (1) preempting reactive crisis decision making by developing deliberate, well-planned measures to reduce pollution and (2) avoiding the confines of command-and-control regulations, as well as poor performance of purely voluntary measures, by institutionalizing a voluntary approach with regulatory backup for bad actors.

Voluntary programs alone have failed to produce significant reductions in agricultural nonpoint source pollution.[29] EPA observes that "[t]he trend in nonpoint source control is towards voluntary approaches as the primary means for BMP implementation with regulatory or quasi-regulatory approaches as backup where voluntary approaches prove inadequate."[30]

Under the planned intervention and microwatershed approach, a state agricultural lead agency and local conservation districts serve as the primary institutions to induce voluntary behavioral change among agricultural producers targeted as significant sources of nonpoint source pollution. The existence of an extensive network of government-supported natural resource agencies distinguishes agriculture from other production sectors of the U.S. economy and provides agriculture with a unique alternative to the type of environmental regulation that has dominated to date. These natural resource agencies include NRCS, the Cooperative Extension Service, state conservation agencies, and local conservation districts. NRCS has traditionally provided technical assistance and some cost-share financing to farmers implementing conservation practices. Under the 1996 Farm Bill, NRCS will be responsible for administering the USDA's consolidated conservation programs, including funding for the EQIP program.[31] Many dairy producers in Erath County have turned to NRCS for assistance in developing required dairy waste management plans. The Cooperative Extension Service has a long history of providing educational information to farmers. Local conservation districts, coordinated by state conservation agencies, provide a county-based organizational framework through which local landowners can access state and federal farm services.

As envisioned in Texas, initial regulatory responsibility for addressing water pollution from agricultural land uses under planned intervention is delegated to the Texas State Soil and Water Conservation Board (TSSWCB), the agricultural lead agency. In conjunction with local conservation districts, TSSWCB cooperates with producers to create and implement water quality management plans and to develop corrective action plans when pollution complaints occur.[32] TSSWCB employs primarily voluntary strategies, underpinned by available cost-share funds, to motivate changes in land-use practices. When "bad actors" refuse to cooperate, TSSWCB refers them to Texas's environmental regulatory agency for enforcement action.[33] This combination of voluntary and regulatory measures provides agricultural operators the opportunity to implement necessary land-use practices to achieve environmental goals within a reasonable time frame, while ensuring that producers who do not comply are not permitted to pollute indefinitely. The adoption of management practices is voluntary only to the extent that producers choose which abatement practices to implement and, within reason, the time frame for implementation. In this manner, planned intervention simply inserts a voluntary, time-and-resources loop into the normally deadline-driven, technology-forcing regulatory process. Regulatory enforcement action is reserved for uncooperative bad actors, after attempts made to induce voluntary compliance fail. This institutional approach, which statutorily links the voluntary programs of the state's soil and water conservation board with enforcement under the Texas Natural Resource Conservation Commission, was enacted in 1993 by the Texas legislature.[34]

In summary, planned intervention envisions close cooperation between environmental regulators, soil and water conservation districts, individual producers, and other local citizens. Properly linking the pollution abatement efforts of natural resource agencies developed during the New Deal with regulatory programs allows policymakers to exploit the strengths of both. By emphasizing a deliberate, well-planned combination of voluntary measures, against a regulatory backdrop, planned intervention feeds into the concept of a watershed approach. And, through employing mechanisms such as stakeholder dialogue to harness evolving priorities, planned intervention feeds into the creation of watershed teams and partnerships.

Facilitating Local Perspectives within Watershed Approaches

Drawing lessons from coordinated resource management

Resource management issues can and often do produce very strong emotional responses among decision makers and the people living within impacted areas. This raises the question of how to ensure these individuals have productive dialogues which enable them to address solving the problem at hand. Federal agencies have gained valuable experience in integrating local perspectives into natural resource problem solving and decision making, which can profitably inform the addition of local perspectives to watershed programs. Through coordinated resource management (CRM), a concept used to resolve specific conflicts involving natural resource management, agency personnel seek to facilitate dialogues among diverse community members with differing or conflicting goals. Facilitating such dialogues requires a viable framework for encouraging cooperation. CRM establishes a level playing field for a variety of landowners residing and/or owning property within a specific area to discuss and reach consensus on the best method for framing a plan to reduce pollution loadings. The Bureau of Land Management, U.S. Forest Service, and the Cooperative Extension Service have used CRM in the management of public land for many years. "The process...has been used successfully to resolve specific resource-use conflicts, such as depredation of agricultural crops by wintering geese near waterfowl refuges and water contamination from dairy farm lagoons and livestock feedlots."[35]

Initiating a CRM effort begins at the local level. "Preferably, a coordinated plan is initiated at the local level by a request from a[n]...organization...that perceives the need for a group-action approach to resolving or averting a local resource problem. A conservation district, for example, might process a request for a coordinated plan because these districts are legal subdivisions of state government with responsibility for land and water conservation."[36] Thus, local soil and water conservation districts can trigger the utilization of CRM-style initiatives in order to formulate watershed plans.

Local conservation districts may then call together a broad range of participants, incorporating representatives of those agencies, organizations, and associations having a direct interest in the resource management issue at hand. CRM recognizes that "the whole community should work together as a team from the beginning to the end of any planning effort. This assures the greatest number of ideas, the widest range of options and the very best courses of action in the interest of the community as a whole."[37] In addition, embracing the full range of actors in a watershed increases the likelihood that planning decisions will be implemented effectively.

Watershed participants meet in a group called the *steering committee* (SC), consisting of approximately 20 management-level representatives. These representatives are decision makers who can speak for their organizations on matters dealing with watershed planning. When technical issues arise, the SC seeks out the assistance of technical review teams (TRTs). "TRTs usually function at the ranch unit, allotment, small watershed, wildlife, wilderness, or similarly sized area. They are composed of interdisciplinary technical experts, landowners, permittee, agencies and interest groups as appropriate for the subjects and issues at hand."[38] A field trip of the local area by watershed participants begins the process of introducing stakeholders to one another and breaking down barriers.

CRM forums, such as the steering committee, allow all involved parties to vent frustration and be heard. Land-use and property rights issues particularly generate very strong emotional response among watershed stakeholders. Within a proper framework, however, this group can come up with diverse, productive options for resolving conflict. A facilitator plays a crucial role in this process. The facilitator is a neutral person who provides structure to the meeting and keeps everyone involved. To promote productive dialogues, the facilitator must structure interactions to allow opportunities for overcoming communication barriers between participants. Lessons from formal mediation principles and CRM establish the framework for guiding a productive discussion. The guiding principles for such a meeting include[39]

1. *Involve the real players in the conflict.* All interested parties must be included in the planning group to ensure productive discussion and implementation of decisions. "To leave some caring interest out is to invite attack."[40]
2. *Identify the problem.* All parties must agree on the real problems and issues. Over time, this will allow hidden agendas to be brought to the surface.
3. *State expectations and objectives.* Each member of the meeting should state what she or he wants from the meeting. Many times members have shared goals, such as preserving the resource. This establishes common ground and a means for moving forward. Where desires differ, the areas of concern are clear.
4. *Analyze the problem.* All aspects of the issue should be discussed. This is the point where solutions can be minimized and options generated.
5. *Make decisions by consensus.* The committee makes all decisions, recommendations, and actions by unanimous agreement. "Any issue not receiving unanimous resolution would be sent back to a working committee for further study or would be tabled."[41] Many find the consensus rule generates trust and responsibility among watershed participants. In working on a CRM project dealing with grazing on public lands, John Weber observed, "The consensus rule has been extremely important to our success story. It has developed

confidence and trust."[42]

6. *Agreement.* All parties must agree on the best management of the resource. This results in formulation of the watershed plan.

Using natural resource building blocks for effective, community-based water quality protection

Citizen participation provides an integral part of the planned intervention and microwatershed approach and TIAER's Lake Waco/Bosque River watershed initiative. Broad citizen support is crucial for achieving and maintaining desired water quality within watersheds since all human activity potentially impacts water quality. Reflecting principles from CRM and experience developed over a 5-year period, TIAER's approach feeds into watershed concepts by employing mechanisms such as stakeholder dialogue to harness evolving principles through two basic forums—constituency committees and microwatershed councils. Citizen participation strategies refined by TIAER promote stakeholder input into the research and policy development for the Lake Waco/Bosque River watershed as a whole.

The constituency process is central to TIAER's research efforts, helping ensure that policy proposals reflect the genuine concerns of direct stakeholders.[43] This process promotes in-person communication between direct stakeholders and elected officials while allowing appropriate input from indirect stakeholders (i.e., agency personnel, professional lobbyists, etc.). Individuals residing within the watershed who have a direct stake in decisions regarding water within the area are nominated to serve on a constituency committee. The committee is chaired by one or several elected officials interested in environmental quality in the watershed and perceived as capable of promoting constructive policy change. Participation of an elected official encourages active citizen participation and connects agencies to the process through accountability. Committee members are informed on development of public programs and research efforts through several means. The committee receives information on existing and proposed research programs that involve water resources, including any ongoing water quality monitoring data. Existing government programs are also presented for committee review. Members are encouraged to give input concerning the present state of water quality and their future expectations for water use. Input regarding the techniques, goals, structure, and focus of government programs for improving water quality is also solicited. In this way, scientific research and agency programs can be made transparent and accountable to the community, and members can identify and raise issues pertinent to the community for researchers and government. Focus group meetings, surveys, and other instruments can assist peer group leaders within the committee to remain abreast of the views of the groups represented. Recommendations made through consensus are encouraged. In this manner, a forum is created for interactive, ongoing communication between research, government, and citizenry.

TIAER has worked with two watershed-based constituency committees to date. The first committee, comprised of milk producers, environmentalists, and a variety of interested local individuals, met for a 2-year period beginning in 1990, to voice its viewpoints and participate in developing alternative policies to address local water quality problems. The committee finalized its work in 1992 and made a series of recommendations to the Texas legislature's Joint Interim Committee on the Environment.[44] The group identified a need for local input into both research and regulation processes and recommended stronger government presence within the watershed. Routine and increased inspections were perceived as important to reduce complaints and avoid selective enforcement. The group believed that TSSWCB, working in concert with regulatory agencies, could initiate innovative strategies to provide incentives for voluntary compliance including funding support, educational support, and technical assistance enabling the agricultural sector to move into full compliance. Last, the committee agreed that all livestock operations should come into compliance with environmental objectives and that operators not voluntarily complying should be subject to traditional regulatory enforcement processes. This bottom-up approach calls for facilitating stakeholder participation into decisions made by researchers and policymakers for addressing water quality problems, and was passed into law by the 73d Texas legislature as Senate Bill 503.[45]

Late in 1995, TIAER began working with a new watershed-wide citizen advisory committee—the Bosque River Advisory Committee—to support the implementation of a microwatershed approach. This group of 20 watershed residents has been nominated by the four elected state representatives and one state senator whose districts overlie the watershed boundaries. These officials also act as committee cochairs. This committee includes representatives of all true stakeholder groups, e.g., dairies, municipalities, riparian landowners, and local environmental interest groups. Responding to local interest, the committee was formed prior to the commencement of microwatershed activities of the Lake Waco/Bosque River initiative. Committee members met, receiving background information and beginning discussions articulating local concerns regarding water quality. To facilitate examination of initial areas of interest, the members formed four subcommittees (1) to assess existing water quality monitoring, (2) to examine in-stream nutrient targets and define success, (3) to state policies and programs for nonpoint source water pollution, and (4) to state best management practices for nonpoint source water pollution.

On the ground level within targeted microwatersheds, local conservation districts will form microwatershed councils to operate within areas identified as presenting nonpoint source pollution problems. Once local conservation districts target priority microwatersheds, they will organize microwatershed stakeholders into groups to collectively discuss pollution problems and assist in development of solutions for reducing loads. These individuals assume responsibility for improving water quality as part of the microwatershed council. Some microwatersheds may be exclusively rural in character; others may involve rural and municipal stakeholders, thereby requiring organizational coordination among more agencies. Periodic meetings over the course of the project will allow council members to develop rapport. Peer pressure by council members, their unique knowledge of local conditions, and their direct motivation to solve perceived local problems will augment voluntary efforts by producers and local conservation districts, and enforcement by water quality regulatory agencies.

To realize viable local solutions for water quality problems, a microwatershed council must be of a size which optimizes communication while encompassing sufficient land areas to produce an effective result. For stakeholders to forge collective, local plans, they must be able to identify pollution sources, and propose and implement solutions. Thus, optimal size can and will vary depending on a number of factors including the types of pollution problems, the number and/or size of water bodies, population density, and available resources. TIAER's approach supposes that a microwatershed will represent a geographic area encompassing no more stakeholders than can realistically meet and interact as a group to resolve problems. Clearly this system favors rural areas. However, application of this definition is conceivably transferable to more populous areas by inclusion of a representative stakeholder forum where direct stakeholders, such as urban or suburban landowners, select representatives.

The goal of consortia meetings is to inform the development of a microwatershed plan that will result in observable reductions to pollution loadings and will prioritize pollution abatement activities. Support from natural resource and environmental regulatory agencies, along with information from biophysical and economic modeling, will assist council members in developing the microwatershed plan. This strategy will implement a bottom-up approach for handling pollution issues. Several groups will participate in council organization and support:

local conservation districts, TSSWCB, and NRCS. These groups, along with TIAER and the local river authority, will cooperate to conduct council meetings. Integrated modeling framework support will be provided by TIAER and its research partners.

These community-based forums offer stakeholders the opportunity to resolve their conflicting positions and voice their views. As professionals in dispute resolution services emphasize, the process of conflict resolution proceeds along well-established lines. The parties need the opportunity to ventilate their frustrations before they become willing to negotiate in earnest. "[E]ncouraging and allowing...ventilation often removes or softens psychological/emotional barriers to compromise."[46] Natural resource conflicts take time to resolve. Many such conflicts take 2 years or more to come to a point of open dialogue and productivity.[47] The opportunity exists to avoid the polarization of conflict by employing consensus-based, community strategies within watershed approaches.

Conclusions

In summary, as expressed in EPA's *Watershed Framework*, the critical ingredient for a successful watershed approach depends upon "people working together to achieve clean and safe water and healthy aquatic ecosystems through comprehensive management practices tailored to the needs within watershed."[48] By utilizing knowledge collected through the National Pilot Project and developed through the planned intervention and microwatershed approach, researchers can actively encourage the reformulation of government institutions and the participation of local citizens in developing policy. While the USDA and EPA retain ultimate responsibility for ensuring that community input informs broader natural resource goals, an important step toward devolving power to local entities can be achieved through incorporating community-based forums into the research and decision-making processes. Thoughtfully structured projects can combine the strengths of federal and state agencies with community members to create a framework capable of addressing the next generation of water quality issues through holistic, landscape-based approaches to watershed management.

References

1. Despite inclusion of section 208 in within the Clean Water Act in 1972, which mandates largely state-run areawide water quality management plans, most state programs produced few results. Pub. L. No. 92-500, p. 2, 86 Stat. 839. Refer to Larry C. Frarey, et al., Conservation districts as the foundation

for watershed-based programs to prevent and abate polluted agricultural runoff, *Hamline Law Review,* vol. 18, no. 2, Winter 1994, pp. 155–156. This eventually led to the passage of section 319 of the Clean Water Act in 1987. 33 U.S.C §1329. Section 319 requires greater efforts toward assessment of nonpoint source pollution and enhances state responsibility to develop watershed strategies to address nonpoint source pollution.

2. Larry C. Frarey, Larry Hauck, Ron Jones, and Nancy Easterling. *Livestock and the Environment: Watershed Solutions,* Texas Institute for Applied Environmental Research (TIAER), Stephenville, 1994, p. 11.
3. Peter Rogers, *America's Water: Federal Roles and Responsibilities,* Massachusetts Institute of Technology, Cambridge, 1993, p. 4.
4. Environmental Protection Agency, *National Water Quality Inventory: 1994 Report to Congress,* EPA 841-R-94-001, Washington, 1996, pp. ES 15–18.
5. Dana A. Rasmussen, Enforcement in the U.S. Environmental Protection Agency: Balancing the carrots and the sticks, *Environmental Law,* vol. 22, pp. 333, 336.
6. John Burt, Ag pollution: A new generation of rules? *Agricultural Outlook,* U.S. Dept. of Agriculture, 1994, p. 21.
7. *Dictionary of Scientific and Technical Terms,* 5th ed., McGraw-Hill, New York, 1994, p. 2153.
8. Environmental Protection Agency, *Watershed Approach Framework—1996.* Draft 11-27-95 appearing in the written materials for The Clean Water Act: New Directions, ABA Satellite Seminar, American Bar Association, p. 58.
9. Ibid., p. 57.
10. Ibid., p. 61.
11. The term *CAFO* has specific regulatory meaning under state and federal laws, but is used here in its general sense, to include all livestock operations which confine animals on limited acreage.
12. Allan Butcher, et al., Livestock and the environment: Emerging issues for the Great Plains, *Proceedings of Conservation of Great Plain Ecosystems: Current Science, Future Options,* Kansas City, MO, 1993, p. 3.
13. National Pork Producers, *1992 Pork Facts,* Des Moines, IA, 1992, p. 14.
14. Butcher, et al., p. 9.
15. United States Department of Agriculture (USDA), *Texas County Grade A Milk Marketings 1980 through 1992,* Carrollton, TX, 1993, table 1.
16. Ibid. Cow numbers were derived using a 305-day lactation period at 50 lb of milk per day.
17. Texas State Soil and Water Conservation Board (TSSWCB), *Management Program for Agricultural and Silvicultural Nonpoint Source Pollution in Texas,* Temple, TX, 1989, pp. 28, 32.
18. The Texas Water Commission merged with the Texas Air Control Board and

other agencies on Sept. 1, 1993, to form the Texas Natural Resources Conservation Commission (TNRCC). TNRCC is currently delegated regulatory enforcement authority for water pollution in Texas.

19. Texas Water Commission (TWC) and Texas State Soil and Water Conservation Board (TSSWCB), *1990 Update to the Nonpoint Source Water Pollution Assessment Report for the State of Texas,* TSSWCB, Austin, 1991.
20. Anne McFarland and Larry Hauck, *Livestock and the Environment: Scientific Underpinnings for Policy Analysis,* Texas Institute for Applied Environmental Research, Stephenville, 1995, p. 1.
21. U.S. House of Representatives, *Review of the U.S. Department of Agriculture's Fiscal Year 1990 Water Quality Initiative: Hearings Before the Subcommittee on Department Operations, Research, and Foreign Agriculture of the Committee on Agriculture,* 101st Cong., 1st Sess. 1991 (hearings held June 21 and 30, 1989).
22. "County Dairymen Ready for Battle," *Stephenville Empire-Tribune,* Sept. 20, 1989. The fines were subsequently reduced by as much as 50 percent in negotiations with the agency; refer to Greg Mefford, "Dairies Agree to Pay Diluted Fines," *Stephenville Empire-Tribune,* Feb. 15, 1990.
23. Mike Kingston, ed., *Texas Almanac 1994–95,* The Dallas Morning News, Inc., Dallas, TX, 1993, p. 329.
24. For a detailed discussion of TIAER's analysis of the relationships between water quality data and land uses, and land practices and soils in the contributing basins above each monitoring site, refer to McFarland and Hauck, *Livestock and the Environment.*
25. Portions of this discussion appeared in Staci Pratt and Ron Jones, Implementing the watershed approach to water quality protection: The importance of microwatersheds, in *Proceedings Innovations and New Horizons in Livestock and Poultry Manure Management,* Austin, TX, Sept. 6–7, 1995, p. 27.
26. McFarland and Hauck, *Livestock and the Environment,* p. 15.
27. McFarland and Hauck, *Livestock and the Environment,* pp. ii–iii.
28. Frarey et al., *Livestock and the Environment.*
29. Logan, Agricultural best management practices: Implications for groundwater protection, *Groundwater and Public Policy,* Ankeny, IA, 1991, p. 6.
30. Environmental Protection Agency, Office of Wetlands, Oceans and Watersheds, *Nonpoint Source Interim Strategy,* Washington, 1994, p. 3.
31. Heather Jones, USDA must define "large" operation, *Feedstuffs,* vol. 68, no. 21, 1996, p. 41; also refer to 16 USC § 3801 *et seq.* as amended by Title III, Pub. L. No. 104-127, 110 Stat. 888, 980-1026 (the Federal Agriculture Improvement and Reform Act of 1996).
32. Texas Agriculture Code Annotated § 201.021 © & (d) (West 1996). 1996. West Publishing, St. Paul, MN.
33. Texas Agriculture Code Annotated § 201.026 *et seq.* (West 1996).

34. Texas Agriculture Code Annotated § 201.026 *et seq.* (West 1996).
35. E. William Anderson and Robert C. Baum, How to do coordinated resource management planning, *Journal of Soil and Water Conservation,* vol. 43, no. 3, 1988. Reprinted in *Coordinated Resource Management Guidelines,* Society for Range Management, Denver, 1993, p. 7-5.
36. Ibid.
37. C. Rex Cleary and Dennis Phillippi, *Coordinated Resource Management Guidelines,* Society for Range Management, Denver, 1993, p. 4-9.
38. Ibid., p. 5-1.
39. Bill Ross, a policy intern at TIAER, aided in the formulation of these guiding principles.
40. Cleary and Phillippi, *Coordinated Resource Management Guidelines,* P. 7-3.
41. C. Rex Cleary, Experimental stewardship—What's happening? *Rangelands,* vol. 6, no. 4, 1984. Reprinted in *Coordinated Resource Management Guidelines,* Society for Range Management, Denver, 1993, p. 4-16.
42. Guy Webster, Old foes working together to help manage public lands, *Reno Gazette Journal,* 1984. Reprinted in *Coordinated Resource Management Guidelines,* Society for Range Management, Denver, 1993, p. 4-14.
43. Frarey et al., *Livestock and the Environment,* 1994, p. 20.
44. Texas Institute for Applied Environmental Research, *Livestock and the Environment: Rethinking Environmental Policy, Institutions and Compliance Strategies,* TIAER, Stephenville, TX, 1992.
45. Acts 1993, 73d Leg., ch. 54 §1, eff. Apr. 29, 1993 codified as Texas Agriculture Code § 201.026 (West 1996).
46. Dispute Resolution Services of Tarrant County, Inc., *Basic Mediation Training Manual,* Fort Worth, TX, 1996, p. 6.
47. James E. Crowfoot and Julia M. Wondolleck, *Environmental Disputes, Community Involvement in Conflict Resolution,* Island Press, Washington, 1990, p. 98.
48. Environmental Protection Agency, *Watershed Approach Framework—1996,* p. 57.

6

Watershed Protection and Coastal Zone Management

Jan Peter Smith, Bruce Kent Carlisle, and Anne M. Donovan
Massachusetts Coastal Zone Management Office, Boston

Introduction

The passage of the federal Coastal Zone Management Act (CZMA) in 1972 demonstrated a recognition by Congress that the coastal zones of the United States needed special attention. Population and development pressures in coastal areas were leading to widespread environmental degradation as well as competition for limited resources. In response, Congress established a national program that offered coastal states financial assistance to develop initiatives to address these problems. Within the first 10 years of the program, most coastal states had developed program plans and established coastal zone management (CZM) offices within state government to work on coastal policy and planning issues. The CZM programs have generally proved to be successful and visible efforts by states to address coastal environmental problems ranging from coastal erosion and water quality to public access and harbor planning.

In 1990 when the CZMA was reauthorized by Congress, an amendment (section 6217) was added that required coastal states to develop nonpoint source pollution control plans. Nonpoint source pollution is diffuse pollution from various land uses, such as agriculture, forestry operations, and roads, that gets carried by rainfall runoff and snowmelt to a surface water body. Part of the intent behind this amendment was to give CZM programs an opportunity to develop nonpoint pollution control plans since the CZM programs had established a solid record in addressing other coastal environmental issues. In recognition of the importance of watershed-based management of nonpoint pollution, the legislation also instructed CZM programs to include inland contributing watersheds in the development of the state plans.

In another important piece of coastal zone management legislation preceding section 6217, Congress created the National Estuary Program (NEP) in 1987 to establish a direct mechanism to protect the nation's estuaries. Managed by the Environmental Protection Agency (EPA), the NEP aims to identify nationally significant estuaries threatened by pollution, development, or overuse and to promote the preparation of comprehensive conservation and management plans (CCMPs) to protect estuary resources. EPA's approach has been to fund estuary programs based on formal applications to EPA from the states. The NEP also focuses on the connection between land use and water quality and the need to include watershed planning as an integral component in the effort to develop comprehensive management plans. Massachusetts is home to two NEPs, the Buzzards Bay and Massachusetts Bays programs, which are administered by the state CZM program.

Spurred by these national legislative directives, the Massachusetts Coastal Zone Management Program developed both a coastal nonpoint pollution control plan as well as two comprehensive conservation and management plans for the Buzzards Bay and Massachusetts Bays NEPs. Using the Massachusetts experience as an example, this chapter discusses how these programs have progressed, what the accomplishments and challenges have been, and what the future role will be for CZM programs in addressing watershed planning concerns.

Impacts on Coastal Waters

Nonpoint source (NPS) pollution affects coastal waters when contaminated rainwater and snowmelt run directly into the ocean or into other coastal waters, such as estuaries and salt marshes. Rainfall and snowmelt that occur many miles inland can impact coastal waters by transporting NPS pollutants to rivers that ultimately run to the sea. Consequently, most land-use activities in coastal watersheds (the geographic area from which water drains into coastal water bodies) can contribute to coastal NPS pollution problems. In Massachusetts, the watersheds affecting the Massachusetts coastal waters includes just over one-half of the state. (See Fig. 6.1) Like most of the coastal regions of the United States, the Massachusetts coastline is the most heavily populated area of the state, further heightening the problem of coastal NPS pollution, since densely populated areas are associated with increased contamination from areas of intensive land use.

Contaminants released into coastal waters because of NPS pollution cause many problems. For example, beaches must be closed to swimming and shellfish beds to harvesting when bacteria levels reach certain thresholds. Nonpoint sources of bacteria include sanitary waste from boats, household pet and farm animal wastes, failed septic systems, and stormwater runoff. In addition, when pesticides and other chemicals are introduced into the marine environment, they may have significant effects on the productivity of coastal habitats, such as estuaries. Because estuaries serve as the breeding grounds for many fish species, commercial fisheries may be affected. Other wildlife and the entire ecosystem may also be impacted either directly or as a consequence.

In Massachusetts, one of the most costly results of coastal NPS pollution is shellfish bed closings. Over the past 15 years, shellfish bed closings have increased dramatically, and many of these closings result from NPS pollution by septic systems and by domestic and farm animals. Because they are filter feeders, shellfish are particularly susceptible to water pollution since they can accumulate contaminants and bacteria in their tissues as they feed. Consequently, the contamination of shellfish with bacteria from human and animal wastes is a serious health threat. Once bacterial levels in coastal waters reach certain levels, the shellfish beds must be closed and harvesting these resource areas is prohibited. Public health concerns may require shellfish bed closings for precautionary reasons in the absence of direct water quality analyses, since the threat of potential contamination from known sources of stormwater runoff of other potential sources of bacteria represents a health concern.

Coastal Nonpoint Source Pollution Control Program

Section 6217 of the 1990 federal CZMA legislation was specifically targeted to improving coastal water quality through the reduction of

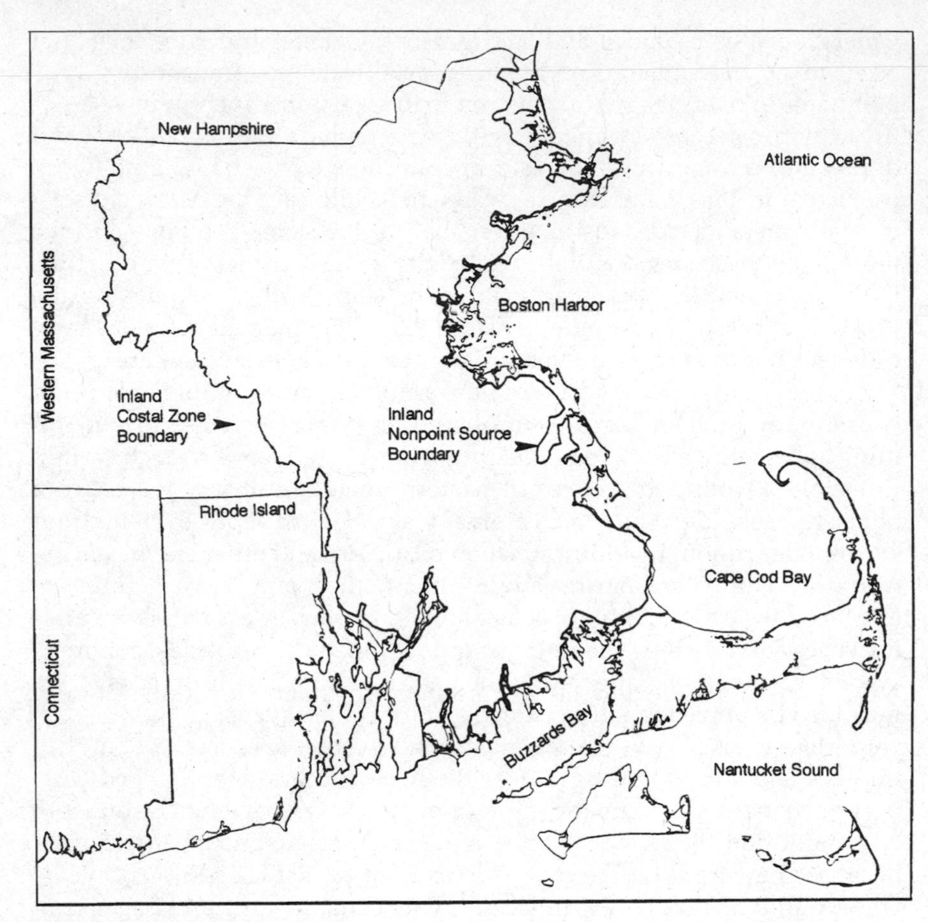

Figure 6.1 Massachussetts

sources of NPS pollution. Under this legislation, states with federally approved CZM programs were directed to develop individual state coastal nonpoint source pollution control programs. A unique aspect of this legislative action was that states were given significant flexibility of develop comprehensive strategies that best suited their individual government structures and that balanced environmental and economic goals within the state. In addition, unlike past legislative attempts at NPS control, the section 6217 coastal nonpoint source pollution control program mandates that states identify the specific enforceable authorities to use if nonregulatory or voluntary approaches are unsuccessful. Finally, state coastal nonpoint source pollution control plans must be approved by the National Oceanic and Atmospheric Administration (NOAA), the federal agency that administers the Coastal Zone Management Act, and the Environmental Protection Agency, the federal agency that administers a separate nonpoint source program under the Clean Water Act.

Federal Guidance

NOAA and EPA developed a guidance document for states to use throughout the planning process entitled *Guidance Specifying Management Measures for Sources of Nonpoint Pollution in Coastal Waters.* The federal guidance outlines all the requirements states must address to have an approvable coastal nonpoint source pollution control plan.

A unique aspect of the section 6217 coastal nonpoint source pollution control plan is that it is technology-based rather than water quality–based. This means that states must implement strategies to prevent pollution and address known causes of pollution, rather than wait until water quality impacts actually occur and then reactively implement strategies to try to correct pollution problems. Under the coastal nonpoint source pollution control program, therefore, the focus is to implement control strategies that have been proved to be effective in reducing the major sources of NPS pollution or to develop innovative strategies to accomplish to same goals.

The technology-based approach is a more direct tactic than the water quality–based approach to effectively address NPS pollution. Because NPS pollution is diverse by nature, and individual sources may actually only contribute a relatively small load of contaminants to coastal waters, the task of identifying and quantifying the exact sources of NPS pollution involves expensive and time-consuming data collection. The technology-based approach relies on existing pollutant loading data to conclude which land-use activities are sources of NPS pollution to coastal waters, thereby avoiding the source identification predicament.

The federal guidance identifies the major land-use nonpoint pollution sources and directs state programs to develop strategies to effectively implement *minimum management measures* to control these sources. The minimum management measures were defined by EPA and NOAA to be those measures that should adequately control NPS pollution under normal circumstances and conditions. For example, for agriculture, keeping grazing animals out of streams is a minimum management measure for controlling stream bank erosion and animal waste contamination. How states ensure that the management measure is implemented is up to them. The guidance also contains requirements for developing and implementing controls beyond those outlined in the EPA and NOAA document. If, after minimum management measures have been implemented, coastal waters still do not meet federal water quality standards (under the Clean Water Act), additional management measures may be necessary. Following on the above example, if fencing for farm animals is not adequate to prevent stream bank erosion and animal waste contamination, then additional measures such as vegetative buffers or other management controls will be necessary to ensure that water quality standards are met. These measures should be targeted directly at reducing the NPS pollution activities that prevent these waters from meeting designated water quality standards, i.e., ensuring the water is safe for drinking, fishing, or swimming.

The suggested management measures contained in the federal guidance can be grouped into two separate control strategies. The first group of management measures are those that are oriented toward planning and source reduction. This type of strategy relies on improving the land-use practices in coastal watersheds. The relationship between land-use practices and NPS pollution has been clearly established. For example, human development of natural areas converts natural land cover, such as forests and fields, to impervious surfaces, such as roads, parking lots, buildings, and other surfaces that do not readily absorb water. After being rendered impervious, rainwater and snowmelt run quickly off these surfaces, transporting contaminants to coastal waters. Implementing land-use strategies that limit impervious area can help minimize NPS pollution problems.

The second group of management measures is structural types of solutions, or best management practices (BMPs), aimed at capturing NPS contaminants before they enter surface waters. An example of a structural BMP is an infiltration basin, which is a depressional structure that is constructed over permeable soils to hold and then infiltrate stormwater runoff.

In addition to these management measures, the federal guidance outlines components that state plans must address:

- *Agency coordination:* States must demonstrate that their coastal nonpoint source pollution control programs are coordinated with both existing state and local water quality plans (which are part of the Clean Water Act nonpoint source program) and state coastal zone management programs. In addition, states should establish ways to improve coordination among state agencies, as well as between state and local officials responsible for habitat protection, land-use programs and permitting, water quality permitting and enforcement, and public health and safety.
- *Technical assistance:* States must develop a plan for providing local governments and the public with technical information and other assistance to help these groups implement any additional management measures that are needed.
- *Public participation:* States must provide opportunities for the public to be involved throughout the development of the coastal nonpoint source pollution control program.
- *Program boundaries:* State coastal zone boundaries must be modified to include inland areas that may contribute to coastal NPS pollution problems. For Massachusetts, this has meant that the eastern one-half of the state that includes all the watershed areas draining to Massachusetts coastal waters is included in the coastal zone boundary.

Nonpoint Source Pollution Control Programs. These categories are:

- *Urban sources:* A wide variety of urban sources of NPS pollution degrade coastal waters. Urban development increases impervious surfaces. Since these surfaces are often covered with oil, trash, animal wastes, and other contaminants, the rainwater and snowmelt that run off these areas pick up and carry these contaminants to surface water bodies. Failing septic systems, which contribute bacteria and chemicals to coastal waters, are another major source of urban NPS pollution. Finally, roads, highways, and bridges also contribute to NPS pollution. Oil, antifreeze, and other contaminants leaked onto these surfaces from motor vehicles, as well as residue left from tires and exhaust, can be washed into surface water bodies.
- *Marinas and boats:* Marinas and boatyards are a source of several types of contaminants, including fuel, cleaning chemicals, paint, and oil. These facilities contribute NPS pollution when these substances are spilled directly into the water or are washed off docks and boats when it rains. Boats are also a source of NPS pollution, including sewage and trash that are purposefully released to the water and gasoline and oil that accidentally leak from engines or bilges.

- *Agricultural sources:* The pesticides and fertilizers applied to cropland become nonpoint sources of pollution when they are washed into waterways by rainfall or snowmelt. In addition, erosion of soil can lead to increased sediment levels in waterways, another type of NPS pollution. Finally, runoff can carry animal wastes from holding pens or grazing fields into waterways.
- *Forestry:* To conduct forestry operations, roads are often built through forest lands. In addition, timber-cutting equipment must be driven across rivers and streams. These operations can increase erosion and release nutrient-rich sediments, contributing to the NPS pollution process.
- *Hydromodification:* Hydromodification refers to channelization (the straightening, widening, or deepening of channels for flood control or navigation), dam construction and dam use, and stream bank and shoreline erosion. Channelization increases water flow rates and changes water flow pathways. This may cause NPS pollution by both increasing erosion rates and increasing the quantity of pollutants reaching downstream sites. Dam construction may increase erosion and sediments problems and may also result in the release of contaminants from construction equipment (such as oil and fuel) which enters the waterway. Dams may cause other NPS pollution problems because sediment and other pollutants build up behind the dam. When the water is released, large levels of these NPS pollutants may be carried downstream. Finally, development near erosion-sensitive stream banks and shorelines accelerates erosion processes beyond natural levels. These activities contribute excessive sediments and other pollutants to the waterways.
- *Wetlands:* Wetlands provide many important environmental and economic benefits. Wetlands help to control flooding, protect the shoreline from storm damage, and provide habitat for commercial fish and shellfish, as well as rare and endangered species. In terms of NPS pollution, wetlands can hold sediments and other contaminants, which can keep these contaminants from reaching coastal waters. When wetland areas are filled or otherwise altered for development, the wetlands no longer serve this function, and NPS pollution problems are increased. In addition, when the natural capacity for wetlands to hold contaminants is surpassed, wetlands release these contaminants to coastal waters. Wetland protection is therefore important to the protection of coastal resources. In addition, wetland restoration (i.e., returning wetlands to their former and more productive natural function, condition, or size) and the construction of artificial wetland should be encouraged to further protect coastal waters.

The Massachusetts Approach

In Massachusetts, the Massachusetts Coastal Zone Management (MCZM) office and the Department of Environmental Protection (DEP) jointly assumed the responsibility for developing and implementing the coastal nonpoint source pollution control program (CNSPCP). Located within the Office of Environmental Affairs (the cabinet-level environmental office in Massachusetts state government), MCZM has developed a role as a policy and planning agency for coastal environmental issues, while DEP is the enforcement agency for environmental regulations. MCZM took the lead role in the effort to develop the program plan, with strong support and cooperation from DEP. From the outset, MCZM and DEP sought and received the critical input from other state agencies and a wide range of environmental and industry groups. The state's focus during this program development process was to protect coastal resources from the impacts of NPS pollution while taking into consideration the needs of those who would be regulated by the program. Working in an interactive forum through a number of different working groups, MCZM's development of the CNSPCP was based on consensus building and flexibility.

Although federal boundary definitions have excluded some inland watersheds in the Massachusetts program (even though these areas drain to coastal waters in other states, including Connecticut and Rhode Island), Massachusetts has made the commitment for consistency purposes to implement the program statewide and not just in its own coastal watersheds. In doing so, Massachusetts took a major step toward addressing all NPS surface water quality issues, not just coastal.

In the early stages of program development, MCZM determined that all the major categories of NPS pollution identified in the federal guidance document existed in Massachusetts. (See Fig. 6.2.) In comparison, some other coastal states claimed that either several of these major source categories were not present in their defined boundary areas or that the source was present but did not pose a threat to coastal water quality. Again, the major NPS source categories are

- Urban sources
- Marinas and boats
- Agricultural sources
- Forestry
- Hydromodification
- Wetlands

Marine Waters Assessment

* Marine Waters Assessed, sq miles

Assessed (12%)

Unassessed (88%)

Total Marine Waters = 1,876 sq miles
Total Assessed = 220 sq miles

* - Relative comparison based on estimate of total marine waters using Division of Marine Fisheries shellfish management areas (see text for explaination)

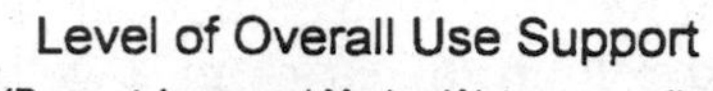

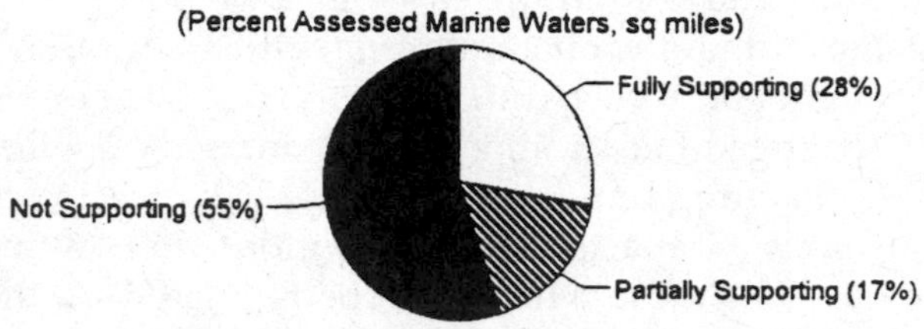

Individual Use	Percent Assessed Marine Waters, sq miles			
	Fully Supporting	Partially Supporting	Not Supporting	Not Rated
Aquatic Life Support	49%	12%	29%	10%
Fish Consumption	4%	0%	21%	75%
Primary Contact-Swimming	44%	43%	10%	3%
Secondary Contact	84%	5%	5%	6%

Figure 6.2 Marine water assessment.

Marine Waters Assessment

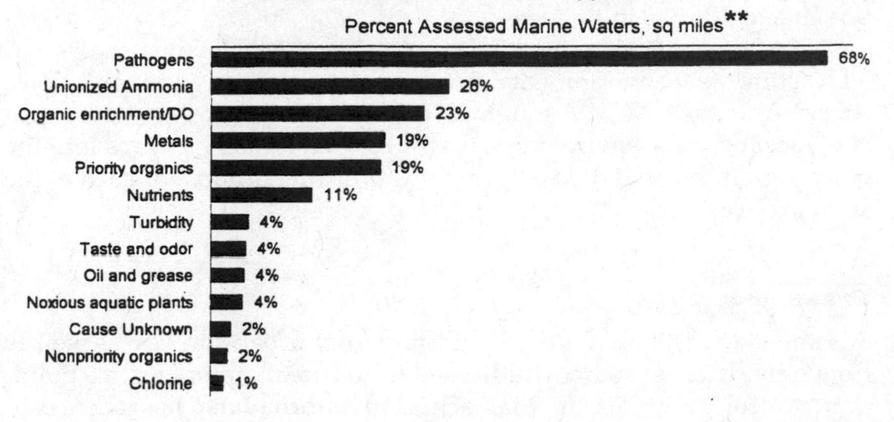

Sources of Non-Support

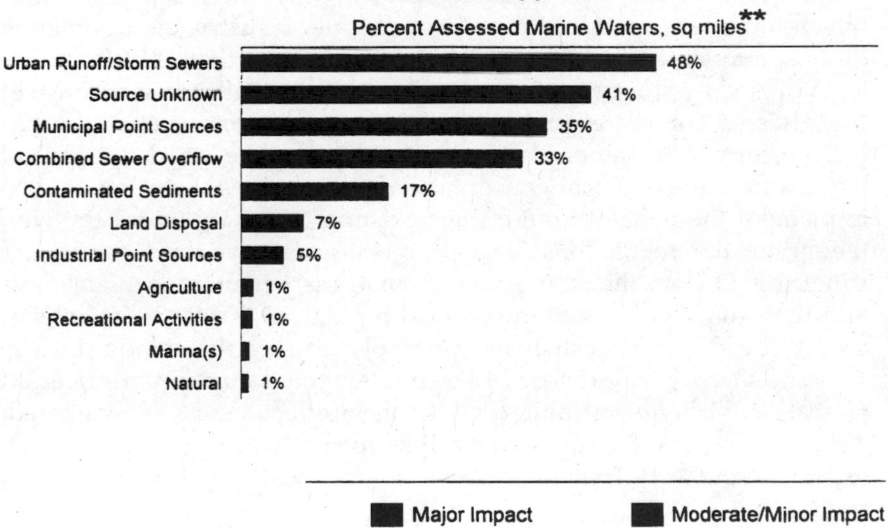

** -Marine Waters Assessed = 220 sq miles

Figure 6.2 (Continued) Marine water assessment.

Other program components that were included in the MCZM approach were interagency cooperation, technical assistance to communities, public participation in the planning process, regulatory changes, and possible legislative initiatives.

MCZM began its planning efforts by conducting a detailed analysis of all existing state environmental programs, legislation, and regulations. At the same time, MCZM sought to engage the appropriate state agencies, local officials, environmental groups and industry groups into the planning process and did so by creating working groups for each of the six source categories.

Urban sources

Stormwater runoff carries NPS pollution from urban sources, as well as from other types of sources addressed in coastal nonpoint source pollution control programs. In Massachusetts, stormwater has been estimated to be the major source of nonpoint source pollution to surface waters. MCZM and DEP initiated a stormwater advisory committee, which included representatives from private industry, environmental groups, and state and federal agencies, to address the pollution problems caused by stormwater runoff. The agencies with the assistance of the advisory committee produced a policy guidance document outlining performance standards and strategies to be used to control stormwater runoff. Local conservation commissions were targeted to implement the policy through their existing authority under the wetland protection regulations. Since problems associated with stormwater runoff result from the sum of many small incremental local land uses and their impacts, the actions of local regulatory boards were considered to be the best mechanism for implementing the standards. The standards have been proposed to be implemented for an interim period as state environmental policy while their effectiveness is evaluated. After 1 to 2 years, the standards will be formally incorporated into the existing regulatory framework for the state water quality protection programs.

One of the major urban sources of NPS pollution in Massachusetts is failing septic systems. DEP, in cooperation with other state agencies including MCZM, revised Title 5 of the state sanitary code. These regulations govern the installation and maintenance of septic systems throughout the Commonwealth. MCZM has worked with DEP to ensure that the requirements of the coastal nonpoint source pollution control program as outlined in the EPA and NOAA guidance document (such as a requirement for regular maintenance and controlling system additives) are met by Title 5.

Another major urban source of NPS pollution is development and construction. MCZM determined that erosion and sediment controls were needed to minimize the impact of new construction on coastal resources. MCZM has worked with the U.S. Department of Agriculture's (USDA's) Natural Resources Conservation Service (NRCS), the federal agency that provides technical and financial assistance to farmers, to develop effective control measures. And a state guidance document directed to sediment and erosion control has been updated to reflect current knowledge of control technologies.

Roads, highways, and bridges are also a significant urban source of NPS pollution. MCZM worked with the state transportation agencies through the Massachusetts Highway Department to develop environmental specifications to accompany the structural design specifications that are in place for all state-funded road work. These environmental specifications include management practices, such as the use of buffer strips and drainage systems that reduce runoff and NPS pollution. Contractors who build, improve, or repair state roads, highways, and bridges are therefore required to meet these environmental specifications. Local highway departments will also soon be given technical assistance so that they can adopt similar specifications.

Finally, Massachusetts agencies are aggressively coordinating their efforts to develop watershed plans. The purpose of watershed planning is to use local zoning and planning tools to institute and coordinate land-use patterns that reduce environmental problems, such as NPS pollution, among all communities within a common watershed. The Massachusetts Executive Office of Environmental Affairs (EOEA) and DEP's Office for Watershed Management have developed a watershed planning approach through a pilot project in the Neponset River watershed.

DEP has divided the 27 watersheds and basins within the state into five separate groups. DEP and an interagency basin team work with the towns and cities to develop and implement watershed management strategies on a five-year cycle. All the watersheds in each of the five groups are on the same year of the cycle. The basin teams develop an outreach effort in year 1, environmental monitoring and data gathering in year 2, information and data assessment in year 3, consistent and coordinated permitting strategies on point source and NPS pollution issues in year 4, and a final evaluation in year 5. Because the permits are effective for 5 years, this creates a continual cycle in which every 5 years DEP returns to review and update all permits within each watershed.

The Neponset River watershed initiative is an excellent example of an integrated watershed management approach to protect and improve water quality in the Neponset River basin. The initiative has attempted to integrate the efforts of both public and private resources to develop a plan to control both point and nonpoint sources of pollution. The approach developed through this initiative was used as a model for other river basins throughout Massachusetts. Consequently, the focus of the initiative is on developing strategies that can be repeated in other areas of the state.

State agencies involved in the Neponset River watershed initiative include DEP, EOEA, MCZM, Department of Environmental Management (DEM), and the Department of Fisheries, Wildlife, and Environmental Law Enforcement, since all have roles in resource protection and/or environmental regulation. DEP has conducted extensive water quality monitoring in the area to determine sources of pollutants and to update water management permitting in the basin. In addition, other state agencies have provided technical assistance to local governments and businesses within the watershed. On the federal level, NRCS is also involved in the Neponset River watershed initiative and provides technical assistance on erosion issues.

The private partners in the Neponset River watershed initiative are led by the Neponset River Watershed Association. This nonprofit group has agreed to bring its members into the initiative, as well as to work with local governments, businesses, and residents within the Neponset watershed to help develop an integrated water quality approach for the river basin.

Together, the public and private partners in the Neponset River watershed initiative are developing comprehensive strategies for protecting water quality. Intermunicipal cooperation to evaluate storm drains for potential illegal sewer connections is one example. The partnership also ensures that all those with a stake in the environmental protection of the Neponset River, as well as the economic health of the region, are included in the initiative.

The Neponset River pilot project provided important experience for DEP with interagency coordination and for coordination with local governments and various nonprofit and citizens' groups. Watershed protection efforts in other basins have greatly benefited and now are modeled on the successes of the Neponset River pilot project.

Marinas and boats

Marinas and boats generate NPS pollution when they are improperly sited, designed, or operated. MCZM determined that the state's Chapter 91 regulations, which govern activities within Massachusetts waterways, can address the primary causes of NPS pollution from boats and marinas since these regulations contain a broad requirement that all state environmental requirements be complied with by any Chapter 91 license holder. To help reduce the sources of NPS pollution, therefore, MCZM has decided to focus its efforts on developing guidance documents that help marinas and harbor masters to implement the Chapter 91 requirements and control NPS pollution. MCZM is also working to provide marinas and harbormasters with the technical assistance they need to meet NPS pollution control requirements. In addition, MCZM has worked to provide grants to communities for the installation and operation of boat pump-out facilities.

Agricultural sources

Unlike many other states, agriculture in Massachusetts is dominated by small, family-owned farms. Consequently, to reduce agricultural NPS pollution problems, the economic realities faced by small farmers must be taken into account. MCZM has worked with the Massachusetts Department of Food and Agriculture (DFA) to develop an NPS pollution control strategy that includes a voluntary compliance component to help minimize economic hardship for Massachusetts farmers. Farmers are expected to develop and implement farm plans that include a prioritized list of NPS pollution control strategies. Technical assistance is provided through NRCS and state cooperative extension services. Farmers making a good-faith effort to follow this process will not be subject to regulatory action when water quality problems are traced to farm operations. Those farmers who do not develop and implement farm plans will be subject to permitting and regulatory requirements. MCZM has sponsored the preparation of an agricultural technical guide to assist farmers in their efforts to develop farm plans. MCZM worked with the Farm Bureau, trade associations for cranberry growers and nurseries, and individual farmers in developing the guidance document.

Forestry

DEM, the agency that manages state parks and forests, recently updated regulations for timber cutting. MCZM determined that the best way to address potential forestry-related NPS pollution was to work with DEM as these regulations were revised. Throughout the revision process, DEM worked with an interagency committee with membership from DEP, the Metropolitan District Commission, the Massachusetts Association of Conservation Commissions, the Massachusetts Association of Conservation Professionals, Massachusetts Audubon, and the University of Massachusetts Forestry Extension, as well as with the timber-cutting and sawmill interests. DEM also held public hearings on the revised regulations and used input from these hearings to refine the regulations as necessary to protect the environment, industry, and public

Once the DEM regulatory revisions were completed, DEM and DEP held workshops for foresters to provide them with the technical assistance they needed to meet NPS pollution controls. In addition, DEM and DEP are holding workshops for local officials to help them implement NPS pollution controls for timber cutting within their jurisdiction. To assist with the implementation of the forestry regulations, MCZM helped DEM develop a best management practices (BMP) manual. This manual provides technical guidance for foresters so that they can effectively comply with regulations and reduce NPS pollution.

Hydromodification

MCZM's analysis found that the Massachusetts Wetlands Protection Program and the Chapter 91 Waterways Program contain specific standards which help to prevent and control NPS pollution impacts from channelization (dredging, flood control, and drainage improvements) and dam building. Strict regulatory requirements are in place to protect against bank, channel, and stream bottom alterations. Erosion and sediment controls are routinely required to prevent impacts associated with construction activities in or near water and wetland resources. In addition, the DEP Office of Watershed Management's basin planning approach helps to determine where surface water quality is being adversely affected by hydromodification. Where problems are detected, DEP works with other agencies and local officials to implement BMPs to restore water quality. MCZM has worked with DEP through these initiatives, as well as with other agencies, to coordinate strategies that address the NPS pollution impacts from hydromodification.

Wetlands

Because the Massachusetts Wetlands Protection Act and its regulations contain extensive mechanisms to control NPS pollution, the act has become one of the most important components of the coastal nonpoint source pollution control plan to protect wetlands from contamination and development. The state's stormwater management policy and performance standards, e.g., rely on the Wetland Protection Program as the first line of jurisdiction and review of projects for proper stormwater control. Massachusetts also has a recently created Wetlands Restoration and Banking Program which actively identifies and promotes wetland restoration efforts throughout the state. In addition, innovative wetland assessment methods are being developed and tested by MCZM, in order to more accurately determine wetlands that are degraded or at risk of degradation from NPS pollution. Finally, MCZM and DEP are working with the Massachusetts Association of Conservation Commissions, a statewide nonprofit organization that works to assist local municipal conservation commissions, to help local governments implement wetland protection measures as well as to pursue the restoration of degraded wetlands and the construction of artificial wetlands. Until recently, wetlands artificially created for stormwater treatment purposes also became subject to wetland protection jurisdiction. New guidance indicates that these wetlands, which are created expressly for the purpose of pollutant removal, should be exempted from regulatory requirements so that maintenance activities can be implemented without regulatory review.

National Estuary Program's Comprehensive Conservation and Management Plans

The goal of the National Estuary Program is to protect and restore the health of estuaries while supporting economic and recreational activities. To achieve this goal, the EPA helps to create local programs by developing partnerships with local government agencies. In Massachusetts, two national estuary programs have been implemented: the Buzzards Bay project and the Massachusetts Bays program. These programs are administered through MCZM and are funded by EOEA and the EPA.

As National Estuary Programs, the Buzzards Bay project and the Massachusetts Bays program are required to develop comprehensive conservation and management plans (CCMPs) for their bays. These CCMPs include management strategies that protect the bays from pollution threats, including NPS pollution. The CCMPs are advisory plans, rather than regulatory documents. Consequently, the cooperation of local governments is needed if the CCMPs are to be successfully implemented.

The goals of the Buzzards Bay project is to characterize and assess water quality problems in Buzzards Bay (a declared estuary of national significance). In 1988, the Buzzards Bay project began the scientific research and coordination with planners, scientists, and local, state, and federal managers that went into developing the final plan. With the completion of the CCMP in 1991, the Buzzards Bay project began working with local governments to implement NPS pollution control strategies as well as other pollution prevention measures.

The Massachusetts Bays program has developed its CCMP for Massachusetts Bay and Cape Cod Bay. This plan aims to institutionalize a planning system to maintain, protect, and restore these valuable resources within the bays. The plan was developed based on extensive scientific research and with strong cooperation from local decision makers. NPS pollution issues, as well as other pollution sources, are also being addressed in this CCMP. Examples of action items contained in the CCMPs include reopening closed shellfish beds and identifying illegal sewer connections to storm drain systems.

Because the focus of the NEPs operates with a watershed perspective, it represents another programmatic effort within MCZM to address water quality issues and actions on a watershed level.

Accomplishments and Challenges

The two most significant challenges for coastal programs in developing nonpoint pollution control plans and working with state watershed initiatives have been (1) to develop the required enforceability for the controls for all the potential nonpoint pollution sources and (2) to coordinate with all the state agencies with varying responsibilities. Additional coordinating efforts have been needed to include federal and local government entities as well as the private and nonprofit sectors and the general public.

Creating enforceable controls for all potential nonpoint pollution sources often required considerable technical and negotiating skills. For example, although agriculture is the largest single contributing source of nonpoint pollution on a national scale, in Massachusetts it represents a minor source (estimated to be 5 percent) of the total pollution to Massachusetts waters. Persuading the agricultural agencies and the industry to accept the need to control on-farm pollution sources with an enforceable farm planning program required considerable negotiation. The lack of extensive data and information on local water quality conditions and impacts hindered to some extent the ability to develop an agricultural pollution control strategy. MCZM emphasized equitableness (that all sources should do their fair share to contribute to the solution) and a clear demonstration that agriculture was a significant source of problems in specific locations (from specific water quality data generated by DEP).

The success of program development and implementation also depends on the ability of coastal zone agencies to persuade other agencies to participate in the program development process. Because MCZM staff with extensive interagency experience were devoted to program development for this initiative and because of a strong commitment by top agency managers, the Massachusetts plan achieved a high level of participation and success by environmental agencies. In addition, the environmental protection programs already in existence in Massachusetts were strong. Consequently, few new regulatory programs were needed to meet the federal requirements contained in the guidance documents. Because there is a general trend in government away from new regulatory efforts and toward the streamlining of existing regulations, Massachusetts had a preexisting advantage for the development and implementation of nonpoint source pollution controls.

An example of the extent of environmental agency "buy-in" in Massachusetts is the Department of Food and Agriculture. Prior to this programmatic effort by MCZM, as an example, the role of DFA focused primarily on the promotion of state agriculture and its products (except for its regulatory role in pesticide regulation). The nonpoint pollution control initiative brought DFA into a closer working relationship with its sister environmental agencies and helped the agency focus more on its environmental protection responsibilities than it had in the past, and this, in turn, has now become the focus of DFA's role in the state watershed initiative. Overall, the level of interagency cooperation and communication has increased due to this MCZM initiative. For CZM programs in other states, interagency cooperation and participation have been the most significant challenge affecting the success of the nonpoint pollution control program.

Future Directions

On a national level, the future role of CZM programs in watershed protection is not clear. While the nonpoint pollution control program was intended to bring coastal zone agencies into the watershed planning process, Congress reauthorized the legislation for the nonpoint pollution control effort in 1995 but did not provide funds to continue state program efforts. The message from Congress continued to be unclear with the continued lack of appropriations in 1996. Ongoing CZM program efforts for nonpoint source pollution and watershed planning are likely to continue, but on a state-by-state basis. In Massachusetts, the coastal program is committed to continue the cooperation and participation begun with the coastal nonpoint pollution control effort and the state's watershed initiative. As policy and planning agencies, the coastal zone agencies have considerable potential to deliver technical information through outreach and educational capabilities and to present considerable opportunities to assist with state watershed programs.

References

Environmental Protection Agency, *Guidance Specifying Management Measures for Sources of Nonpoint Pollution in Coastal Waters,* EPA 800-B-92-002, EPA Office of Water, Oceans, and Wetlands, Washington, 1993.

Environmental Protection Agency and U.S. Department of Commerce/National Oceanic and Atmospheric Administration, *Coastal Nonpoint Pollution Control Program: Program Development and Approval Guidance.* EPA and NOAA, Washington, 1993.

Massachusetts Coastal Zone Management, *Coastal Nonpoint Pollution Control Plan,* Massachusetts Coastal Zone Management and Department of Environmental Protection, Boston, 1995.

7

Groundwater Aspects of Watershed Management

Dominique Brocard

The watershed approach to environmental management frequently concentrates on surface water and surficial features including land use. In many cases, however, groundwater is an integral part of the water regime at the scale of the watershed. Including groundwater in the watershed approach is particularly germane for situations in which the boundaries of the watershed and related aquifer(s) approximately coincide and for aquifers qualified as sole-source aquifers, namely, aquifers which provide the majority of the water supply requirements of an area (Belfit and Cambareri, 1996; Jehn and Paque, 1996). It is appropriate to consider groundwater at the watershed scale in regard to both water quantity and quality. Groundwater quantity depends on its replenishment (recharge), which is affected by land use; and in many cases, groundwater and surface water are closely tied, principally for surficial aquifers. Then groundwater withdrawals have a direct impact on stream flow, and excessive withdrawal can have significant environmental or water supply implications downstream. Relative to water quality, the main aspects in which a watershed approach is justified are nitrate loadings and salinity intrusion. Localized groundwater contamination, due to improper disposal of hazardous substances, e.g., is best addressed at the scale of the contamination plume, although broader impacts on the watershed clearly need to be assessed: Many groundwater contamination problems affect surficial aquifers which are hydraulically connected to surface bodies of water. Thus groundwater contamination can represent pollution sources for surface waters, and groundwater remediation needs to account for possible effects on surface waters. For example, extensive groundwater extraction can significantly lower the water table and the water level in connected ponds and wetlands.

Groundwater Hydrology

Groundwater is an integral part of the hydrologic cycle (Fig. 7.1), resulting mainly from the deep percolation of precipitation. Groundwater occupies spaces between or inside soil particles or bedrock fractures. Regions of subsoil saturated with groundwater and in which groundwater can move are *aquifers*. In *water table* (or *unconfined*, or *phreatic*)

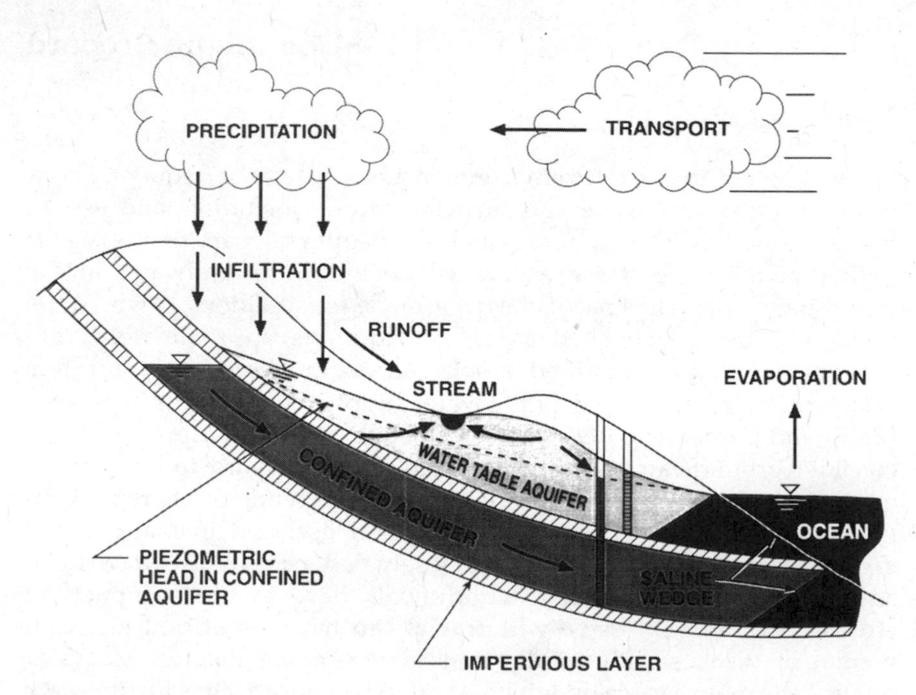

Figure 7.1 Hydrologic cycle and groundwater.

aquifers, the saturated zone extends up to the water table, at the elevation where water would stabilize in a well open to the aquifer. This height defines the *piezometric head*. Just above the water table is a *capillary fringe* (whose thickness can range from 2 to 5 cm in coarse sand to 2 to 4 m in clay) which is largely saturated and into which water is drawn by capillary forces. This region can often be neglected in groundwater analyses. Above the capillary fringe, in the *vadose zone*, the soil usually contains water, but below saturation. *Confined* aquifers are sandwiched between impervious formations, and the piezometric head is above the top of the aquifer. The confining layers can be semipervious, allowing some groundwater flow between aquifers, albeit at reduced rates.

Groundwater is *recharged* (replenished) by percolation of rainwater or snowmelt in unconfined aquifers, or unconfined portions of otherwise confined aquifers. The rate of recharge, measured in meters per year (or other dimensionally equivalent units) depends on the precipitation, soil type, slope, and land use. In particular, development frequently is often accompanied by an increase in the degree of imperviousness of a watershed, which increases stormwater runoff and decreases aquifer recharge. Groundwater flows (usually slowly) from areas of high piezometric head to areas of low piezometric head, eventually discharging into streams, lakes, and seas. Groundwater texts should be consulted for further information on the mechanics of groundwater flow and the many aspects not covered in this short treatment (Freeze and Cherry, 1979; Bear, 1979).

Groundwater discharge is typically the main provider of stream base flows, during dry weather periods. Many small ponds and wetlands are expressions of the water table in areas where land elevation is below the water table or piezometric head. In those areas, significant ecological impacts can result from changes of groundwater levels, e.g., due to withdrawals. Thus, groundwater is an essential component of watershed hydrology.

Groundwater Quantity—Mass Balance

As shown above, groundwater is an essential element of the water mass balance of most watersheds. The extents of the surficial watershed and related aquifer(s) are not always identical, however, complicating matters somewhat. In glacial valley provinces, such as in New England, individual aquifers exist in each valley system, with less permeable, unconsolidated formations on the hillsides. For those situations, groundwater divides often coincide approximately with the topographical highs defining the extent of the watershed, so that the aquifer and watershed have approximately the same footprint. For those situations, the aquifer and watershed mass balances have common elements. For example, an increase in stormwater runoff due to development translates to an approximately equal reaction in aquifer recharge (neglecting possible changes in evapotranspiration) and base flow. The resulting changes in stream flow may be minimal on a long-term basis (e.g., more than a year), but significant short-term changes can result, with increases of peak flow and reduction of summer base flows. In other geologic provinces, aquifers and watersheds can have significantly different extents, and their interaction is more complex.

Thus, on a case-by-case basis, the extent, geology, and regime of the aquifer(s) interacting with the surface water bodies in the watershed must be identified. This identification can usually be based on previous geologic investigations. In the United States and many other countries, aquifer delineations have been conducted by the relevant agencies (e.g., U.S. Geological Survey). The regime of the aquifer relates to the recharge areas, flow patterns, and discharge points or areas.

The different elements of an aquifer mass balance and means of estimating them are briefly described below. For illustration, units of different parameters are listed in terms of meters and days except for empirical formulas requiring specific units.

Precipitation recharge

This component is equal to precipitation P minus runoff RU minus evapotranspiration ET filtered by the storage STO in the root zone of plants. Precipitation recharge can be estimated by conducting a root zone water budget, typically with a time step of 1 month. An example of root zone mass balance is shown in Table 7.1. The water storage STO in the root zone at the end of each month is equal to the storage at the beginning plus infiltration (precipitation minus runoff) minus evapotranspiration minus recharge.

Precipitation can be obtained from meteorological stations. Runoff can be estimated using the Soil Conservation Service (SCS) method with the following formula (SCS, 1964; McCuen, 1982):

$$Q = \frac{(P - 0.2S)^2}{P + 0.8S} \qquad \text{with} \qquad S = \frac{100}{CN} - 10$$

where Q = excess precipitation = runoff per unit land area, in, P = precipitation, in, S = potential maximum retention, in, and CN = curve number, which is dependent on land use, antecedent soil moisture, and hydrologic soil group. Values of the curve number can be obtained from

Table 7.1 Root Zone Water Budget Example

Soil type:	Fine sandy loam	Field capacity	0.2
		Wilting point	0.08
Vegetation:	Deeply rooted	Root depth	1000 mm
Root zone:		Max. storage STOMAX	200 mm
		Min. storage STOMIN	80 mm

	Precipitation (incl. snowmelt) P	Runoff Q	Infiltration INF	Potential evapotranspiration PET	Storage at beg.	Storage at end STO	Evapotranspiration ET	Groundwater recharge RE
Jan.	5	2	3	0	200	200	0	3
Feb.	18	5	13	0	200	200	0	13
Mar.	51	9	42	0	200	200	0	42
Apr.	65	13	52	28	200	200	28	24
May	92	19	73	68	200	200	68	5
June	75	12	63	112	200	151	112	0
July	66	6	60	135	151	80	131	0
Aug.	96	5	91	119	80	80	91	0
Sept.	165	23	142	84	80	138	84	0
Oct.	121	12	109	46	138	200	46	1
Nov.	89	9	80	10	200	200	10	70
Dec.	30	2	28	0	200	200	0	28
Annual total	873	117	756	602			570	186

All depths in millimeters.

tables in the previously mentioned reference. This analysis must be conducted on a storm-by-storm basis and runoff totalized monthly. The infiltration INF, including depression storage, which is subject to evapotranspiration, is INF = $P - Q$.

The storage in the root zone at the end of each month STO_i is the storage at the end of the previous month STO_{i-1} plus infiltration minus evapotranspiration and aquifer recharge:

$$STO_i = STO_{i-1} + INF - ET - RE$$

The maximum storage (or retention) capacity STOMAX is equal to the depth of the root zone multiplied by the field capacity of the soil. Average values for different soils and vegetation are listed in Table 7.2 (Thornthwaite and Mather, 1957; ASCE, 1990).

The aquifer recharge is equal to the excess storage over the retention capacity at the end of the month

$$RE = STO_{i-1} + INF - ET - STOMAX$$

Evapotranspiration is the fraction of the *potential* evapotranspiration PET which can be satisfied by moisture available in the root zone of plants growing in the area. Potential evapotranspiration depends on the vegetation type, air temperature, and time of year and can be estimated by using the Blaney-Criddle formula (SCS, 1970) among others (ASCE, 1990)

$$PET = (0.0173T^2 - 0.313T)kd$$

where PET = potential evapotranspiration, in, T = average air temperature, °F, k = crop factor, and d = fraction of annual hours of daylight occurring during the month. Alternatively, tabulated values as a function of location and month can be used (Carter). The full PET is realized when sufficient water is brought by infiltration and/or available in storage. However, plants can extract water from storage only up to a lower limit of extractable water, also referred to as the *wilting point*. These depend on the soil, and average values are listed in Table 7.2 in terms of fraction of the root zone; i.e., the minimum storage STOMIN is the lower limit listed in Table 7.2 multiplied by the root zone depth. In summary, ET can be estimated by using the following equations; alternatively, ET can be calculated by using empirical formulas as a fraction of PET depending on the soil moisture content.

$$\text{If} \quad STO_{i-1} + INF - PET > STOMIN \qquad \text{then} \quad ET = PET$$

$$\text{If} \quad STO_{i-1} + INF - PET < STOMIN \qquad \text{then} \quad ET = STO_{i-1} + INF - STOMIN$$

Table 7.2 Root Zone Depths and Retention Capacities of Different Soils and Crops

Soil type:	Fine sand	Fine sandy loam	Silt loam	Clay loam	Clay
Field capacity:	0.12	0.20	0.30	0.34	0.36
Lower limit of extractable water (wilting point):	0.04	0.08	0.15	0.19	0.21
Root zone depth, m					
Shallow-rooted crops (spinach, peas, beans, carrots)	0.50	0.50	0.62	0.40	0.25
Moderately deep-rooted crops (corn, cotton, tobacco, grain)	0.75	1.0	1.0	0.80	0.50
Deep-rooted crops (alfalfa pastures, shrubs)	1.0	1.0	1.25	1.0	0.67
Orchards	1.5	1.67	1.5	1.0	0.67
Closed mature forest	2.5	2.0	2.0	1.6	1.17

In wet areas, the monthly calculations can be started at the end of the wet season, assuming that the root zone storage is full: STO_1 = STOMAX. The validity of this assumption is verified if the same is found at the end of the year. Otherwise, an iterative approach is needed in which different values of STO_1 are successively tried, the correct one being that duplicated at the end of the year.

Stream recharge

Stream recharge is the infiltration from a stream to the groundwater in areas where the piezometric head in the aquifer is below the stream stage and the stream is a losing stream. The discharge rate per unit length of the stream Q_s' (m^2/day) can be estimated from flow measurements across two stream transects. If the flow at the downstream section is lower than the upstream flow by ΔQ, then $Q_s' = \Delta Q/L$, where L is the distance between the two sections. Another means of estimating the stream discharge is by using a seepage meter (Fig. 7.2). Estimates from seepage meters, however, are only representative of the immediate area of the meter location and, therefore, can be quite variable. Several meters in

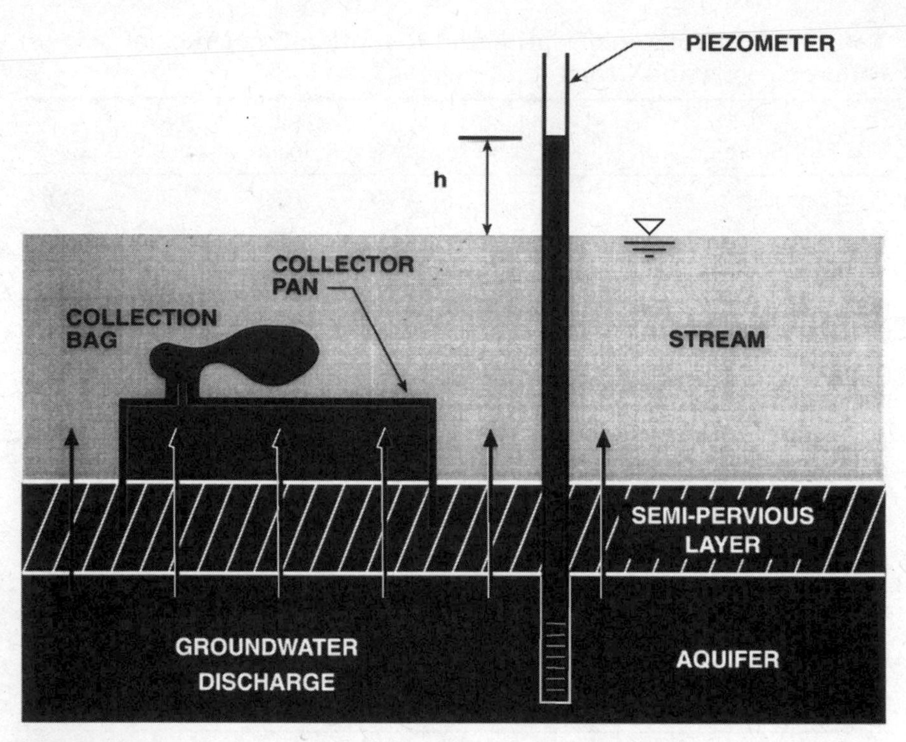

Figure 7.2 Seepage meter to estimate groundwater discharge to stream.

zones with different streambed characteristics are recommended. In general, the rate of stream discharge is proportional to the difference in elevation between the stream stage and aquifer head ΔH(m). To permit estimates under different conditions, the stream leakage can be characterized by a leakance C (m/day) such that $Q_s{}' = C\,\Delta H$. Most computer models of groundwater flow characterize stream discharges in this fashion.

Septic system recharge

In areas not served by sewers, septic systems provide recharge to surficial aquifers. This component is approximately 90 percent of the indoor water use, which can be gauged by the winter water consumption.

Leakage from other aquifers

For aquifers separated from other aquifers by semipervious boundaries, leakage between the aquifers can occur. The leakage flow rate is proportional to the difference in piezometric head between the aquifers, with a proportionality constant depending on the conductivity and thickness of the separating layer. This constant can be estimated from pumping tests, but it is generally difficult to determine with accuracy.

Water supply withdrawals

This includes municipal as well as private water wells and irrigation wells. Records of municipal wells are usually available, and withdrawals from private wells for indoor use can be determined from the number of persons serviced. Records of irrigation withdrawals are frequently not available.

Ocean and sea discharge

Coastal aquifers typically ultimately discharge into the ocean or sea. Estimation of this component of the groundwater balance must often rely on calculation of groundwater flow per unit width Q' from measured hydraulic gradients $\overline{V}H$ (m/m) and estimated aquifer transmissivity T (m^2/day), using Darcy's law: $Q' = T \times \overline{V}H$.

Change in storage

If the aquifer is in equilibrium and the mass balance represents average conditions, the quantity of groundwater stored in the aquifer is constant, on an annual basis. Changes in storage, however, occur due to wet and dry years as well as changes in withdrawals. These changes can result in changes of other flow inputs and outputs which may take many years to stabilize. Therefore, it is important to take the change of storage into consideration in aquifer mass-balance exercises.

Verification

It is often not possible to accurately estimate all the components of the groundwater balance of an aquifer. To adjust those estimates, development of a groundwater flow model of the aquifer can be very valuable. The estimated flow inputs and outputs are specified in the model at their appropriate locations, and the model calculates piezometric heads, which can be compared to measurements at monitoring wells or piezometers. An example of a computer model grid for the Los Osos Valley aquifer in California and associated mass balance are shown in Fig. 7.3 and Table 7.3 (Yates and Wiese, 1988).

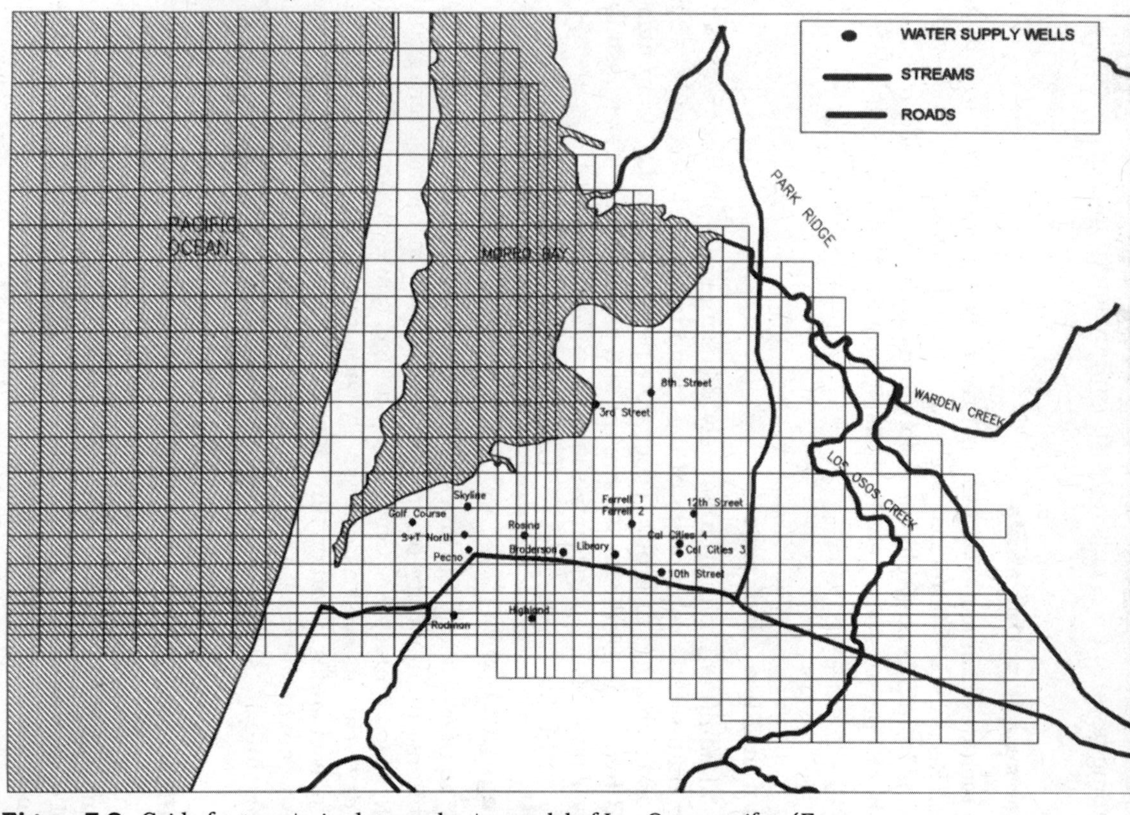

Figure 7.3 Grid of computerized groundwater model of Los Osos aquifer. (*From Yates and Wiese, 1988.*)

Table 7.3 Mass Balance for Los Osos Valley Aquifer, acre-feet for 1986

Budget item	Inflow	Outflow	Net flow
Rainfall recharge	2530	0	2530
Los Osos Creek seepage	690	640	50
Leakage from bedrock aquifer	420	0	420
Ocean discharge	0	590	−590
Domestic water use			
Public well pumpage	0	2220	
Private well pumpage	0	210	
Septic percolation	1550	0	
Irrigation return	330	0	−550
Agricultural water use			
Pumpage	0	970	
Irrigation return	400	0	
Phreatophyte transpiration	0	200	−570
Spring flow	0	360	−360
Net = change in storage			730

Seawater Intrusion

An issue closely related to the groundwater mass balance is seawater intrusion into coastal aquifers. Because seawater is denser than freshwater, seawater wedges intrude into freshwater aquifers near shorelines. As groundwater movements are generally slow, there is often surprisingly little mixing of the saline and freshwater, and a sharp interface exists. A schematic situation is depicted in Fig. 7.4, with an expression for calculating the length of the wedge. Another useful result is the Ghyben-Hertsberg relationship, which relates the depth of the interface below sea level Z and the water table elevation (or piezometric head for confined aquifers) h under static equilibrium conditions: $Z = h(\rho/\Delta\rho)$, where ρ = freshwater density ($\approx$ 1 kg/L) and $\Delta\rho$ = density difference between freshwater and saline water ($\approx$ 0.025 kg/L for seawater). For these numerical values, $Z \approx 40h$. Measurements of the water table elevation (or piezometric head) relative to sea level thus provide an estimate of the depth to the seawater interface.

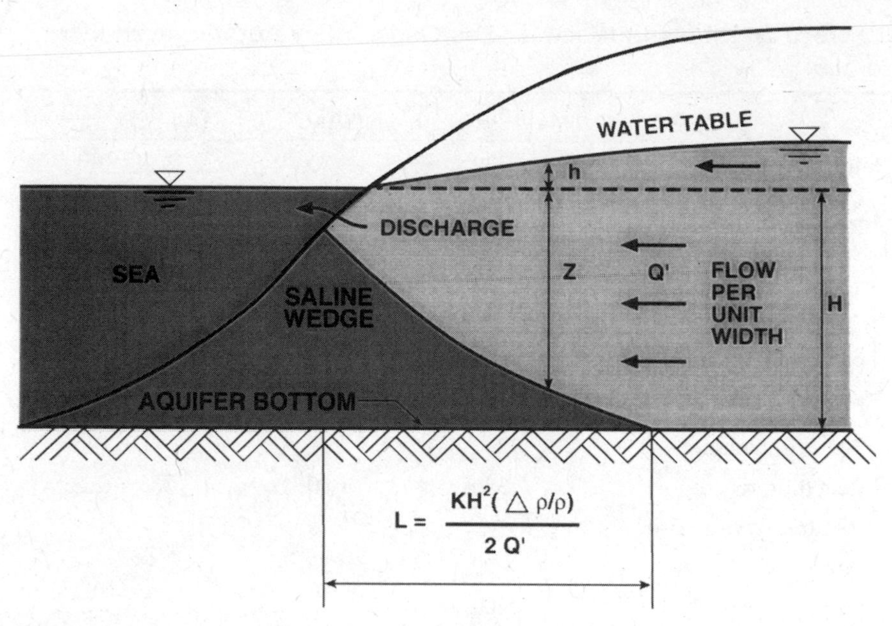

Figure 7.4 Saline wedge intrusion in freshwater aquifer.

A crucial parameter in the equation provided in Fig. 7.4 for the length of the seawater wedge is the groundwater discharge per unit length (perpendicular to shore) Q'. Increased groundwater withdrawals in the watershed reduce this flow and increase groundwater intrusion. As a result, the elevation of the saline water interface rises, and wells withdrawing freshwater from the upper portion of the aquifer may start ingesting a mixture of seawater and freshwater. Seawater ingestion may then be alleviated by reducing the withdrawal rate of the contaminated wells and compensating for the shortfall by installing other shallow wells. This approach, however, has obvious limitations.

The simple formula provided in Fig. 7.4 is only a first step in the analysis and control of seawater intrusion. Further information on the mechanics of seawater-freshwater interaction is provided in groundwater texts (Bear, 1979). Computer models have also been developed to simulate seawater intrusion (SUTRA, MOCDENS).

Nitrate Loading Analyses

Nitrate is a groundwater contaminant which is well suited to the watershed approach because it is commonly introduced by numerous independent and distributed sources which act cumulatively relative to the concentration in public water supply wells. Nitrate loadings to groundwater occur from such sources as precipitation, septic systems, fertilizers, and animal wastes. Part of these loadings occur in the form of ammonia (NH_3) and nitrite (NO_2), which are rapidly oxidized to nitrate (NO_3). Nitrate is of concern because of the health hazards associated with excessive consumption, including methemoglobinemia, particularly in infants (blue baby syndrome), and the potential formation of carcinogenic nitrosamines in the human body by interaction with other nitrogen compounds with nitrite, itself generated by bacterial action on nitrate. Because of these potential health effects, the Environmental Protection Agency (EPA) has set a maximum contaminant level (MCL) of 10 mg/L for nitrate nitrogen in water supplies.

Nitrate does not decay and is not retarded by adsorption to soil particles; therefore, it is entirely and readily transported by groundwater flow, including that generated by water supply wells. The nitrate concentration in the water withdrawn by a well is simply the flow-weighted average of the concentrations in the water reaching the well from different directions. Therefore, at steady state, the nitrate concentration in a well C (mg/L) is equal to the loading W (mg/day), divided by the well flow rate Q_w (L/day): $C = W/Q_w$. The loading is that occurring in the recharge area of the pumping well, i.e., the area within which percolating precipitation eventually reaches the well. Different states in the United States have different names and definitions for this area, which depend on the aquifer characteristics, well flow, and precipitation. For example, in Massachusetts, the Department of Environmental Protection defines the recharge area of a well in terms of a zone II, which is "the area of an aquifer that recharges a well (the land surface which overlays that part of the aquifer that recharges a well) under the most severe and recharge conditions that can be anticipated. It is bounded by the groundwater divide that results from pumping the well and the contact of the edge of the aquifer with less permeable materials such as till and bedrock (310 CMR 24.00)." This definition is adapted to the dominant glacial valley nature of the aquifers in the state, and very detailed procedures are set forth for the delineation of the zone II.

For a given well, a nitrogen loading analysis can be performed given a delineation of the recharge area of the well and an inventory of the different sources within this area (Frimpter et al., 1990). The loadings from the different sources (expressed in mass per unit time) can be estimated based on load factors such as presented in Table 7.4. This analysis can be used to determine if a well is likely to develop nitrate concentrations in excess of the regulation values at a point in the future, given existing and proposed sources. Indeed, it may take several years or even decades

Table 7.4 Nitrate Loading Factors for Typical Sources

Source	Unit	Flow, gal/day per unit	NO_3-N concentration, mg/L	NO_3-N loading, lb/yr per unit
Domestic septic system	Person	60–70	30–35	5.5–7.5
Restaurant	Seat	35	35–40	3.7–4.3
Campground	Person	25	35–40	2.7–3.0
Office building	Person	10–15	35–40	1–1.8
Shopping center	Employee	60	35–40	6.4–7.3
Lawn fertilizer	1000 ft^2			0.6–1.0

for all the sources in a well's recharge area to be felt at the well. Based on this information, educated restrictions can be set on land use in the area. This approach is based on the existing or proposed public wells in an area. Another regulatory methodology is based on the concept that each new development should not introduce more than its allowable share of nitrate to the aquifer, accounting for existing and possible future development in the watershed. In this approach, the flow-weighted concentration of all flows recharging the aquifer from this particular development is limited. A maximum of 5 mg/L is frequently applied to provide a factor of safety. An example of nitrogen loading analysis for a proposed development is presented in Table 7.5.

The loading from atmospheric sources can be estimated from measurements of nitrate concentrations in groundwater in areas with no anthropogenic sources. This concentration is lower than the nitrogen concentrations measured in precipitation, as much of this nitrogen is taken up by plants before reaching the groundwater table. Recharge from impervious areas such as roads, parking lots, and buildings, however, benefits from much lower nitrogen uptake. Depending on the recharge mechanism (e.g., infiltration basin) a loss factor of 0.1 to 0.3 can be assumed.

Table 7.5 Example of Nitrate Loading Analysis

XYZ Development	Input data	Groundwater recharge, ft^3/yr	Nitrogen loading, lb/yr	Nitrogen concentration, mg/L
Rainfall characteristics				
Average annual rainfall, in	45			
Average nitrate concentration, mg/L	0.5			
Septic systems				
Number of residences	14			
Number of persons per residence	6			
Wastewater volume per person, gal/person/day	65			
Average nitrate concentration in septic effluent, mg/L	35			
Average groundwater recharge characteristics		266,430	582.1	35
Impervious land contribution				
Impervious land area, acres	6.3			
Runoff coefficient	0.5			
Fraction of runoff recharged	1			
Nitrate removal fraction in infiltration basin(s)	0.3			
Average groundwater recharge characteristics		514,553	11.2	0.35
Natural land contribution				
Natural land area, acres	20			
Fraction of precipitation recharged	0.5			
Nitrate removal fraction of natural land	0.9			
Average groundwater recharge characteristics		1,633,500	5.1	0.05
Lawns				
Lawn area, acres	3.2			
Fraction of precipitation recharged	0.5			
Unit nitrate loading from fertilizer, $lb/1000\ ft^2/yr$	0.8			
Average groundwater recharge characteristics		261,360	111.5	6.84
Total site		2,675,843	709.9	4.25

References

ASCE. 1990. *Evapotranspiration and Irrigation Water Requirements.* American Society of Civil Engineers. Manuals and Reports on Engineering Practice, no. 70.

Bear, 1979. *Hydraulics of Groundwater.* McGraw-Hill, New York.

Belfit, G. C., and T. C. Cambareri. 1996. A Groundwater Lens Based Strategy for Water Quality Protection on Cape Cod. *Proceedings of Watershed '96.* Water Environment Federation, Baltimore, MD.

Carter, D. B. *Basic Data and Water Budget Computation for Selected Cities in North America.* Earth Science Curriculum Project, Reference Series RS-8. Prentice-Hall, Englewood Cliffs, NJ.

Freeze, A. R., and J. A. Cherry. 1979. *Groundwater.* Prentice-Hall, Englewood Cliffs, N. J.

Frimpter, M. H., J. J. Donahue, and M. V. Rapacz. 1990. *A Mass-Balance Nitrate Model for Predicting the Effects of Land Use on Groundwater Quality.* U.S. Geological Survey, Open File Report 88-493. Denver, CO.

Jehn, P. J., and M. Paque. 1996. Groundwater, Source Water Protection and the Watershed Approach. *Proceedings of Watershed '96.* Water Environment Federation, Baltimore, MD.

McCuen, Richard H. 1982. *A Guide to Hydrologic Analysis Using CSC Methods.* Prentice-Hall, Englewood Cliffs, NJ.

SCS. 1964. *National Engineering Handbook,* sec. 4: Hydrology, Soil Conservation Service. Washington, DC.

SCS. 1970. *Irrigation Water Requirements.* Soil Conservation Service, Washington, DC. Engineering Division, Technical Release 21.

Thornthwaite, C. W., and J. R. Mather. 1957. Instructions and Tables for Computing Potential Evapotranspiration and the Water Balance. Drexel Institute of Technology, *Laboratory of Climatology,* vol. 10, no. 3.

Yates, E. B., and J. H. Wiese. 1988. *Hydrogeology and Water Resources of the Las Osos Valley Ground-Water Basin—San Luis Obispo County, California.* U.S. Geological Survey, Water Resources Investigations Report 88-4081. Denver, CO.

8

Role of Highways in Watershed Planning

Richard A. Moore
P.E., Vice President
Rizzo Associates, Inc.

Henry L. Barbaro
Project Manager
Massachusetts Highway Department

Introduction

Highways crisscross the country and play a role in defining both the physical and water quality characteristics of most watersheds. This chapter presents an approach to deal with three of the most common water quality concerns related to highways: sodium from deicing activities, conventional pollutants from stormwater runoff, and spills from vehicular accidents. Since there are few new highways being built today, this chapter focuses on existing conditions, specifically those in the Hobbs Brook and Stony Brook watersheds in eastern Massachusetts. These watersheds provide the drinking water for the city of Cambridge, Massachusetts.

The Watershed

The city of Cambridge draws most of its 15 million gallon per day (Mgal/day) public water supply from the Hobbs Brook and Stony Brook reservoirs. The watersheds for these reservoirs comprise approximately 23.7 mi^2 in Lexington, Lincoln, Weston, and Waltham, as shown in Fig. 8.1. The characteristics of the reservoirs and their watersheds are summarized in Tables 8.1 and 8.2.

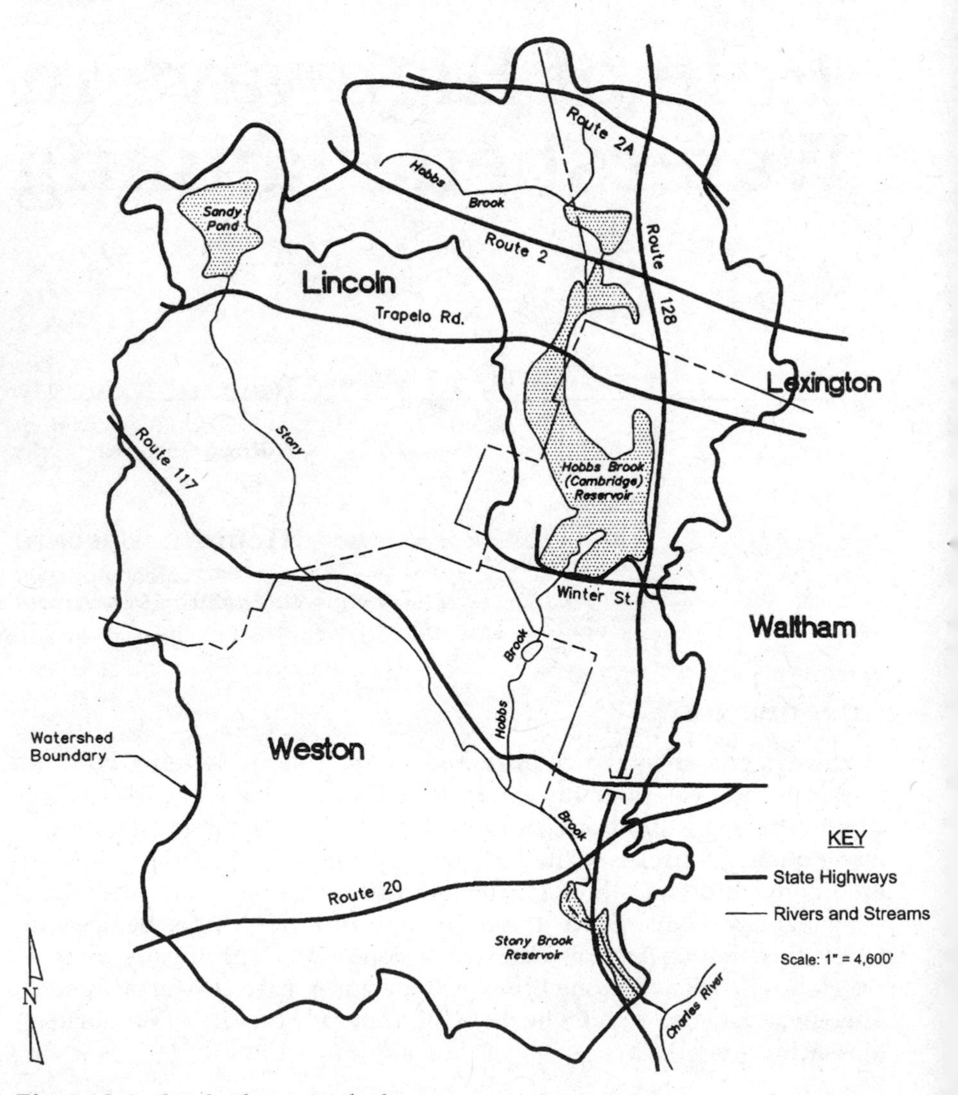

Figure 8.1 Cambridge watershed.

Table 8.1 Reservoir Characteristics

Characteristics	Hobbs Brook	Stony Brook
Drainage area, mi^2	6.8	16.6
Surface area, acres	580	63
Normal storage volume, Mgal	2,700	345

Table 8.2 Combined Watershed Characteristics

Land use	Area, acres	Percentage of total
Open space and forest	9,320	61
Residential	4,620	30
Transportation	390	3
Commercial and industrial	860	6
Total	15,190	

Five major roadways are within the watersheds. They include route 128 (Interstate 95), route 2, route 2A, route 20, and route 117. The Massachusetts Highway Department (MHD) is responsible for maintenance and deicing of approximately 90 lane mi of roadway in the watershed. The local roadways are maintained by the individual communities.

While all five Mass Highway roadways contribute urban runoff to the reservoir system, routes 128 and 2 are directly adjacent to the reservoirs and have direct stormwater discharges to them. Within the watershed, a total of 127 storm drain outfalls serve route 128, and 33 outfalls serve route 2. In order to identify and rank those Mass Highway storm drain outfalls with the greatest potential for impacting water quality in the reservoirs, a ranking system was applied to the known outfalls serving route 128 and route 2 as illustrated in Table 8.3.

Table 8.4 summarizes the rankings assigned to the outfalls serving routes 128 and 2. In all, 58 outfalls, or 36 percent of those ranked, fall into the A category. Figure 8.2 indicates the general location of the different outfall types, ranking A through D.

Table 8.3 Outfall Ranking System

Rank	Distance to reservoir, ft
A*	<100*
B	100–500
C	500–1000
D	>1000

*Also includes outfalls that discharge directly to perennial brooks.

Table 8.4 Ranking of Outfalls within the Cambridge Reservoir Watershed

Ranking	Route 128	Route 2	Total	Percent of total
A	49	9	58*	36
B	16	6	22	14
C	19	0	19	12
D	43	18	61	38
Total	127	33	160	100

*Includes 12 outfalls that discharge directly to tributary brooks.

The Issues

The presence of highways in the Hobbs Brook and Stony Brook watersheds raises the following three typical issues.

Issue 1: Sodium from deicing operations

Sodium is contained in almost all foods and is a necessary part of the human diet. However, chronic excessive consumption of sodium has been related to heart and kidney ailments. Although sodium in drinking water typically accounts for less than 2 percent of a person's total daily intake, the American Heart Association in 1957 proposed a limit of 20 mg/L of sodium in drinking water for those persons on a "strict" diet of 500 to 1000 mg/day.

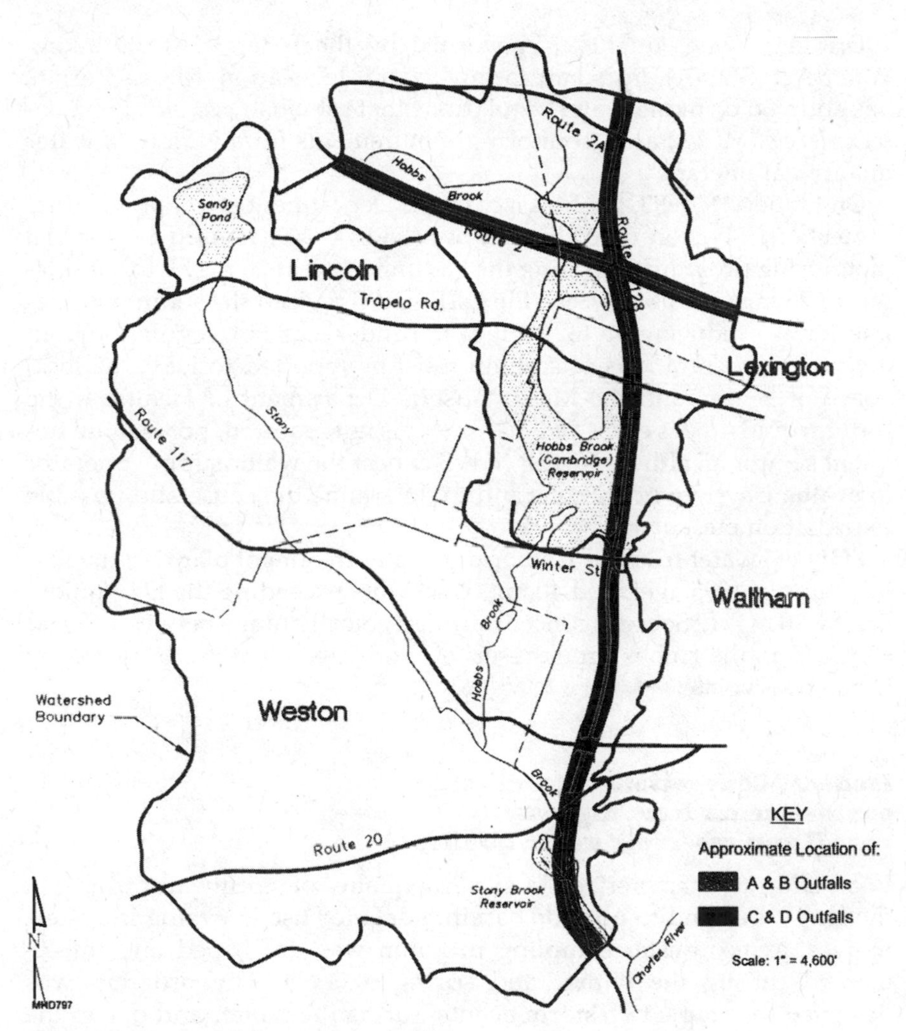

Figure 8.2 Outfall locations by ranking.

Drinking water quality is governed by the federal Safe Drinking Water Act (SDWA). Pursuant to this federal legislation, Massachusetts has adopted drinking water regulations that establish *maximum contaminant levels* (MCLs) and monitoring requirements for a variety of water quality parameters.

On January 1, 1993, the Massachusetts Department of Environmental Protection, Division of Water Supply (DEP/DWS) revised its sodium monitoring program, changing the sodium MCL of 20 mg/L to a guideline of 28 mg/L. In May 1994, the DEP/DWS revised the sodium guideline again, reducing it to 20 mg/L. Under current regulations, all detected concentrations of sodium must be reported to DEP, the local board of health, and the Massachusetts Department of Health. If the sodium guideline is exceeded, treatment is not required, nor is more frequent sampling, although DEP/DWS expect the water supply operator to evaluate system operations and to determine the cause and possible remediation measures.

Finished water from the Cambridge water treatment plant for the first 6 months of 1995 averaged 40 mg/L sodium, exceeding the DEP guideline by 20 mg/L. Sodium concentrations typically range between 40 and 60 mg/L in the Hobbs Brook reservoir and 20 to 40 mg/L in the Stony Brook reservoir.

Issue 2: Conventional stormwater contaminants from highway runoff and reservoir water quality

In an effort to characterize the existing quality of stormwater runoff in the study area and to provide baseline data for use in evaluating alternatives, a stormwater sampling program was developed and implemented during the winter and spring of 1994. The program was designed to sample two storm events during the winter and one in the spring. This plan was to collect data for comparison with published data from other sites around the country.

Table 8.5 presents a summary of the stormwater pollutant *event mean concentrations* (EMCs) obtained from this field sampling effort, compared with data from other highway and urban runoff investigations. Figure 8.3 shows the sampling locations. In general, the quality of stormwater in the study area is similar to that of the reference data (i.e., Federal Highway Administration and Nationwide Urban Runoff Program) for typical highway and urban runoff. Deicing practices and snowmelt can result in elevated concentrations of some pollutants.

Table 8.5 Stormwater Runoff Quality—Study Samples and Reference Data

Pollutant	Storm no.	Cambridge Watershed Study Monitoring Station* HB-1	HB-2	HB-3	FHWA, 1990† Normal runoff	Snow wash-off	NURP 1983† urban runoff
Chloride (mg/L)	1	4200	1500	2700	13	400	—
	2	42	29	340		5600	
	3	71	8	250			
TSS (mg/L)	1	490	340	100	93	204	100
	2	350	160	110			
	3	120	58	15			
Oil and grease (mg/L)	1	16.0	11.0	7.9	10	—	10–15
	2	9.1	3.5	2.3			
	3	13.0	2.5	0.5‡			
TKN (mg/L)	1	3.5	4.7	4.1	1.48	1.94	1.50
	2	0.8	1.0	1.0			
	3	7.2	4.4	1.2			
NO_3 (mg/L)	1	0.320	0.670	0.880	0.660	0.680	0.680
	2	0.140	0.270	0.650			
	3	2.300	1.200	0.490			
Total phosphorus (mg/L)	1	0.080	0.240	0.020	0.293	0.570	0.330
	2	0.350	<0.010	0.170			
	3	0.480	0.410	0.040			
Copper (μg/L)	1	90	50	60	39	91	34
	2	60	20	10			
	3	70	50	10			
Lead (μg/L)	1	200	110	80	234	549	144
	2	130	60	50			
	3	50	50	50			
Zinc (μg/L)	1	530	310	520	217	420	160
	2	290	80	200			
	3	740	140	70			

*Concentrations listed are from composite samples. Runoff from storm 3 at the Station HB-1 highway ramp provided a grab sample only. Grab samples are listed in place of the composite for this sample.

†Results for the concentrations listed are median values for median site in data set. Individual EMCs range from about 0.3 to 3.0 times value shown.

‡Detection limit.

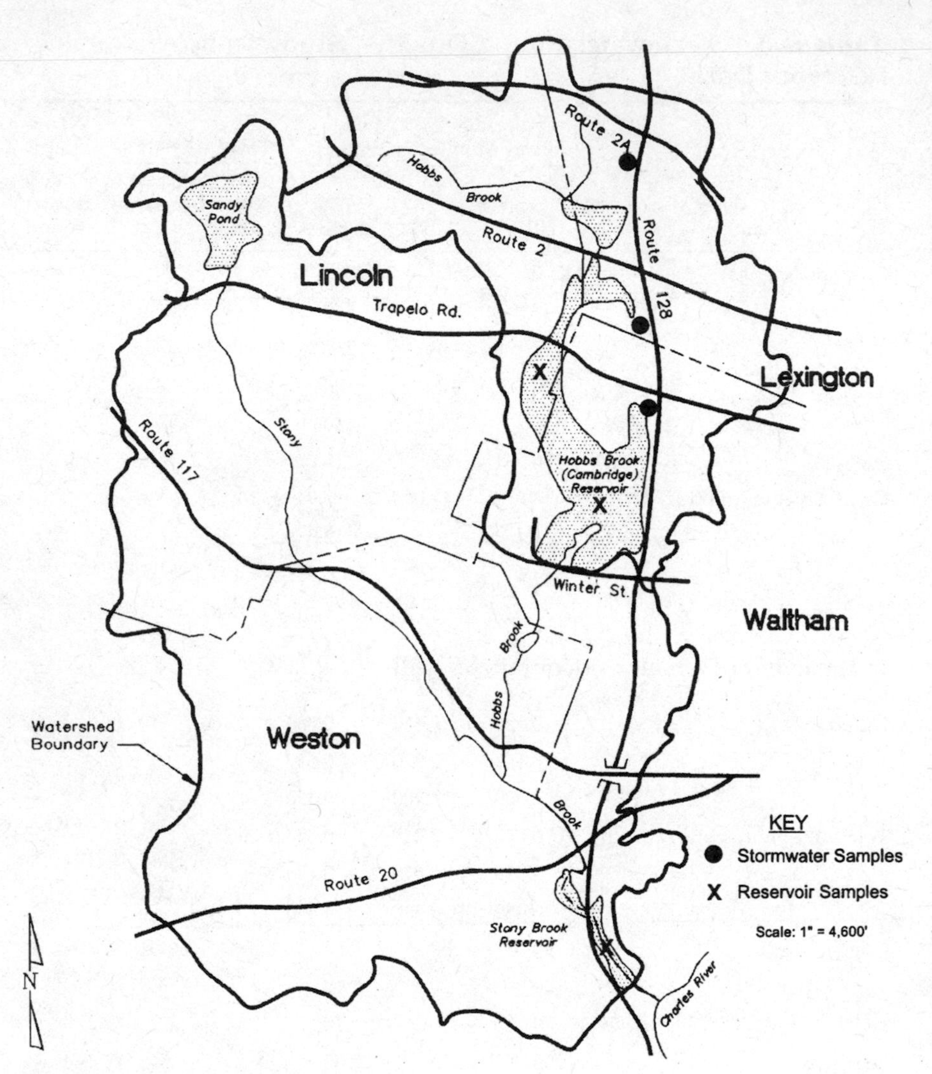

Figure 8.3 Cambridge watershed.

Except for sodium, there are limited water quality data describing surface water conditions in the reservoirs and no data for the tributary streams. Three 1994 sampling events shown in Table 8.6 are typical of the limited available data in Hobbs Brook and Stony Brook reservoirs. The sampling locations are shown in Fig. 8.3. In general, based on the data that are available, the water quality in the reservoirs is good and does not show adverse impacts from conventional highway contaminants. As noted above, there is no information describing stormwater impacts on the stream tributary to the reservoirs.

Issue 3: Spills

Fortunately, there have not been any major transportation spills on the highways in the Hobbs Brook and Stony Brook watersheds. There have been small spills, generally less than 100 gal, that in several cases have entered the drainage system. In such cases, emergency response teams have contained and cleaned up the material before it reached the reservoir. However, as noted above, 56 drainage outlets are within 100 ft of the reservoirs or perennial streams, creating a threat to the water supply in the event of a major uncontained spill.

The Solutions

The Mass Highway has developed plans to address each of the three issues noted above. The plans were selected based on the ability to address each issue, other environmental impacts, costs, and operational considerations including safety. The plans are described below.

Management of sodium

The Massachusetts Highway Department has operated a reduced salt program within the Hobbs Brook and Stony Brook reservoir watersheds since the winter of 1986–1987. Starting in 1986–1987, the Mass Highway used a 1:1 mixture of sand and a premix of sodium chloride (rock salt) and calcium chloride at a ratio of 4:1. During ice and snow events, the entire mixture was applied at a rate of approximately 240 lb/lane mi to the 90 lane mi of highway maintained by the Mass Highway within the Cambridge reservoir watershed. Application rates sometimes were increased on the interchanges when required by severe storm events.

Prior to 1986, the Mass Highway applied straight sodium chloride to these roadways at a rate of 400 lb/lane mi per application. From the

Table 8.6 Reservoir Quality Sampling Data

	Site: Location Date-Time:		Hobbs Brook Reservoir: RES-1 Sample 1 4/20/94-12:35		Sample 2 6/23/94-10:50		Sample 3 8/25/94-13:05	
	Sampling location:		Surface	17N Depth	Surface	10N Depth	Surface	9N Depth
Parameter	Units	MDL	Result	Result	Result	Result	Result	Result
Dissolved sodium	mg/L	0.5	71	72	63	65	66	66
Chloride	mg/L	10	130	130	130	130	130	130
Total dissolved solids	mg/L	5	270	280	260	260	280	260
Total suspended solids	mg/L	5	ND	ND	ND	ND	ND	ND
BOD_5	mg/L	5	5	ND	2.3	ND	ND	ND
COD	mg/L	7	21	21	20	19	15	10
Nitrate	mg/L	0.1	0.39	0.41	0.28	0.29	ND	ND
Total Kejhdal nitrogen (TKN)	mg/L	0.05	0.27	ND	0.81	0.4	0.36	0.92
Ortho-phosphate	mg/L	0.01	ND	ND	ND	ND	ND	ND
Chlorophyll	mg/m^3		5.61	4.01	0.84	ND	2.06	2.30
Oil and grease	mg/L	*	ND	ND	ND	ND	ND	ND
Total petroleum hydrocarbons	mg/L	0.5	ND	ND	ND	ND	ND	ND
Polycyclic aromatic hydrocarbons	mg/L	‡	ND	ND	ND	ND	ND	ND

Dissolved metals								
Chromium (hexavalent)	mg/L	0.02	ND	ND	ND	ND	ND	ND
Total cyanide	mg/L	0.005	ND	ND	ND	ND	ND	ND
Cadmium	mg/L	0.005	ND	ND	ND	ND	ND	ND
Copper	mg/L	0.01	ND	ND	0.02	ND	ND	ND
Lead	mg/L	0.05	ND	ND	ND	ND	ND	ND
Zinc	mg/L	0.01	0.02	0.02	0.02	ND	ND	ND
DO†	mg/L	NA	10.0	3.0	6.7	4.9	A	A
Conductivity†	Fmhos	NA	370	—	300	300	410	410
Temperature†	°C	NA	12	14	27	26	23	22
pH†	Su	NA	6.2	6.2	7.2	7.2	7.5	7.5

Table 8.6 (*Continued*) Reservoir Quality Sampling Data

	Site: Location: Date-Time:		Hobbs Brook Reservoir RES-2 Sample 1 4/20/94-14:40		Sample 2 6/23/94-13:25		Sample 3 8/25/94-14:10	
	Sampling location:		Surface	16N Depth	Surface	7N Depth	Surface	6N Depth
Parameter	Units	MDL	Result	Result	Result	Result	Result	Result
Dissolved sodium	mg/L	0.5	69	71	64	64	67	68
Chloride	mg/L	10	130	130	130	130	140	130
Total dissolved solids	mg/L	5	270	270	260	260	260	250
Total suspended solids	mg/L	5	ND	ND	ND	ND	ND	ND
BOD_5	mg/L	5	3.9	3.5	ND	ND	ND	ND
COD	mg/L	7	26	23	20	15	ND	ND
Nitrate	mg/L	0.1	0.41	0.4	0.29	0.29	ND	ND
Total Kejhdal nitrogen (TKN)	mg/L	0.05	0.32	ND	ND	0.33	0.96	1.3
Ortho-phosphate	mg/L	0.01	ND	ND	ND	ND	ND	ND
Chlorophyll	mg/m^3		3.2	2.4	0.801	ND	2.98	3.05
Oil and grease	mg/L	*	ND	ND	ND	ND	ND	ND
Total petroleum hydrocarbons	mg/L	0.5	ND	0.7	ND	ND	ND	ND
Polycyclic aromatic hydrocarbons	mg/L	‡	ND	—	ND	ND	ND	ND

Dissolved metals								
Chromium (hexavalent)	mg/L	0.02	ND	ND	ND	ND	ND	ND
Total cyanide	mg/L	0.005	ND	ND	ND	ND	ND	ND
Cadmium	mg/L	0.005	ND	ND	ND	ND	ND	ND
Copper	mg/L	0.01	ND	ND	ND	ND	ND	ND
Lead	mg/L	0.05	ND	ND	ND	ND	ND	ND
Zinc	mg/L	0.01	0.02	0.01	0.01	ND	ND	ND
DO†	mg/L	NA	7.5	4.2	6.5	5.6	A	A
Conductivity†	Fmhos	NA	380	—	370	370	410	410
Temperature†	°C	NA	12	10	25	24	24	23
pH†	Su	NA	6.5	6.5	7.2	7.2	7.6	7.6

Table 8.6 (*Continued*) Reservoir Quality Sampling Data

	Site: Location: Date-Time:		Stony Brook Reservoir RES-3 Sample 1 4/20/94-9:30		Sample 2 6/23/94-8:50		Sample 3 8/25/94-9:10	
	Sampling location:		Surface	18N Depth	Surface	5N Depth	Surface	12N Depth
Parameter	Units	MDL	Result	Result	Result	Result	Re sult	Result
Dissolved sodium	mg/L	0.5	25	24	38	38	57	58
Chloride	mg/L	10	47	48	74	75	110	110
Total dissolved solids	mg/L	5	140	130	190	180	250	260
Total suspended solids	mg/L	5	ND	ND	ND	ND	ND	ND
BOD_5	mg/L	5	3.4	7.8	2.4	2.4	ND	ND
COD	mg/L	7	30	24	24	25	ND	ND
Nitrate	mg/L	0.1	0.6	0.61	0.6	0.6	0.19	ND
Total Kejhdal nitrogen (TKN)	mg/L	0.05	0.3	0.38	0.36	0.49	0.40	0.40
Ortho-phosphate	mg/L	0.01	ND	ND	ND	ND	ND	ND
Chlorophyll	mg/m^3		1.98	10.1	0.817	1.2	2.48	1 .84
Oil and grease	mg/L	*	ND	ND	ND	ND	ND	ND
Total petroleum hydrocarbons	mg/L	0.5	ND	0.7	ND	ND	ND	ND
Polycyclic aromatic hydrocarbons	mg/L	‡	ND	ND	ND	ND	—	—

Dissolved metals								
Chromium (hexavalent)	mg/L	0.02	ND	ND	ND	ND	ND	ND
Total cyanide	mg/L	0.005	—	—	ND	ND	ND	ND
Cadmium	mg/L	0.005	ND	ND	ND	ND	ND	0.006
Copper	mg/L	0.01	ND	ND	ND	0.04	ND	0.03
Lead	mg/L	0.05	ND	ND	ND	ND	ND	0.014
Zinc	mg/L	0.01	0.01	0.01	ND	ND	ND	0.23
DO†	mg/L	NA	10.6	3.2	5.2	4.8	3.5	1.8
Conductivity†	Fmhos	NA	170	—	350	350	370	—
Temperature †	°C	NA	12	9	22	22	23	22
pH†	Su	NA	7.2	7.2	7.0	7.0	7.5	7.5

— Not applicable or not tested for
ND Not detected
MDL Mean detection limit
A Measurement not taken due to equipment failure.
*All samples laboratory-analyzed except where indicated.
†Field measurement, taken by Rizzo Associates
‡Variable detection limits for specific PAHs analyzed by GC/MS method.

winter of 1970–1971 through 1985–1986, the average annual sodium chloride application rate in the watersheds was 23 tons per lane mi. Since 1986, the annual sodium chloride application has been reduced to an average of approximately 9 tons per lane mi. This is a 60 percent decrease in the sodium chloride application rate within the watershed. However, due to other potential sources of sodium chloride not accounted for or underestimated, measurable decreases of sodium concentrations within the reservoirs have not yet been realized.

As the 60 percent reduction in sodium chloride application rate suggests, the Mass Highway has already accomplished a great deal in the management of deicing practices. However, other alternatives that could further reduce the amount of sodium used warrant consideration by both the Mass Highway and municipalities within the watershed. These fall into two categories: alternative deicing practices and diversion of sodium-laden water out of the watershed. Each is discussed below.

Alternative deicing practices. The primary purpose of evaluating alternative deicing methods is to reduce the amount of rock salt (sodium chloride) and therefore the resulting sodium load to the reservoir system. Over the past 20 years, a variety of alternatives have been evaluated and tested throughout North America. The most practical include using rock salt more effectively or using alternative chemicals.

Rock salt has long been the preferred deicing chemical of highway departments, due to both its low cost and its effectiveness at breaking and preventing the ice-pavement bond at most commonly encountered winter storm temperatures.

Before we discuss practices to minimize rock salt use, it is important to understand its role in the deicing operation. Rock salt prevents and/or breaks the formation of the ice-pavement bond. For both chemical and physical reasons, ice bonds tenaciously to the porous surface of highway pavement. Since water expands when it freezes, water-filled pores in the pavement become tightly wedged pockets of ice, which serve as anchors for the layer of ice that follows. Even chemical energy has a difficult time penetrating the pore-filled pockets of ice that are the foundation of the ice-pavement bond.

Practices utilized to reduce total rock salt use, while at the same time maintaining safe operating conditions during ice and snow events, are reviewed below.

1. *Application of a calcium chloride brine solution prior to storm onset.* A strong brine solution applied to a roadway pavement prior to the onset of ice-forming conditions will act to prevent ice formation. Ice cannot form when the pavement pores are filled with strong chloride brine. That strong brine is approximately 20 percent heavier than water, so it

is not easily displaced from the pavement pore spaces. If applied early enough, the chemical will act in tandem with the heat of hydration and the heat generated by vehicular tire-pavement friction in preventing the formation of the ice-pavement bond. The heat melts the ice, and the brine depresses the freezing point of the resulting solution.

2. *Prewetting solid calcium chloride during application.* Prewetting rock salt with liquid sodium chloride or liquid calcium chloride provides the necessary solution to "quick-start" the dissolution of the rock salt. Prewetting also reduces the quantity of rock salt wasted due to overscattering, wind loss, traffic action, and premature plowing.

If a strong brine solution is applied prior to the solid rock salt application, there may be no need to prewet the solid rock salt. The brine solution may provide the necessary solution to both dissolve the rock salt and prevent scattering. One convention is to prewet solid rock salt and apply at an average rate of 150 lb/lane mi on the main line and up to 300 lb/lane mi on the ramps depending on weather conditions.

3. *Tailoring rock salt application to weather conditions, equipment, and personnel limitations.* Tailoring rock salt application rates to weather conditions involves estimating the workload that is to be "assigned" to the rock salt and using what is known about its ability to perform under the given or expected conditions. The first step in tailoring the rock salt application rate is to determine the amount of rock salt required to penetrate any accumulated snow.

The ideal rate of rock salt application is obviously influenced by the amount of existing ice pack and snow on the road, the anticipated snowfall between treatment, the pavement temperature, predicted ambient temperature and precipitation, and traffic. Current practice in Massachusetts generally dictates a constant 120 to 240 lb/lane mi deicing chemical. This is considered reasonable.

To fine-tune application rates to weather conditions, an accurate accounting procedure must be in place to monitor and document the materials applied as well as to provide information for ongoing program adjustments. The spreader routes, material applied, and weather and roadway conditions are documented at each depot, by Mass Highway. These data are routinely entered into a computer database.

4. *Maximizing plowing operations.* An important step in minimizing the use of rock salt is to maximize the use of snowplows. Ideally, the chemical deicer should be used to accomplish only what the plows cannot. Plows cannot break or prevent the formation of the ice-pavement bond. That task is accomplished by deicing chemicals.

Maximizing the effectiveness of rock salt application is accomplished by plowing off as much as possible of the snow and ice before chemical application, and by synchronizing each subsequent plowing and salting application with the effective time spans of the previous application. This is accomplished through careful monitoring of the roadway surface and by calculating the timing of the next plowing application, a standard practice in Massachusetts.

5. *Eliminating rock salt applications below 20°F.* At temperatures below 20°F ice is not likely to form at the pavement surface, primarily because there is very little moisture available to freeze. Actual field conditions will also depend on traffic, pavement, and air temperatures and precipitation. Straight rock salt application at temperatures below 20°F likely serves little purpose, both because no ice is forming and because there is no moisture available to dissolve the rock salt. At these temperatures, rock salt offers only traction, a benefit better served by sand. Hence, rock salt use may be curtailed or eliminated at these temperatures.

These five practices have been implemented either wholly or partly by the Mass Highway. The agency continues to consider and evaluate these five practices which have the potential for reducing the average annual sodium chloride applications within the reservoir watersheds. In addition to more effectively using rock salt, alternative chemicals are available. The list of alternative chemical deicers is relatively long and includes calcium chloride (already mentioned in liquid form), sodium chloride, and calcium chloride "premix," calcium magnesium acetate (CMA), urea, glycol, magnesium chloride, sodium formamides, and sand. Sand is also included in the list of alternative deicers as it is often used in conjunction with, or as a direct substitute for, many of these compounds.

As shown in Table 8.7, certain compounds are inappropriate deicing alternatives for use within the reservoir watersheds. These include urea, glycol, magnesium chloride, and sodium formamides. CMA is also not considered appropriate because it would promote eutrophication in the reservoirs. This leaves calcium chloride as the preferred alternative to sodium chloride.

Calcium chloride used in a brine solution as described above has been tested over the last several years by the Mass Highway. Using calcium chloride brine solution at the beginning of the storm followed by applications of prewetted sodium chloride could reduce the average annual sodium chloride loading to the reservoirs. Mass Highway has tested the use of liquid $Ca\ Cl_2$ for both antiicing and deicing within the watersheds.

Diversion of sodium out of the watershed. Despite the efforts being made by the Mass Highway since 1986 with regard to reducing sodium use in the watersheds, the concentrations have not significantly changed in the surface waters of the Hobbs Brook and Stony Brook

Table 8.7 Alternative Chemical Deicer and Sand

Deicer	Cost	Availability	Corrosive	Environmental impacts	Effectiveness/notes
Premix 4:1 (sodium chloride and calcium chloride)	$80/ton	Good	Yes	Low to medium	Roadway may appear wet or slippery to motorist for some time after the storm, if overapplied.
Sodium chloride (rock salt)	$32/ton	Good	Yes	Medium to high	Effective deicer at temperatures above 20°F.
Calcium chloride	$175/ton as flake	Good	Yes	Low	Effective at lower temperatures than sodium chloride (< 20°F). Attracts moisture from the air and hastens dissolving and melting. Supplied in bags and requires extra handling. Roadway may appear wet or slippery to motorist for some time after the storm, if overapplied.
Liquid calcium chloride	$0.50/gal	Good	Yes	Low	Not selective at melting or penetration (e.g., undercutting and disbandment). Good for larger temperature range for effectiveness (above 5°F). Accelerates solution of sodium chloride.

Table 8.7 (*Continued*) Alternative Chemical Deicer and Sand

Deicer	Cost	Availability	Corrosive	Environmental impacts	Effectiveness/notes
Calcium magnesium acetate	$1000/ton	Good	No	Low to medium	Poor deicing ability at lower than 25°F and limited ability to penetrate ice. May need to use 20–70% more material than sodium chloride. Not effective at low temperatures. Promotes eutrophication in surface waters and depletes oxygen in groundwaters.
Sand	$6/ton	Good	No	Medium	Costly to clean up. Has limited applications under specific conditions. Does not address deicing.
Urea	$200–250/ton	Good	No	Medium to high	Urea is a fertilizer rich in nitrogen. Promotes eutrophication in surface waters and adds nitrogen to groundwaters. Has a limited melting point of 25°F. Expensive. Primarily used at airports.
Glycol	$3.50/gal	Good	No	Low to high	Effective at low temperatures, although high BOD can cause DO deficit in surface water. Commonly used at airports. Noncorrosive.
Magnesium chloride	$420/ton	Good	Yes	Low to medium	Works well at low temperatures (0°F). Wicks moisture from air and hastens dissolving and melting but can result in pavement being wet for extended time.
Sodium formamides	$800/ton	Poor	Yes	Low	Contains sodium, is very expensive and not readily available.

reservoirs. Because of this an alternative was considered that consisted of collecting, diverting, and treating sodium-laden runoff with a discharge either back into the Stony Brook reservoir or downstream to the Charles River.

The system needed to implement this alternative would be extensive and would include the following:

- 12.5 mi of 30-ft-wide snow berms
- 9 mi of storm drain 12 to 42 in in diameter
- 3.5 mi of force main to 6 to 26 in in diameter
- 8 pumping stations
- 4 in-line storage facilities and an advanced wastewater treatment plant

Figures 8.4, 8.5, and 8.6 show the typical drainage system modifications, 30-ft-wide snow berm design, and overall drainage system layout, respectively.

The effect of this construction on highway operations and adjacent properties would be significant. Of equal significance would be the impact on wetland resources, which is quantified below:

- Bordering vegetated wetland = 520,000 ft^2
- Bank = 9400 ft
- Land under water = 38,400 ft^2

The combined impacts of costs in excess of $50 million and the serious environmental consequences do not make this a viable alternative to manage sodium in the Hobbs Brook and Stony Brook reservoirs.

Monitoring and modeling to predict sodium concentrations. To better understand the possible causes for the persistently high sodium concentrations in the reservoirs, the Mass Highway has undertaken two companion programs. One includes monitoring and the other modeling. The Mass Highway has funded the installation of eight monitoring stations in the watersheds. The stations have been installed and are operated by the U.S. Geological Survey (USGS). The stations continuously record flow and specific conductance. The goal is to collect enough data to better identify the sources of sodium in the watersheds.

As a companion to the monitoring program, a mass-balance water quantity and quality model has been developed for the reservoir system. The data obtained in the monitoring program will be used to calibrate the model and make water quality predictions. Because of the lack

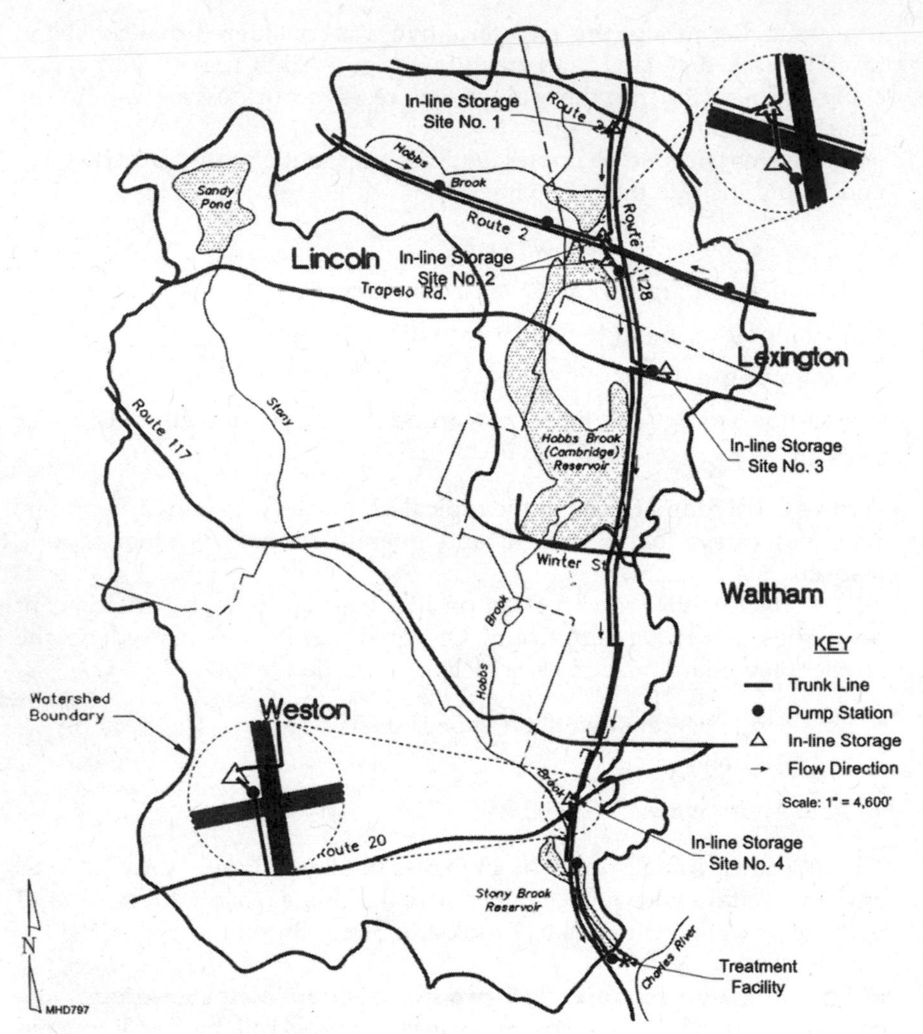

Figure 8.4 Proposed drainage system.

of site-specific data, a number of assumptions were made in constructing the model and developing input data, as described below.

The model is designed to track both quantities of water and masses of sodium throughout the Cambridge watershed system. Figure 8.7 provides a schematic diagram of this system. In general, both precipitation, which falls over the entire watershed, and sodium move downstream from the Hobbs Brook reservoir watershed to the Stony Brook watershed, and finally to Fresh Pond reservoir, about 10 mi away. Therefore,

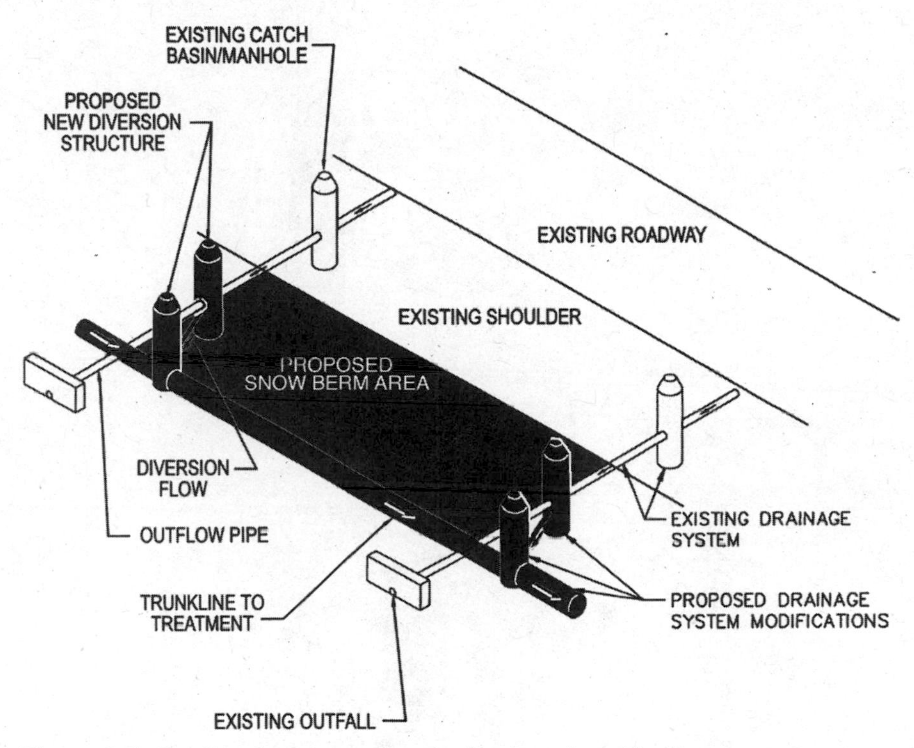

Figure 8.5 Existing drainage system and proposed modifications.

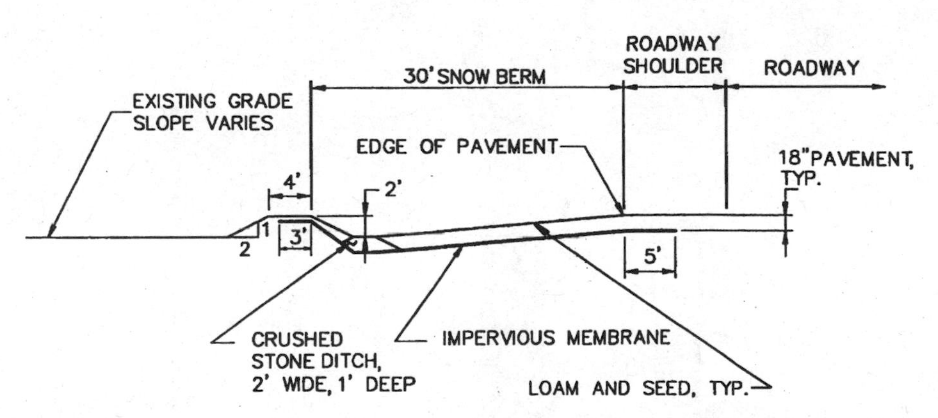

Figure 8.6 Typical level section of proposed snow berm.

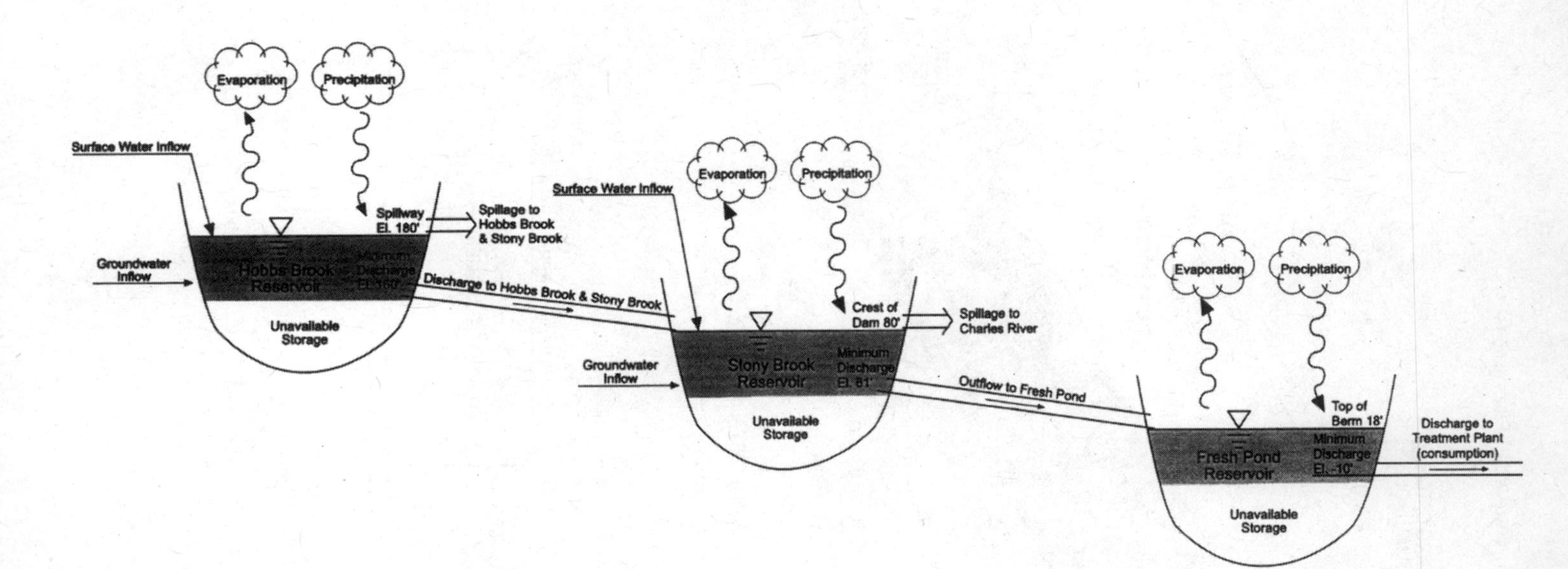

Figure 8.7 Schematic diagram of the Cambridge reservoir system.

the two major inputs to the model are precipitation and sodium used for deicing.

Sodium moves throughout the system in both surface water and groundwater. As a result, the time of travel of both the water and the sodium is a function of the hydraulic and hydrogeologic characteristics of the watershed. Because the watershed is not uniform everywhere with regard to the hydrogeologic characteristics of the subsurface materials and because of different hydraulic characteristics of the three reservoirs, the watershed is divided into 11 separate subdrainage areas and 3 separate reservoir models. Each subdrainage model area (also referred to as *area*) and each reservoir model are modeled separately and linked together on a monthly time step, as both water and sodium are tracked from the subdrainage areas to the reservoir models.

Each of the 11 areas is assigned a unique set of values of the variables that appears to best represent generalized local hydrogeologic characteristics (based on limited data however). The hydrodynamic dispersion and time of travel of the sodium in groundwater are simulated separately within each area.

Figure 8.8 is a schematic diagram of the conceptual model that indicates the movement of water and sodium throughout the Cambridge watershed system. As seen from the diagram, water enters the system through precipitation (P) falling on land and on the reservoirs. The precipitation either infiltrates (INF) into the ground or runs off (R) as surface water flow to a nearby stream, swamp, or pond. The infiltrating water enters the groundwater (GWD), which flows downhill and discharges either to a local stream or directly to the reservoir. The stream flows to either the Hobbs Brook reservoir or Stony Brook reservoir. Once in the Hobbs Brook reservoir, the surface water is released through either spillage (RS) or discharge (RD) and travels to Stony Brook reservoir via Hobbs Brook. Surface water in the Stony Brook reservoir is released through spillage to the Charles River or discharged to Fresh Pond reservoir. Water can also leave the system through evaporation (E), transpiration (T), and discharge (D) from Fresh Pond.

Sodium is input to each of the subdrainage areas based on the total estimated annual mass of sodium load to each area. The total estimated sodium loads are determined by adding sources of sodium within each area. These sources include the sodium that is applied to both municipal and private roadways and parking lots, and the Mass Highway roads as deicing salt. The total estimated load for each source is calculated by first determining or estimating an average annual application rate (i.e., tons per lane mi) for each source and then applying that rate to the total source area within the subdrainage area (i.e., total lane mi). The total annual sodium load is added to the subdrainage areas during the three winter months of January, February, and March. In practice, salt can be applied as early as November and as late as April.

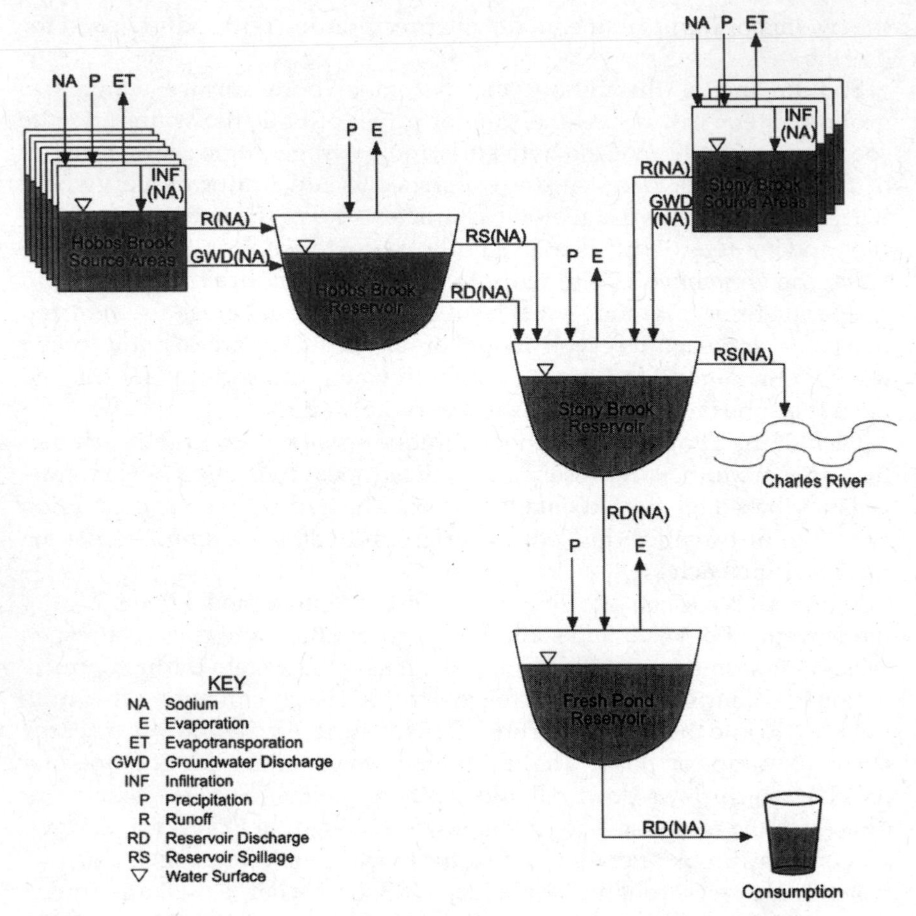

Figure 8.8 Schematic diagram of conceptual hydrogeologic sodium model.

Once added to the subdrainage areas, the sodium is dissolved and travels in the surface water and groundwater. Sodium in surface water discharges directly into the reservoir during the next time step. The sodium modeled in groundwater is subjected to dispersion, and its travel time is a function of the local hydrogeologic conditions. In effect, the sodium is modeled in the groundwater as individual plumes or slugs that may eventually overlap as they discharge to the nearest body of surface water over time.

The total reservoir water volume is recalculated each month, based on the monthly water balance. The total mass of sodium in the reservoir is also recalculated each month with input from both surface water runoff and groundwater discharge, and output from spillage and discharge.

The total sodium mass in the reservoir is divided by the total reservoir water volume to determine the reservoir sodium concentration.

The sodium loading and distribution between runoff and groundwater were based on data from Mass Highway and municipal salt application data and estimates of loadings from private sources such as industrial and commercial parking lots and driveways.

The sodium load is assumed to be distributed between the runoff and groundwater based on the estimated runoff coefficient. The mass of sodium in the runoff is discharged directly into the downstream reservoir during the next monthly time step. The sodium mass in the infiltrating groundwater is added to the initial sodium mass in groundwater storage. At the end of each month, the total sodium mass in the groundwater is recalculated along with a new groundwater concentration. This assumes complete mixing for groundwater subdrainage areas. However, a time delay is established by delaying the discharge of the sodium mass into the surface water. The mass of sodium in groundwater discharged into the surface water is a function of the initial sodium mass in storage, the current mass loading, the sodium hydrodynamic dispersion, and the groundwater time of travel.

The amount of hydrodynamic dispersion is directly related to the geologic material and the time of travel prior to discharge. As a result, the model simulates plug flow and allows for mixing between plugs or sodium plumes.

The sodium time of travel is directly linked to the advective transport mechanism of groundwater flow. No attempt is made to simulate any potential retardation or degradation of sodium. Hydrodynamic dispersion of sodium occurs as it travels along with the groundwater. Hydrodynamic dispersion causes the sodium mass to spread out over time, resulting in a portion of the sodium mass discharging both ahead of and behind the advective front.

Figure 8.9 and Fig. 8.10 illustrate a theoretical distribution of the total sodium discharge mass percentages over time within till and bedrock and sand and gravel subbasins respectively. These plots indicate the effect of dispersion and the difference in travel time with respect to the surficial geology and the average distance to a discharge point. The initial sodium mass is discharged over a much longer period within the till and bedrock area than within the sand and ground area.

Figure 8.11 illustrates the theoretical sodium loads in groundwater and the corresponding mass discharge as well as the resulting sodium mass in storage over time in a till and bedrock subbasin. The figure illustrates how the increased sodium loading in the watershed leads to an increase in discharge mass.

Both total system storage volumes and historical water quality for 1970 through 1992 were input into the model. Figure 8.12 illustrates the

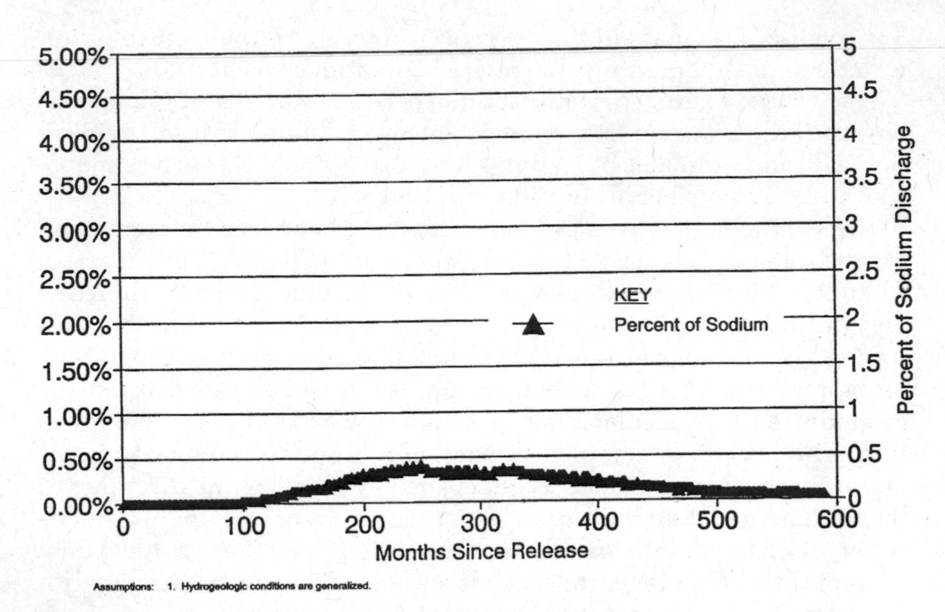

Figure 8.9 Simulated sodium discharge curve, area HB-2.

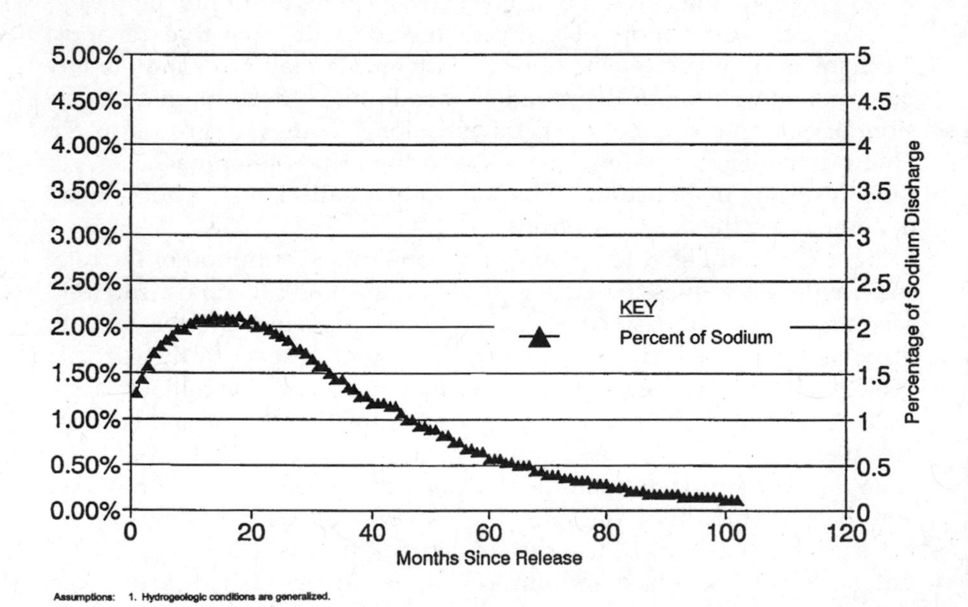

Figure 8.10 Simulated sodium discharge curve, area HB-7.

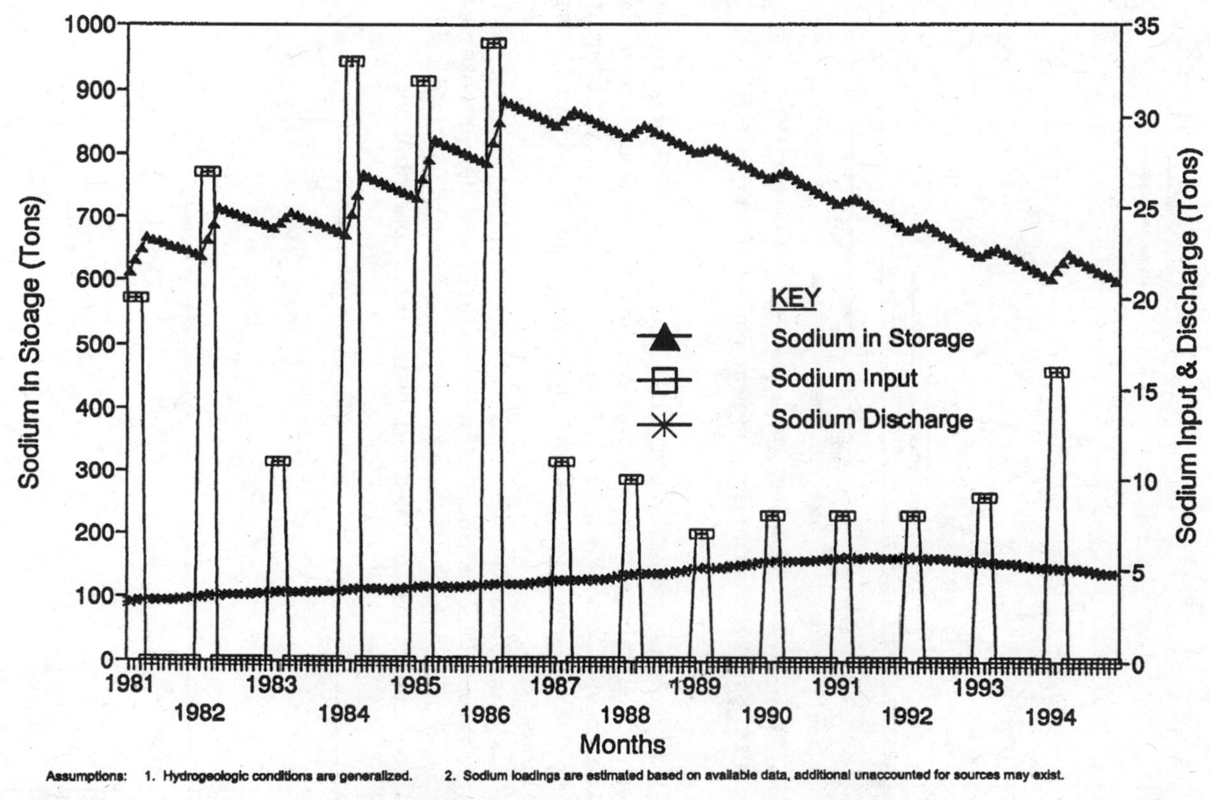

Figure 8.11 Simulated sodium balance in groundwater, area HB-3.

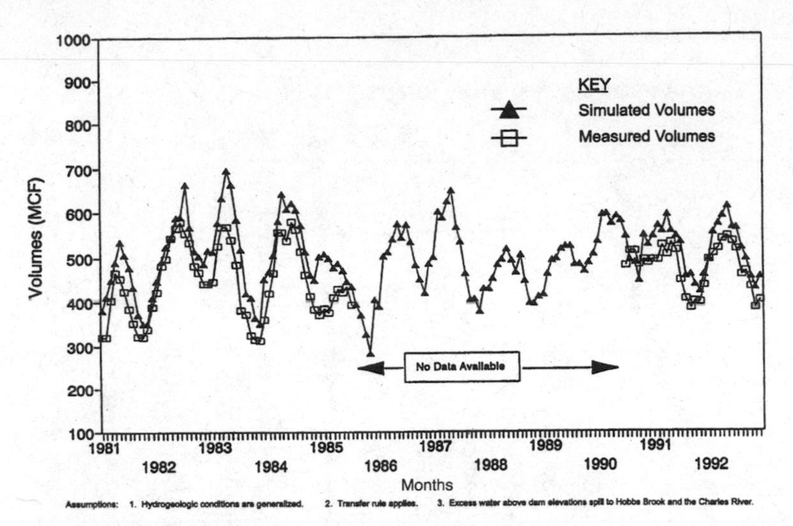

Figure 8.12 Simulated and measured system volume, 1981 to 1993.

final total volume calibration results. This graph indicates the fit between the simulated volumes and the historical volumes over the period of record from 1981 to 1993. A statistical analysis of the comparison indicates a mean error of 4 percent between simulated and historical volumes.

The model is calibrated to historical sodium concentrations in the reservoirs, as shown in Figs. 8.13 and 8.14 for Hobbs Brook and Stony Brook reservoirs, respectively. Where data are available, these graphs illustrate a good correlation between the simulated and the measured concentrations.

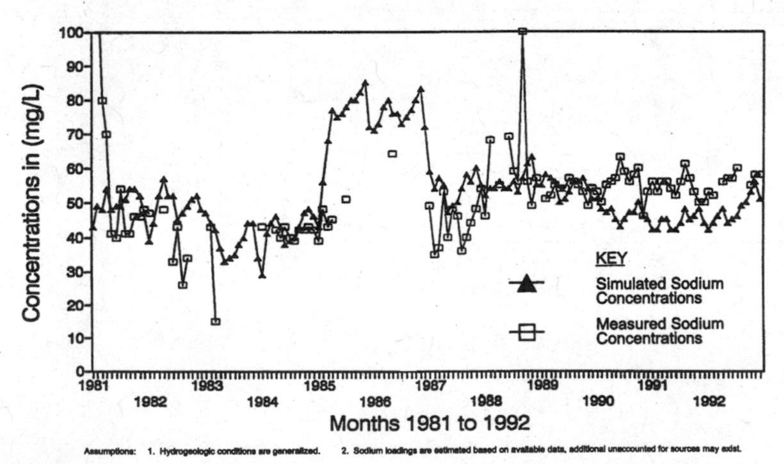

Figure 8.13 Simulated and measured sodium concentrations for Hobbs Brook reservoir.

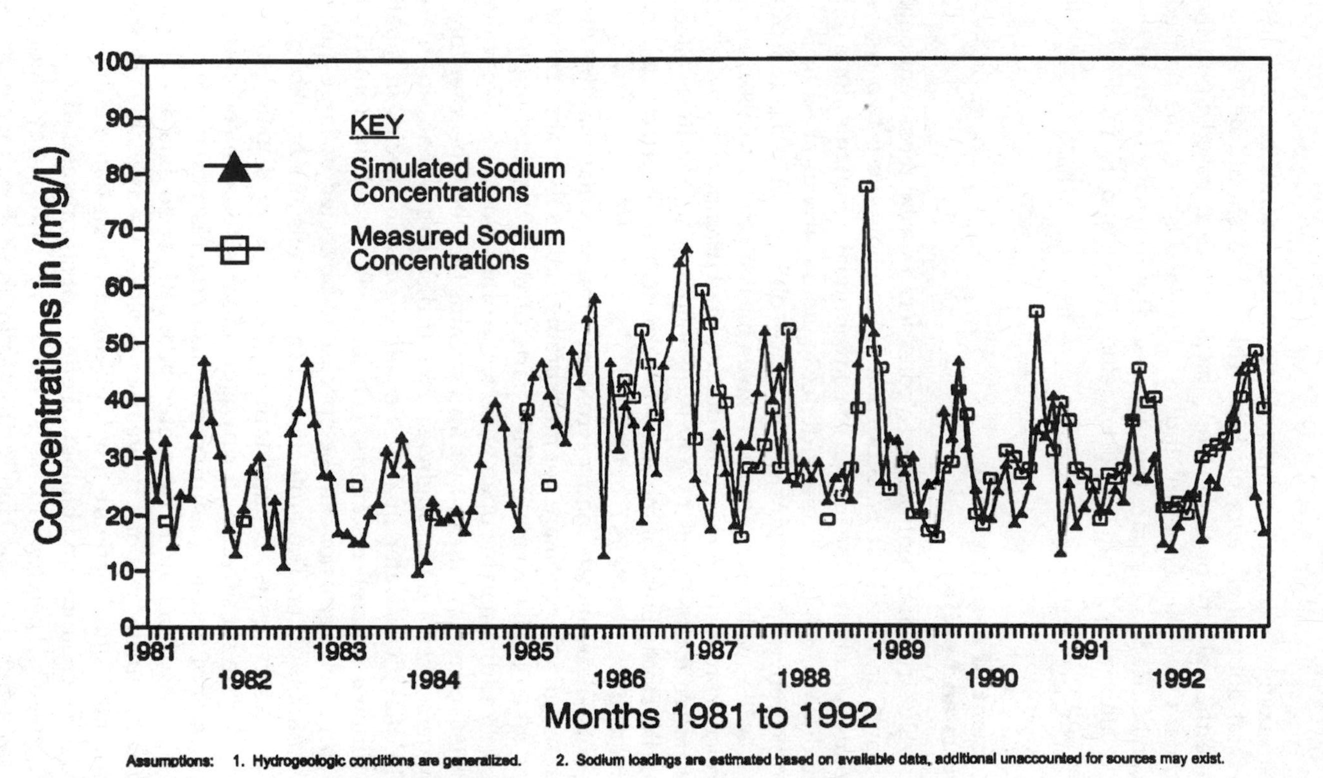

Figure 8.14 Simulated and measured sodium concentrations for Stony Brook reservoir.

Prior to using the model, however, the model should be validated with several additional years of data. These data should be used to determine the validity of the model and to revise it based upon the actual data. Once this is completed as described above, the revised model can be used to compare the water quality impacts of each of the proposed alternatives with respect to the others and to modify the model as appropriate.

Management of conventional pollutants and spills

Smaller-scale structural controls were considered for both conventional contaminants and spills. It is important to combine these two issues because an evaluation of the impact of conventional pollutants alone did not provide convincing enough evidence to justify structural controls.

In terms of conventional highway pollutants, solids, metal, and petroleum products are the greatest concern. Even given the concern for solids, metals, and petroleum products, there is no evidence in the currently available water quality data (see Table 8.6) that these types of pollutants are negatively affecting either the Hobbs Brook or the Stony Brook reservoir. In fact, overall, the water quality in the reservoirs is quite good.

Since the solids, metals, and petroleum products that may be discharged in highway runoff are not measurable in the reservoirs, a closer look was taken to determine if other parts of the watershed might be impacted. In addition to the open waters, the other most sensitive water environment consists of the brooks and streams that feed the reservoirs. These waters receive direct highway runoff discharges that can intermittently alter their water quality. Sensitive aquatic species in small streams, unlike in large reservoirs, cannot easily avoid the impact of stormwater discharges, nor can the species be protected by large quantities of dilution water. The protection of the streams and brooks tributary to the reservoirs is an important part of watershed management because they are sensitive to stormwater impacts and can serve as an early indicator of potential future problems in the reservoirs.

While many streams are in the watershed, little or no data describing water quality during wet or dry weather periods are available. This is not uncommon since there are few well-documented data nationwide describing the chemical and biological effects of stormwater discharges to small streams. Therefore, to assess the impact of highway runoff in the stream segments in the Hobbs Brook and Stony Brook watersheds, an existing statistical model was used.

The model, referred to as the *probabilistic dilution model* (PDM), was originally developed for the Environmental Protection Agency (EPA) under the Nationwide Urban Runoff Program (NURP) and later was adopted by the Federal Highway Administration (FHWA) as a methodology for evaluating the significance of highway runoff.

This model is intended as a screening tool to assess if potential water quality and biological problems may exist. Also, if problems are suspected, the model results can be used to focus the direction of additional data collection efforts necessary to better define actual local conditions in the watershed.

When stormwater runoff from a highway discharges to a flowing stream, the runoff mixes with the stream flow, producing a concentration that is a function of the following four parameters:

- Stream flow
- Stream concentration
- Runoff flow
- Runoff concentration

The in-stream concentration produced by the runoff, after mixing with the receiving waters, can thus be calculated to assess whether it represents a potential threat to the aquatic habitat. The PDM model used to perform this analysis, illustrated schematically in Fig. 8.15, employs a procedure that computes the distribution of in-stream concentrations

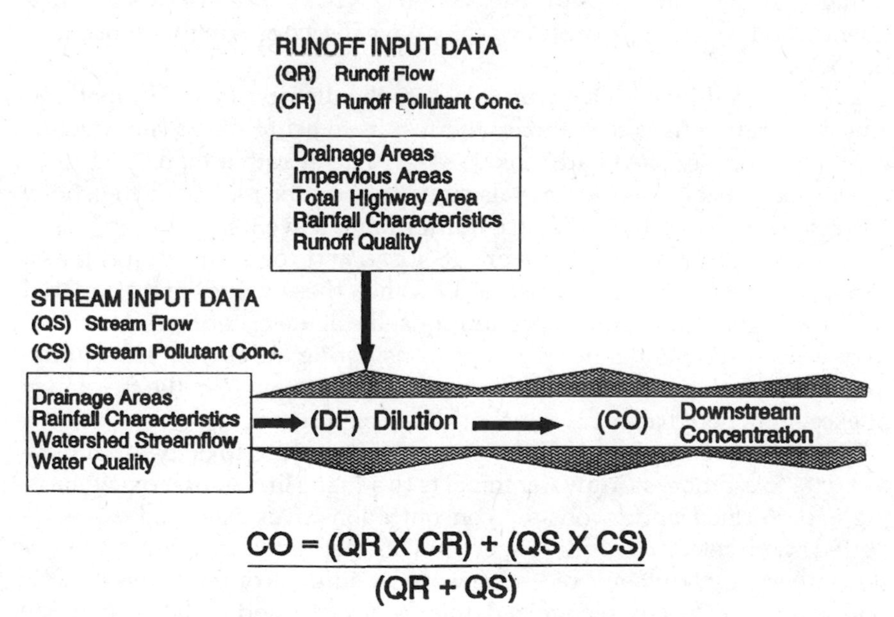

Figure 8.15 Highway runoff pollutant stream impact model.

directly from the statistical input parameters. A more detailed description of the method is provided in the FHWA document *Pollutant Loadings and Impacts from Highway Stormwater Runoff*, volume 1: *Design Procedures*.

It is important to note that the model provides only a statistical prediction of in-stream conditions—in this case based on limited local data. This situation is not unusual since the extensive local database needed to develop robust statistics in all four input parameters is typically not available. Therefore, the use of the model relies heavily on literature values. When it does, the model should be used as a screening tool.

Model Applications. There are many individual locations where highway runoff is discharged to streams in the watersheds. In application of this model, three areas representing different drainage area sizes, land uses, and percentages of highway runoff were used. The results for the most significantly impacted area are shown in Fig. 8.16 and Fig. 8.17.The interpretation of these results is not straightforward, and little regulatory guidance addresses the complex impacts that result from stormwater runoff.

1. *TSS and Oil and Grease.* The distribution of TSS and oil and grease concentrations is not considered to be of concern. Short-term suspended solids concentrations over 1,000 mg/L could be tolerated by the aquatic community. These are predicted to be exceeded on average less than once every 10 years. With respect to oil and grease, average concentrations of 25 mg/L are also predicted to be exceeded less than once in 10 years. Well-operated oil/water separators typically have discharges in this concentration range.

2. *Copper and Zinc.* One way to assess the distribution of copper and zinc concentrations is to compare the results to acute freshwater aquatic criteria for metals (EPA Gold Book, 1988). The acute rather than the chronic criteria have been used for metals since they better represent the short-term exposure conditions that occur after rainfall events.

The acute criteria for copper are 18 μg/L and for zinc 120 μg/L at a water hardness of 100 mg/L as CaCO. When these values are compared with the highest in-stream concentrations, the acute criteria are predicted to be exceeded several times per year. Considering the current EPA guidelines that allow several combined sewer overflows per year, the frequency of exceeding noted above is not considered significant.

There are questions as to the merits of using acute toxicity laboratory test data to estimate stormwater impacts to aquatic life. Acute toxicity testing is performed under constant concentrations over extended exposure periods with selected sensitive species. This methodology does not represent either actual dynamic receiving water conditions or the native aquatic populations. EPA has recognized this deficiency and is beginning the

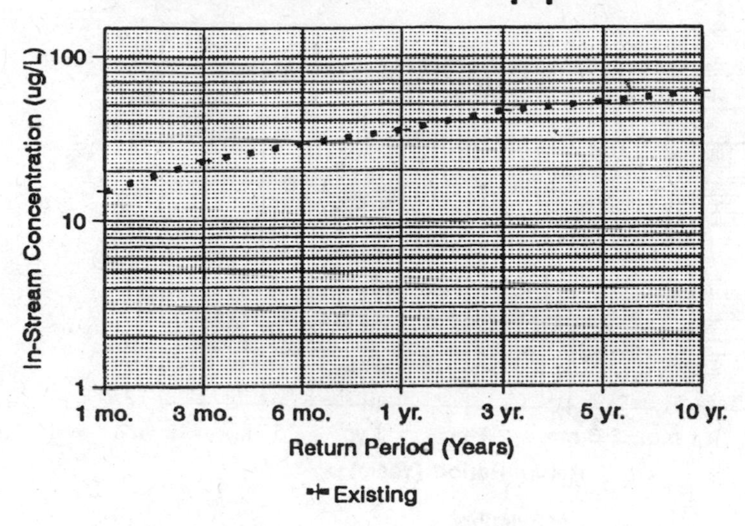

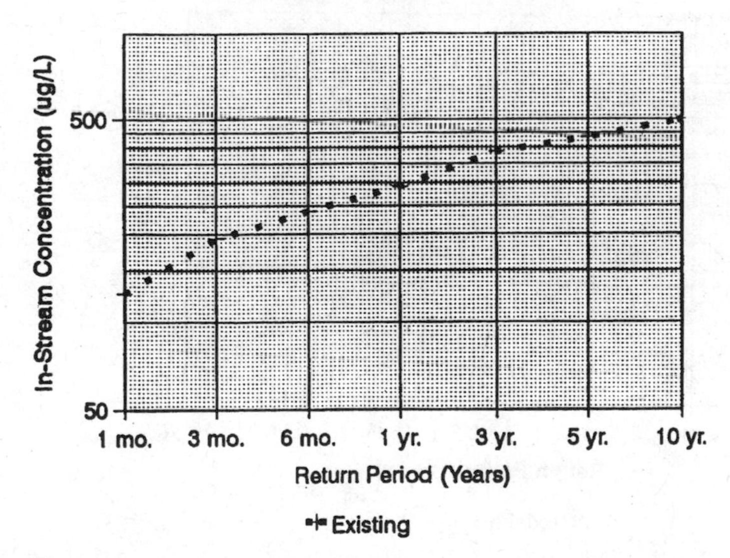

Figure 8.16 Simulated existing in-stream concentrations due to highway runoff.

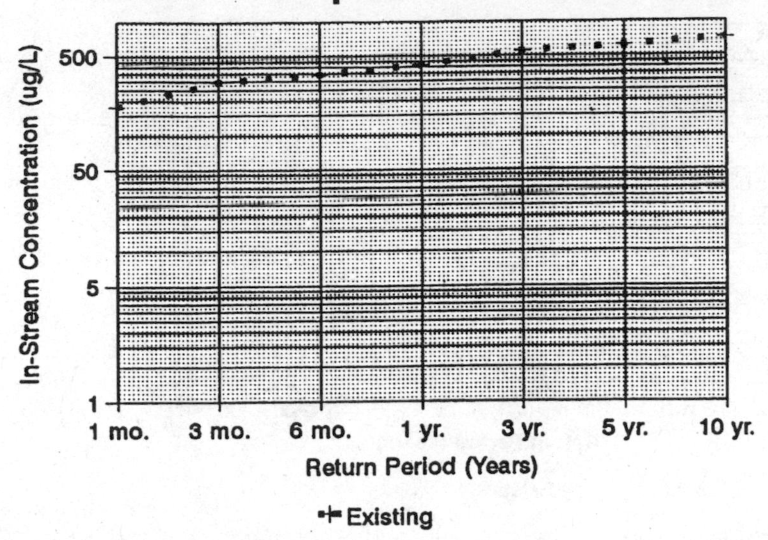

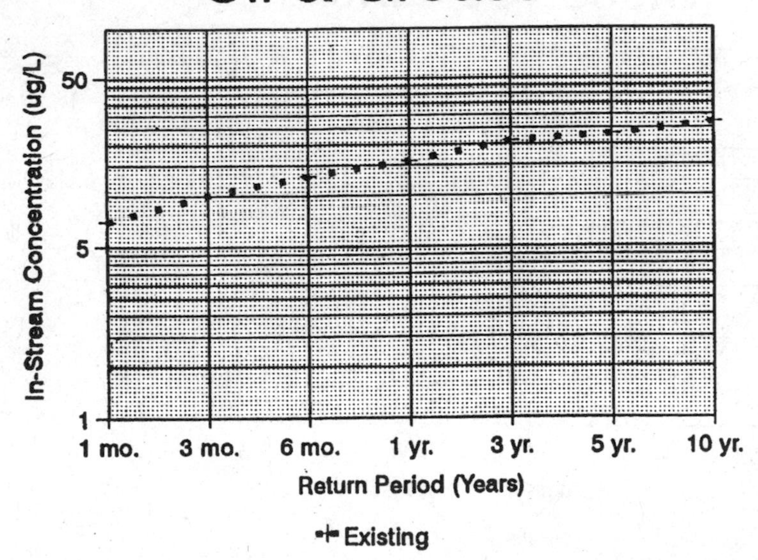

Figure 8.17 Simulated existing in-stream concentrations due to highway runoff.

process of revising and updating data on toxicity levels based on current scientific data.

In addition to water quality criteria, there has been an effort to use bioassays to assess the impact of stormwater discharges. In this process, instead of focusing on a single parameter such as copper or zinc, the effect of the entire discharge is observed on sensitive aquatic species in a laboratory environment. Typically sensitive species are introduced into a discharge sample at 100 percent strength and at diluted concentrations while their mortality is observed over time. The water quality of the discharge is also measured, and an attempt is made to correlate water quality to mortality to the extent it occurs.

In summary, there is contradictory information regarding how serious the impact of stormwater runoff is on stream habitat. When taken together, however, the data suggest that direct highway discharges could at times stress stream habitat due to metals, solids, or oil and grease loading. This information together with the need to manage spills provides the rationale for considering small-scale structural controls.

Structural controls. There is not good agreement concerning when specific structural controls should be used to control highway runoff. However, two general guidelines can help in their selection. First, discharges should be located as far from open water as possible, thus maximizing buffer zones to allow for natural settling, filtering, and dissipation of stormwater velocities and time to react to spills. Second, discharge controls are most appropriately considered for high-quality receiving waters such as drinking water supplies.

To implement these guidelines, it is necessary to select where buffer zones and/or discharge controls should be provided and which controls should be used. (Figs. 8.16 and 8.17). As previously noted, 56 of the 187 tributary outfall locations have been classified as type A, signifying a distance of less than 50 ft to the edge of the mean high water along the Hobbs Brook and Stony Brook reservoirs or their tributary streams, or discharge to perennial brooks or tributaries that maintain a constant base flow and discharge directly to the reservoirs.

By virtue of their location, the 56 outfalls of primary concern need to be physically relocated to increase the buffer to open water. To accomplish this, these outfalls were reduced to 16 groups. For each group, new "low-flow" collection systems, as previously illustrated in Fig. 8.5, would consolidate the individual discharges to one location. The low-flow collection system would be designed to divert flows up to those produced by a 3-month, 24-h storm to a structural control. This design criterion would allow for the capture of at least 90 percent of the annual runoff. Excess runoff from large storms would continue to discharge to the reservoirs via existing outfalls.

In addition to the collection and diversion of stormwater discharges, a suitable discharge control needs to be selected. There are a number of possible controls, with the most effective generally including a storage component to slow down the runoff, to allow for the settling of solids and associated contaminants, and to trap spills. Slowing the runoff also allows for the control of floatables and the prevention of short-circuiting and resuspension of captured solids. Such controls can be designed as one of the following types:

- Extended detention basin
- Wet pond
- Retention basin
- Constructed wetland
- Grit and particle separators

Best management practices. The following best management practices (BMPs) are designed to provide additional control of sand, sodium, conventional pollutants, and potential spills of toxic or hazardous materials from the highway systems. Figure 8.18 shows the general locations of each of the 16 structural controls.

1. *Continuation of operations and maintenance of existing highway drainage structures and roadways.* The maintenance of existing highway drainage structures and roadways prevents excess sand and debris from reaching the local wetlands, waterways, and reservoirs. It also provides proper highway drainage and prevents water from ponding on the highway surface. The highway should be swept clean, and catch basin and manholes should be cleaned out on a regular basis. The current practice of sweeping the roadway and cleaning out catch basins once per year in the early spring should be sufficient.

2. *Emergency Response Maps (ERMs).* These maps are important to spill response planning as they indicate the layout of the highway system and the route of stormwater runoff throughout the drainage system to specific outfalls. During a spill response, a response team can refer to these maps to determine the potential routes of a spill through the highway drainage system. The team can then take measures to isolate and contain the spill and prevent the discharge of hazardous materials into the surrounding environment.

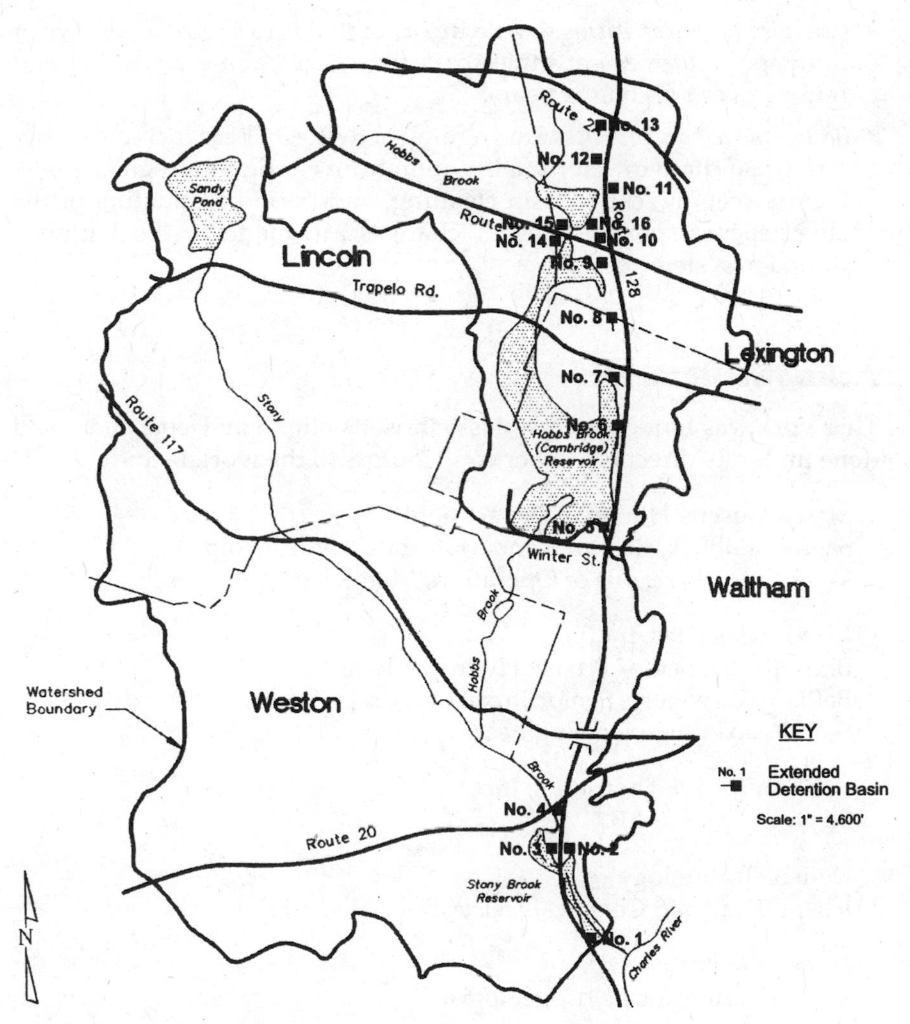

Figure 8.18 Proposed structural controls.

Summary. The components of the recommended plan include the following:

- *Deicing practices* include applying a calcium chloride brine at the beginning of the storm and afterward, the "smart" use of sodium chloride, calcium chloride, and sand coordinated with plowing. Monitor the effectiveness and make adjustments as needed.

- *Extended detention basins* will be incorporated into future roadways or drainage improvement projects as they occur along route 128 and route 2 over the coming years.
- *Best management practices* are ongoing activities. These include a long-term quantitative water quality monitoring program, regular pavement sweeping, catch basin cleaning, and periodic updating of the emergency response maps as changes are made to the highway drainage system.

Acknowledgments

This work was funded by the Massachusetts Highway Department and done under its direction. Other contributors to the work include

Massachusetts Highway Department
Samuel Pollock, Research and Materials, Hydrogeologist
Mark Cain, Divisions of Operation, Maintenance Engineer

Rizzo Associates, Inc.
Brian Butler, Senior Project Hydrogeologist
Bethany Eisenberg, Senior Project Hydrologist
Sean Reardon, Project Engineer

Fay Spofford & Thorndike, Inc.
Steve Chapman, P.E.

De-Icer Technology
Joseph Fedosoff, Chemical Deicer Specialist

Horsley & Witten, Inc.
Michael Frimpter, Hydrogeologist*

DISCLAIMER: The views, opinions, and/or findings contained in this chapter are those of the authors and do not necessarily reflect the official view or policies of the Massachusetts Highway Department. This chapter does not constitute a standard, specification, nor regulation.

9
Watershed Management in U.S. Urban Areas

David R. Bingham
Metcalf & Eddy, Inc.
Cleveland, OH

Introduction

Urban and urbanizing areas can pose significant risk to health of local water bodies and can jeopardize the beneficial use of these water bodies. These areas usually include metropolitan regions that consist of a heavily developed urban core surrounded by a residential, suburban zone. The urban core, or the city, is characterized by a dense concentration of commercial and industrial land uses interwoven by a complex system of streets, parkways, and interstate highways. The surrounding residential areas, or the suburbs, are characterized by a mix of residential neighborhoods, apartment complexes, and commercial developments connected by streets and less traveled secondary roads. The outer fringes of these suburban areas are often undergoing rapid development, converting agricultural or forested areas into parking lots and shopping malls. In addition to the effect of urbanization on the surface of the land, cities and suburbs are underlain by a complicated network of sanitary sewers, combined sewers, and storm drains that collect wastewater and runoff for eventual discharge to local water bodies.

The nature and character of a water body are greatly influenced by its watershed. A watershed is the land area that drains water to a specific water body such as a river, lake, or wetland. Depending on the water body, a watershed can contain numerous tributaries and their associated subwatersheds. The characteristics of a watershed such as precipitation, topography, soil type, and vegetation along with the human use of the watershed directly affect the health and usefulness of a water body. Therefore, when one is developing a program or strategy for restoring and protecting the beneficial uses and overall health of a water body, an approach that considers the entire watershed will have the greatest chance of success.

The urban watershed poses the greatest challenge to water resource professionals because of the complex mix of environmental stressors that act on urban water bodies. Both quantitative and qualitative water resource issues must be addressed in the urban watershed.

Water quantity problems. Due to the predominance of paved surfaces and buildings in the urban core, runoff quantity and peak discharge rates are extremely high compared to those in predevelopment or natural conditions. Figure 9.1 depicts the runoff discharge rates for pre- and postdevelopment conditions for a typical storm. In addition to the existing hydrologic conditions in the urban core, suburban areas under constant urbanization can see dramatic increases in surface runoff and significant physical damage to sensitive aquatic habitats. As illustrated in Fig. 9.2, increasing the proportion of impervious surfaces decreases the infiltration of precipitation into the ground and the evapotranspiration of water from the surface due to direct evaporation and transpiration from plant surfaces. The resultant increase in surface runoff from these development watersheds will likely change the natural hydrology by

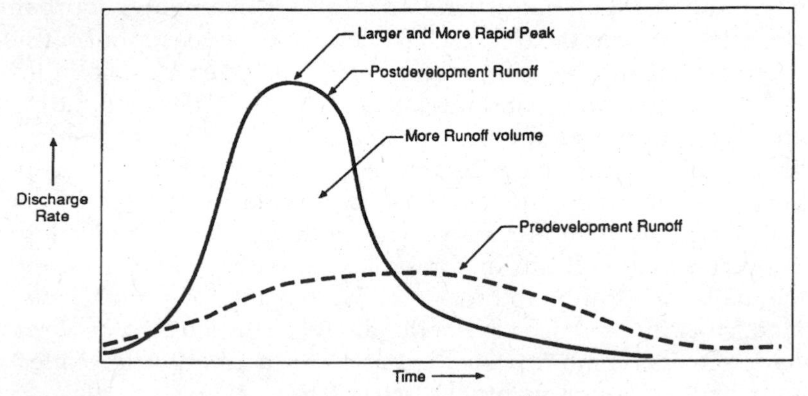

Figure 9.1 Pre - and postdevelopment hydraulics.

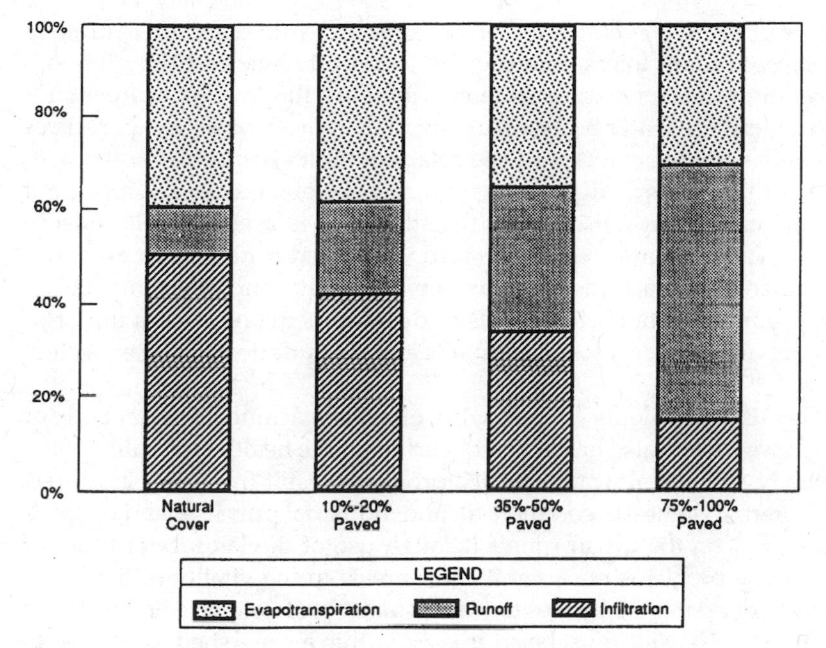

Figure 9.2 Typical changes in runoff flow resulting from development.

- Increasing peak discharges
- Increasing the volume of storm runoff
- Decreasing the time for runoff to reach a stream
- Increasing the frequency and severity of flooding
- Reducing stream flow during periods of prolonged dry weather
- Increasing runoff and stream velocities during storm events

Each of these changes in the natural hydrology of a watershed can result in physical and chemical changes that directly affect the biological health and beneficial uses of the water body. As peak runoff rates increase, pollutant transport and total pollutant loading to the water body can increase. In addition, dramatic increases in stream flow during rainfall events result in greater erosion and channel-scouring problems. Increased erosion results in greater downstream sediment deposition. These sediments can carry and deposit nutrients, metals, and other pollutants.

Changes in the natural hydrology of a watershed can affect the physical habitat of the stream and its riparian zone. Increased stream velocities can result in channel widening and stream bank erosion. The decrease in the infiltration capacity of the ground surface along with the loss of buffering vegetation due to stream bank damage results in increased pollutant loads entering the water body. In addition, the loss of vegetation in the riparian zone can increase the levels of direct sunlight reaching the water body, resulting in increased water temperatures and changes to the structure of the resident biological community.

Research has shown that watershed imperviousness is an important indicator of water resource impairment. There is a strong relationship between increased imperviousness and watershed runoff, reduced water quality, increased warming, decreased biodiversity, and other impacts on water resources. At fairly low levels of development (10 percent impervious and above), water resource impacts and degradation can be expected.

Water quality problems. The quality of water reaching the water body or an urban watershed also has a direct bearing on the health and utility of the water body. In general, water quality problems result from people's direct use of water for domestic, commercial, and industrial purposes and people's indirect effect on the urban runoff through use of the land. Both types of water quality problems pose significant threats to the vitality of the water body and formidable challenges to water quality professionals. Both types of water quality problems must be addressed within a watershed to succeed in improving and protecting the health and usefulness of the water body.

While the water quality problems that can potentially result from domestic, commercial, and industrial wastewater discharges are great, these sources can generally be addressed through traditional wastewater collection and treatment technologies and practices. While the performance of these collection and treatment systems is quite reliable during dry weather conditions, their reliability during rainfall events can be questioned. In many urban watersheds, combined sewer overflows (CSOs) and sanitary sewer overflows (SSOs) can result in significant water quality degradation and must be addressed as part of any watershed management program. In general, the quality wastewater discharges from public wastewater collection and treatment systems in urban watersheds is predominated by pollutants associated with common sanitary sewage. These pollutants include oxygen-demanding material, usually measured by biochemical oxygen demand (BOD); nutrients, with greatest concern placed on phosphorus and nitrogen; and pathogens, represented by indicator organisms such as fecal coliform.

Industrial wastewater can reach the water body of the urban watershed directly as a result of the discharge of wastewater from dedicated treatment plants or indirectly through public wastewater collection and treatment systems. These industrial wastewaters can result in discharges of metals and organic pollutants that can cause aquatic toxicity and can accumulate in sediments.

Urban runoff can also be a significant source of water pollutants. When rainfall falls on the generally impervious urban land surface, contaminants that have accumulated on these surfaces are quickly washed directly into water bodies. Table 9.1 presents a list of urban runoff pollutants.

Overview of the planning process

The watershed management planning process that has evolved in the United States has been shaped largely by various federal and state regulatory requirements. Each of these requirements has traditionally focused on controlling one type of pollution source (municipal wastewater treatment plants, combined sewer overflows, urban stormwater, etc.) or protecting one type of water resource (lakes, coastal waters, drinking water supplies, etc.). Based on these various regulatory requirements, current watershed management planning practices have four major components:

- *Determine existing conditions.* Analyze existing watershed and water resource data, and collect additional data to fill gaps in existing knowledge.
- *Quantify pollution sources and effects.* Utilize assessment tools and models to determine source flows and contaminant loads, extent of impacts, and level of control needed.
- *Assess alternatives.* Determine the optimum mix of prevention and treatment practices to address the problems of concern.
- *Develop and implement the recommended plan.* Define the selected system of prevention and treatment practices for addressing the pollution problems of concern, and develop a plan for implementing those practices.

The watershed management planning process that guides urban watershed management consists of nine steps which generally proceed in the following order:

1. Initiate program.
2. Determine existing conditions.
3. Set site-specific goals.
4. Collect and analyze additional data.
5. Refine site-specific goals.
6. Assess and rank problems.
7. Screen BMPs.

Table 9.1 Summary of Urban Runoff Pollutants

Category	Parameters	Possible sources	Effects
Sediments	Organic and inorganic	Construction sites	Turbidity
	Total suspended solids (TSS)	Urban/agricultural runoff	Habitat alteration
	Turbidity	CSOs	Recreational and aesthetic loss
	Dissolved solids	Landfills, septic systems	Contaminant transport
			Navigation/hydrology
			Bank erosion
Nutrients	Nitrate	Urban/agricultural runoff	Surface waters
	Nitrite	Landfills, septic systems	Algal blooms
	Ammonia	Atmospheric deposition	Ammonia toxicity
	Organic nitrogen	Erosion	Groundwater
	Phosphate		Nitrate toxicity
	Total phosphorus		
Pathogens	Total coliforms (indicator)	Urban/agricultural runoff	Ear/intestinal infections
	Fecal coliforms (indicator)	Septic systems	Shellfish bed closure
	Fecal streptococci	Illicit sanitary connections	Recreational/aesthetic loss
	Viruses	CSOs	
	E. coli	Boat discharges	
	Enterococcus	Domestic/wild animals	

Organic enrichment	Biochemical oxygen Chemical oxygen demand (COD) Total organic carbon (TOC) Dissolved oxygen (DO)	Urban/agricultural runoff demand (BOD) CSOs Landfills, septic systems	Dissolved oxygen depletion Odors Fish kills
Toxic pollutants	Toxic trace metals	Urban/agricultural runoff	Bioaccumulation in food chain organisms and potential toxicity to human and other organisms
	Toxic organics	Pesticides/herbicides Underground storage tanks Hazardous waste sites Landfills Illegal oil disposal Industrial discharges	
Salts	Sodium chloride	Urban runoff Snowmelt	Vehicular corrosion Contamination of drinking water Harmful to salt-intolerant plants

8. Select BMPs.
9. Implement plan.

Setting and refining watershed management goals are one aspect of planning that parallels the step-by-step process listed above. As illustrated in Fig. 9.3, a watershed management planning process will likely be initiated based on some very general watershed-wide goals. As the planning process proceeds, data and information are collected, and goals can be refined and made more specific. These more specific goals can better guide the selection of management and control measures within the watershed. Finally, long-term monitoring can be designed based on these goals to measure program success.

Types of goals. The two main types of watershed management goals are water resource- and technology-based goals. Water resource-based goals are based on receiving water standards which consist of designated beneficial uses and criteria to protect these uses. For example, water resource-based goals may relate to uses such as "opening half of the currently closed shellfish beds." They also may consist of more specific pollution reduction goals, such as lowering the trophic state index or reducing the amount of oxygen-demanding substances in a lake. In addition, water resource-based goals can place numerical limits on the concentrations of specific pollutants. Further, examples of water resource-based goals include no degradation, no significant degradation, and meeting water quality standards. As a defining characteristic of water resource-based goals, success in meeting such a goal is determined by the condition of the water resource.

In contrast, technology-based goals require specific pollution prevention or control measures to address water resource problems. They can be very general, such as "Implement minimum technologies for CSO control," or very specific, such as "Implement runoff detention at 50 percent of the industrial sites in a watershed." A community might be able to determine the effectiveness of implementing these goals without conducting future water quality monitoring. With most technology-based goals, implementing the control measure is presumed to be adequate to protect water resources. Monitoring, however, is still essential after implementation to gauge the program's effectiveness and to see if the desired environmental results are being achieved.

The types of goals set by a community usually depend on the natural or political forces driving watershed management and the public's level of knowledge about the affected water body. If a community undertakes a watershed management program because it has lost a resource (e.g.,

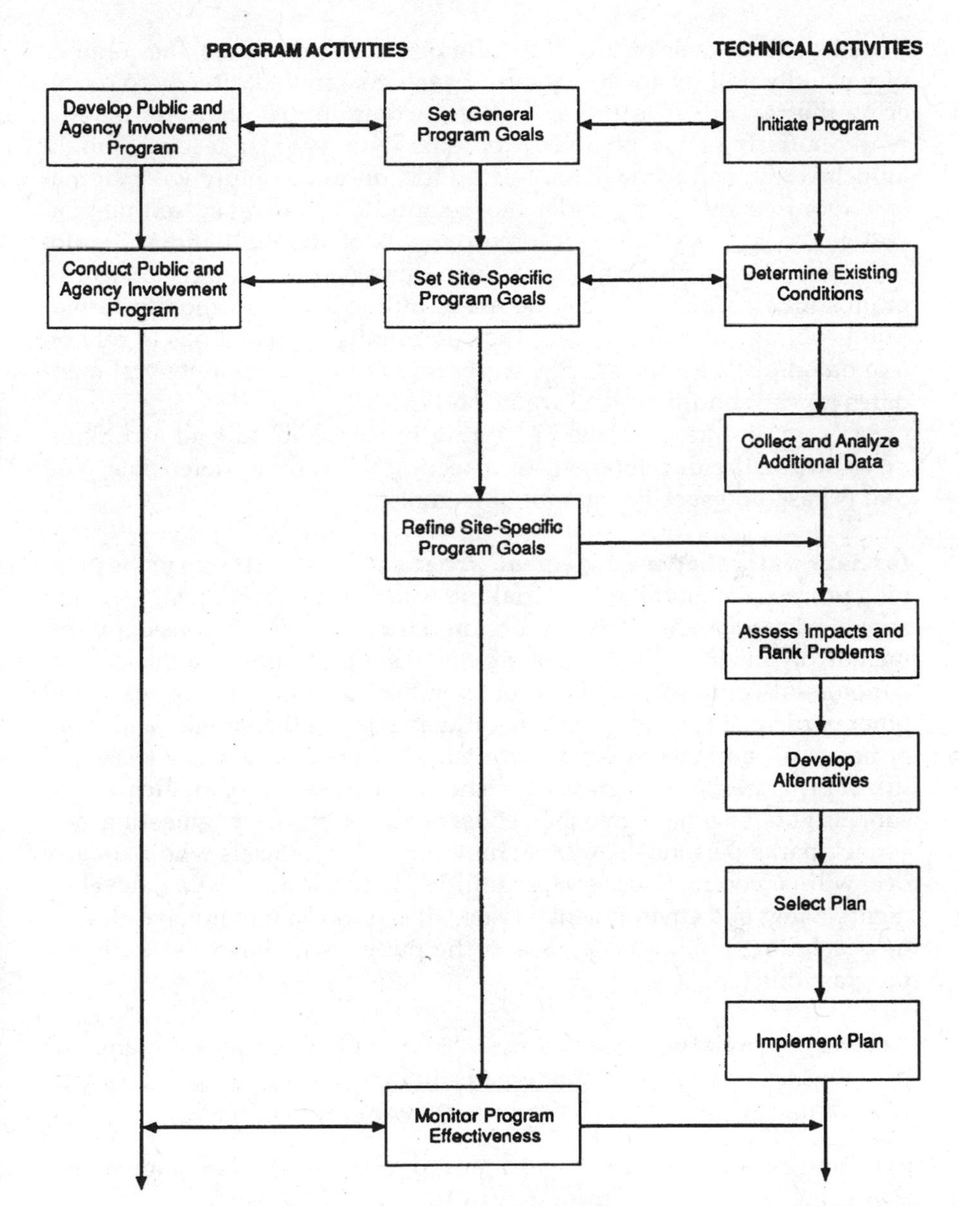

Figure 9.3 Example of watershed planning process.

closed shellfish beds or loss of fishing or swimming areas), the community usually will set a water quality-based goal linked directly to recovering the resource. On the other hand, communities that are not currently suffering from obvious problems with a water resource might launch watershed management programs only to comply with regulatory requirements. Even under these conditions, however, setting general goals, such as "Meet the requirements of the regulations," is not only possible, but also important. Even this general goal directs the program's focus, which then can be made more specific as more information is obtained. In these cases, the community typically has to rely on mandated goals for the specific water body of concern or general mandates for the condition of all water bodies.

Each step in the watershed planning process should lead a community toward the development of a technically sound, defensible plan that is based on specific, quantifiable goals.

Initiate watershed management program. As a first step in the planning process, communities undertaking watershed management planning should develop an overall program structure. Early considerations include organizing a program team; establishing communication, coordination, and control procedures for members of the planning team and other participants; identifying tasks and estimating the number and types of personnel and other resources for each task; and scheduling tasks. For urban programs, the program team should be made up of municipal personnel; public works personnel; conservation officials; engineering personnel; parks personnel; and planning and other officials who regularly deal with or control issues such as utilities, land use and zoning, development review, and environmental issues. It is important to involve all entities, including political officials and the public, who have a stake in the program outcome.

Determine existing conditions. After initiating the program, the planning team must develop a greater understanding of existing watershed characteristics and water resource conditions in order to

- Define existing conditions pertinent to the watershed management
- Identify data gaps
- Maximize use of existing available information data
- Organize a diverse set of information in a usable way

The required research is typically done by gathering existing available watershed information (e.g., environmental, infrastructure, municipal, and pollution source information), as well as receiving water data (e.g., hydrologic, chemical, and biological data, and water quality standards and criteria). This information can be obtained from various data-based mapping resources as well as federal, state, and local agencies. The information can then be used to develop watershed maps; to determine water, sediment, and biological quality; and to establish the current status of streams. The program team can organize the information into a coherent description of existing conditions and determine gaps in knowledge. In this way, the existing conditions of the watershed and receiving waters can be defined.

Collect and analyze additional data. Even under the best circumstances, communities usually will not have all the required information to describe adequately a program area's existing conditions. The program team, therefore, might have to gather additional information through field investigation and data collection. With this additional information and existing data, the program team can evaluate more fully the existing conditions of the watersheds and water resources of concern. Given the cost and time involved in data gathering, the program team will have to weigh the benefits of additional data collection against using limited funds for plan development and implementation. If additional data are required, a plan to gather these data must be developed. The plan should include an assessment of available staffing and analytical resources; identification of sampling stations, frequencies, and parameters for sampling and analysis; development of a plan to manage, analyze, and interpret the collected data; and analysis of available or needed financial resources.

Assess and rank problems. Once sufficient data have been collected and analyzed, they can be used to assess and rank the pollution problems. Based on data gathered in earlier steps, the team will need to develop a list of criteria to assess problems. These criteria are used in conjunction with water quality assessment methods and models to determine current impacts and future desired conditions.

Having determined the problems of concern, the project team can rank these problems to set priorities for the selection and implementation of pollution prevention and control measures. The emphasis on ranking of resources and problems is very important. This concept assumes that focusing resources on targeted areas or sources enhances water resource improvement. Further, it assumes that demonstrating water resource benefits increases public support of watershed management programs as citizens become more closely attuned to overall water quality goals.

Screen best management practices. Once the water resource problems have been prioritized, specific water resource problems and their sources can be addressed. The program team should complete a list of various pollution prevention and treatment practices and review them for their effectiveness in solving the prioritized problems. While the team initially faces a large number of potential practices, obviously inappropriate practices are eliminated in this step based on criteria such as the primary pollutants removed, drainage area served, soil conditions, land requirements, and institutional structure. Following this initial screening, the program team will have a list of potential practices to be evaluated further.

Select best management practices. During this step, the program team investigates the list of potential pollution prevention and treatment practices developed from the previous step to determine which to include in the plan. More specific criteria should be used for analyzing these potential practices than during the initial screening. To make the final selection, the program team must use the analytical tools developed during the ranking and assessment of problems, as well as decision factors such as cost, program goals, environmental effects, and public acceptance. As with the initial screening step, these evaluation criteria depend on established priorities. Generally, the selection process yields a recommended system of various pollution prevention and treatment practices which together address the pollution sources of concern. Availability of required resources to implement the practices is a major consideration. If needs and resources do not match, the community might have to adjust its expectations of what realistically can be accomplished.

Implement plan. After choosing control practices, the program team moves from planning to implementation, which often occurs through a phased approach. Inexpensive and well-developed practices can be implemented early in the program as pilot or demonstration studies; and these results might influence further implementation. Given the added requirements of implementation, operation, and maintenance, the original program team might expand to include members with more construction experience. Also, funding sources are needed for initial capital expenses and continuing operation and maintenance costs. Nonstructural practices must be implemented, and the team must arrange for the detailed design and construction of structural practices.

During this step, program responsibilities must be clearly delineated. All involved entities must be familiar with and accept their role in implementing and enforcing the plan. Continuing activities also should be clearly defined, and monitoring schedules should be set to determine the program's effectiveness in meeting its goals. Maintenance programs should be developed so that structural practices continue to operate as intended. Finally, the community should be aware of available technical assistance that could help throughout implementation of the plan.

The remainder of this chapter presents each major step in the planning process in greater detail.

Watershed inventory

An important early step in the watershed planning process is to develop an inventory of the factors pertinent to watershed management. This inventory supports watershed planning by

- Providing a basis for establishing and reassessing water resource protection and improvement objectives
- Identifying pollutants of concern and related effects on water resources
- Providing a base map for locating pollution sources and controls
- Defining areas of concern where pollutant loadings pose a high environmental or public health risk and where source control efforts should be focused
- Providing information for development of water quality models, if needed
- Planning, designing, and implementing BMPs
- Evaluating postimplementation improvements and beneficial use attainment
- Identifying areas of good water quality and high value to focus protection efforts

The watershed inventory has two major components:

- Watershed description, which characterizes the land-based information and the "causes" of water resource problems
- Receiving water description, which characterizes the receptors of the watershed sources and their effects

The watershed description defines the watershed area and its subwatersheds and further identifies pertinent geographic and environmental features (e.g., land use, geology, topography, and wetlands), infrastructure features (e.g., sewerage and drainage systems), municipal data (e.g., population, zoning, regulations, and ordinances), and potential pollution source data (e.g., in-stream sediments, landfills, underground tanks, and point source discharges). The receiving water description provides water resource information for water bodies affected by the watershed, which can include any type of receiving water (e.g., rivers, streams, lakes, and estuaries) and its sediment and biota as well as groundwater.

The first step in describing a watershed is to delineate the watershed and smaller watersheds or subwatersheds within it. Once the watersheds are delineated on a base map, additional features and data for each watershed and/or subwatershed are compiled. Pertinent information is presented in Table 9.2. The watershed description is organized by data type (i.e., environmental, infrastructure, municipal, and potential sources and existing BMPs). Each data type has its own section with a narrative description of each data subtype supported by appropriate tables and/or maps.

In addition to a watershed description, a receiving water description should be prepared, which includes the types of water resource data that should be sought, sources of data, and methods to summarize and analyze existing receiving water conditions. Many watersheds have multiple receiving waters, such as tributaries, larger rivers or estuaries, or lakes; in many cases, adding groundwater to this list could be useful. Effective identification and use of available water resources data could reduce inventory costs significantly by reducing needed sampling and analysis. In addition, review of historical water quality data provides a basis for

- Establishing and reassessing goals
- Documenting the type and extent of water resource impacts
- Identifying data gaps that should be addressed with a sampling program
- Identifying priority areas and major pollution sources
- Quantifying pollutant loads
- Documenting impairment or loss of beneficial uses and water quality standard violations

After the water resource data have been gathered, a receiving water description is developed. This description includes summaries of the data collected, organized by data type (i.e., physical/hydrologic, chemical, biological, and water quality standards and criteria). Each summary includes a narrative description outlining the information gathered for each data type. This information is presented in a way that indicates existing data gaps and priorities for addressing those gaps.

Expending resources early in the watershed planning process to

Table 9.2 Watershed Inventory Data

Watershed data	Receiving water data
Environmental	**Source inputs (flow and quality)**
Topography	CSO
Land use	Storm water
Recreational areas	Other NPS
Soil and surface/bedrock geology	**Receiving water**
Vegetation	Physiographic and bathymetric data
Natural resources	Flow characteristics
Temperature	Water quality data
Precipitation	Sediment data
Hydrology	Fisheries data
Infrastructure	Benthos data
Roads and highways	Biomonitoring results
Storm drainage system	Federal standards and criteria
Sanitary sewer (and combined sewer) system	State standards and criteria
Treatment facilities	
Other utilities	
Municipal	
Population	
Zoning	
Land ownership	
Regulations and ordinances	
Municipal source control BMPs	
Potential sources/BMPs	
Landfills	
Waste handling areas	
Salt storage facilities	
Vehicle maintenance facilities	
Underground tanks	
NPDES discharges	
Pollution control facilities	
Retention/detention ponds	
Flood control structures	

locate as much existing information as possible is cost-effective in the long term, because it helps maximize use of existing information, minimize data collection costs, and avoid overlooking of important data resources. The information, having been gathered and analyzed, has to be examined to determine existing knowledge gaps. If necessary information is unavailable, the need and extent of added data collection must be determined.

Data collection and analysis

After the watershed inventory is complete and existing data and information have been collected and analyzed, existing conditions can begin to be evaluated. Some program goals can be refined at this point, and some priority problems in the watershed may start to emerge. However, it is likely, if not certain, that additional watershed and water resources data will be needed. The challenge for the community will be to determine how much new data will be needed. Care must be taken to strike a balance between using valuable resources for monitoring and data analysis and using those resources to implement management and control measures after the planning is complete.

Based on the existing data collected during the watershed inventory and the watershed management goals that have been refined based on these existing data, the community must develop a set of critical questions to be answered before the planning process can proceed. It will be these key questions that will determine the amount and characteristics of the additional data to be collected. Therefore, the formulation of these key questions must be given considerable time and effort by the watershed management planning team. The challenge is not to ask every question that remains. Rather, the challenge is to ask the minimum number of questions that must be answered. Also, if asked correctly, these questions will clearly define the required monitoring and data analysis that must be completed to provide the watershed management planning team with a solid foundation for its management program.

The monitoring and data analysis needs will be determined based on the watershed inventory and the program goals refined based on the analysis of existing data and information. In general, monitoring and data analysis needs will reflect several objectives: assessing existing conditions, refining problem identification, calculating pollutant loads, providing data for modeling, and addressing important pollution sources and water resources.

Assess existing conditions. If existing data are not sufficient to establish current dry or wet-weather conditions, additional data are needed. Dry-weather sampling of water resources could include areas affected by urban wet-weather sources and areas upstream of, and therefore not influenced by, the urban wet-weather source in the watershed. It might also include sampling of dry-weather base flows and continuous point sources entering the water resource. In addition to water sampling, sediment and biological sampling is particularly useful for determining a water resource's relative health. Also, sampling of habitats, wildlife, soils, and other components of the watershed might be required to establish existing conditions.

Wet-weather sampling can be used to determine runoff pollutant concentrations and to observe their effects. Wet-weather sampling is critical in urban planning because significant source loadings occur in wet weather. Sampling of runoff and measurement of flow in both sources and receiving waters during a storm can be used to determine the variability of runoff volumes and pollutant loads and to assess receiving water impacts for a particular storm. Results from sampling of receiving waters during storms can be used to evaluate the effects of storm water runoff on ambient water quality, violations of water quality standards, and the effects of storm water on beneficial uses. Other types of wet-weather observations could be useful to assess flow paths, ponding, areas of erosion, and other wet-weather conditions in the watershed.

Refine problem identification. Data collection programs might focus on collecting the additional information needed to identify problems clearly, such as pollutant sources and water resource impacts, that first were identified during the existing-conditions assessment. These data can provide the basis for source identification, problem assessment, and BMP selection. Data collection for problem identification could again involve dry- or wet-weather sampling of sources, receiving waters, or watershed factors.

Calculate pollutant loads. Flow concentration data from sources of pollutants collected in dry or wet weather, as appropriate, can be used to estimate pollutant loadings and to identify priority pollution sources and watersheds. Pollutant loadings may be estimated by using numerous methods ranging from simple to complex. These estimates can be used to evaluate event or annual pollutant loadings from the watershed, assess resource impacts, and select appropriate BMPs.

Provide data for computer models. Computer models can be used as predictive tools to assess problems and the potential benefits of alternative pollution prevention and control strategies. Quantitative models that are calibrated and verified using data from site-specific sampling programs can be used to estimate impacts of future pollution loadings anticipated under potential control strategies. Models quantify pollutant loads as well as assess impacts on receiving waters or other watershed components. These models often require particular types of input data that might have to be collected. These typically involve dry- or wet-weather source flow and concentration data, but can also include other specialized parameters. For example, data on sediment oxygen demand in the receiving water might be needed if dissolved oxygen modeling is a primary concern, or physical and chemical characteristics of street surface solids might be tested if pollutant buildup and washoff dry wet-weather are to be simulated.

Address important pollution sources or resource areas. The monitoring program might need to focus on known or suspected major pollution sources, to supplement available data, and to confirm the existence of pollutant loading from a source. Pollution sources could be either point or nonpoint sources expected to be of particular importance to the program. The monitoring program also might need to focus on critical resource areas. Natural resources that could warrant special consideration for sampling include shellfish beds, wildlife sanctuaries and refuges, wetlands, coral reefs, spawning grounds, recreational fishing areas, bathing beaches, and drinking water resources.

Data collection programs

Development of a data collection program depends on numerous factors. The program should have clear objectives. Design of the data collection program also depends on factors such as the size and nature of the watersheds and receiving waters. The plan must take into account available funding, resources, and schedule constraints.

Designing the data collection program. Since data collection programs are site-specific and varied, providing detailed guidance on what should "typically" be done is not realistic. The major considerations in design of a data collection program are parameter selection, sampling station selection, and frequency of data collection.

Selection of parameters. Parameters to be measured during the sampling program should be selected based on the review of existing conditions; the program's overall goals; the specific objectives of the data collection program; and the requirements of local, state, and federal regulations. For example, most state water quality standards have numerical limits for indicator bacteria levels in waters intended for swimming and boating. If local beaches are threatened by bacterial contamination from storm water or CSOs, bacteria sampling needs to be included in the program.

Given the long list of potentially important parameters, site-specific considerations and the key questions that need to drive the selection of parameters to be tested. The most common pollutant categories are solids, oxygen-demanding matter, nutrients, pathogens, and toxic substances; temperature, dissolved oxygen, turbidity, and specific conductivity can be included as indicator parameters to support specific assessments of pollution sources and receiving waters. It is also important to characterize particle settling velocities, particle diameters, and dissolved and nondissolved chemical fractions for use in evaluating runoff treatability and pollutant routing in the watershed and receiving waters.

In addition to the source and receiving water quality parameters outlined above, sediment samples may be analyzed for physical and chemical parameters, such as grain size distribution, organic content, total organic carbon (TOC), nutrients, metals, petroleum products, polychlorinated biphenyls (PCBs), and other parameters. As pollutants are partitioned between the dissolved and particulate phases, sediment chemistry reflects the portion of the particulate-bound pollutants that settle. These pollutants can, through other physical and chemical mechanisms, be introduced into the water column. Sediment chemistry can indicate potential pollution problems caused by the sediments, such as the release of metals and other pollutants into the water column and the depletion of overlying dissolved oxygen (DO) as organic matter is broken down by microorganisms.

The sediment characteristics reflect the long-term effects of intermittent and variable urban discharges. In fact, it is easier and more cost-effective to test sediments and plant and animal populations in the affected areas than to conduct sampling of the intermittent pollution sources and receiving water responses. The existing substrate and communities integrate the cumulative effects and can be characterized rapidly since they do not vary extensively.

Sampling of aquatic biota involves collecting biological species from the water column and sediments to determine the species diversity, dominance, and evenness. This process can include sampling for plankton, periphyton, macrophyton, macroinvertebrates, and fish and determining the number and density of populations in the water resource. In addition, physical habitat indicators, such as substrate and plant types and conditions, are useful reflections of pollution impacts. As with sediment, these habitats reflect the long-term effects of the intermittent urban impacts. These effects might be subtle and take a long time to occur, depending on the nature of the transport mechanisms and receiving water body.

Toxicity test sampling can be used to determine the relative toxicity of receptors that might be receiving contaminants. Toxicity test results also provide information on the relative degree of chronic and acute toxicity, which again reflects the period of exposure of organisms to toxic effects.

Selection of sampling stations. Sampling stations should be selected strategically so that data collected from a limited number of stations satisfy multiple sampling objectives. The major types of sampling are watershed-based (source sampling) and water resource-based (receiving water and aquatic ecosystem sampling).

Source sampling. Wet-weather generated discharges (e.g., storm water, CSO, and nonpoint source) can contribute large pulses of pollutant load and could constitute a significant percentage of long-term pollutant loads from urban and suburban areas. Wet-weather sampling can be used to characterize runoff from these discharges, determine individual pollutant source and total watershed loadings, and assess the impact on receiving waters. Pollution sources, tributaries, or entire watersheds can be ranked by total pollutant load and prioritized for implementation of pollution prevention and control measures.

In selecting a site for urban sampling during wet weather, the following criteria should be considered:

- *Discharge volume:* Select sites that constitute a significant portion of the flow to the water body.
- *Pollutant concentrations:* Based either on historical information or on land use or population density, select sampling sites to quantify representative or varying pollutant load sources.
- *Geographic location:* Select sites that permit sampling of flows from major subwatersheds or tributaries to permit isolation of pollutant sources.
- *Accessibility:* Select sites that allow safe access and sample collection.
- *Hydraulic conditions:* Utilize existing flow measurement devices, such as weirs or gauging locations, or sample where hydraulic conditions are conducive to manual or automated flow measurements.

Sampling should also include dry-weather flows from storm drains or other structures to determine if they result from illicit connections or from groundwater infiltration. The magnitude of these dry-weather discharges determines the need to identify and remove these illicit connections.

Water resource sampling. For the impact of urban watersheds to be assessed, the water quality of receiving waters during normal dry-weather periods should be known. Water quality data collected during dry-weather conditions provide a basis of comparison to data collected during wet-weather conditions. These data are also needed to quantify dry-weather pollutant transport from tributaries and groundwater flows. If existing data are not sufficient to characterize current conditions, stations should provide good spatial coverage within the receiving waters. Based on initial sampling results, the number of stations potentially could be reduced. For example, if initial sampling results show that a particular stream within a watershed is of high quality, then sampling coverage of this stream could be reduced. Additional stations could be added in response to expected changes in land use (such as high-density development projects), which might affect water quality. Critical stations, however, such as those that previously indicated water quality violations, need to be maintained. Also, use of existing stations from other programs should be maximized.

Wet-weather sampling stations should be located to assess impacts of significant urban runoff pollutants and major storm drain systems and CSO outfalls. Additional stations may be sampled within tributaries affected by storm water, CSO, or other discharges and land-use types of particular concern.

Other general site selection criteria for receiving waters include

- History of available data
- Easy accessibility
- Safety of personnel and equipment
- Entry points of incoming sources or tributaries
- Adequate mixing of sources or tributaries
- Straight reaches, rather than bends

Sediment sampling. Sediments in receiving waters affected by urban sources integrate the long-term effects of dry- and wet-weather discharges because of their relative immobility. Grab samples can be taken to indicate historical accumulation patterns. Sampling sites could be distributed spatially at points of impact, upstream (or downstream) reference sites, areas of future expected changes, or other areas of particular interest. Selection of specific locations is subject to accessibility, hydraulic conditions, or other aforementioned criteria.

Biological sampling. Benthic or bottom-dwelling organisms are affected both by contaminants in the water column and through contact or ingestion of contaminated sediments. The type, abundance, and diversity of these benthic organisms thus can be used to investigate the presence, nature, and extent of pollution problems. Comparisons of areas upstream and downstream of a suspected pollution source require that sampling locations have similar bottom types, because physical characteristics affect both the chemical composition and the habitat requirements of organisms.

Regional data or indices might be available for comparisons with local site conditions to determine whether an ecosystem is stressed. Such data provide a reference for comparison and might suggest appropriate habitat types or areas to sample in determining the level of pollution impact.

Frequency of data collection. The frequency of data collection significantly affects program cost and should be determined judiciously based on the need for sufficient data to develop statistically valid conclusions. Wet-weather runoff sampling is often limited to several events and selected representative subwatersheds because of the large resource requirements and high costs. Data must then be extrapolated to other similar subwatershed areas and used to calculate storm-related pollutant loading for an entire watershed. Depending on the area's size and number of subwatersheds, and on financial resources, adequate characterization of different subwatersheds might require a phased approach. Areas of greatest concern are sampled first, with subsequent sampling to characterize other areas based on a watershed priority sequence. Given the cost of such sampling, collection of sediment and ecosystem data that integrate the long-term effects of urban runoff may be fruitful since the data on sediment and ecosystems are relatively stable and do not need to be characterized as frequently.

For water resource monitoring, the sampling schedule should account for seasonal climatic changes as well as seasonal land-use activities, such as fertilizer application in spring, or road deicing activities in winter, that might influence water quality. In temperature areas with pronounced seasonal changes, monitoring stations are usually sampled at least seasonally. This is especially important for sampling of aquatic biota. For characterization of urban sources, several sampling events are ordinarily scheduled during worst-case conditions: in spring during snowmelt and heavy rains when runoff and contaminant transport are significant, or during summer conditions when stream flow is low, receiving water dilution is minimal, and contaminant concentrations are potentially highest. In addition, the relatively high temperatures in summer can affect aquatic biota, as well as reduce the capacity of water to maintain high DO levels and stimulate bacterial metabolism, placing additional demand on oxygen supplies in the water column. This scenario represents worst-case conditions in areas that experience organic and nutrient enrichment. In areas with fairly constant climate, less emphasis is placed on seasonality, with perhaps more attention placed on land-use activities.

Impact assessment

Watershed impact assessments can address a wide range of issues, including

- The types of urban runoff pollution in the watershed
- The extent to which these pollution sources adversely affect resources

- The institutional needs and constraints in addressing the problems
- The goals and objectives established for the watershed

Criteria to conduct these assessments can be developed for each of these major issues. Example criteria, such as those listed in Table 9.3, can be evaluated qualitatively or quantitatively. Pollutant source and resource assessments tend to be more quantitative, making use of source and receiving water data and models. Institutional goals and objective assessments tend to be qualitative, requiring substantial judgment by the watershed planner.

Pollutant source assessments. Pollutant source assessments address the type, magnitude, and transport mode of pollution sources (existing or

Table 9.3 Criteria for the Assessment of Watershed Management Impacts

Pollutant source
Type of pollutant
Pollutants typically associated with the source
Source magnitude and pollutant loading
Transport mechanisms to the resource (i.e., direct pipe, overland flow, groundwater)
Wet-/dry-weather trends
Resource
Existing use of the affected resource (type, status, and level of use)
Designated or desired use of the affected resource
Type and severity of impairment
Relative value of resource affected
Institutional
Available resources and technologies
Understanding of problems and opportunities
Appraisal of potential for solving the identified problem
Implementability of controls
Applicable regulations
Multiagency responsibilities
Funding sources and limitations
Public perception
Goals and objectives
Water-resource goals (water-use objectives)
Technology-based goals
Land-use objectives
Objectives of planner and sponsor

potential) in a watershed. These assessments are frequently aimed at quantifying the source flows and pollutant loads under various conditions.

Computer modeling is valuable in quantifying the flows and loads of pollution sources needed for pollution source assessments. Models can be used to estimate source strengths as well as to evaluate the effectiveness of proposed corrective measures or BMPs. Models available for urban runoff assessments vary widely in complexity, ranging from simple estimation techniques to sophisticated and expensive computer models. Information on urban and nonurban models is available from the literature and from agencies that sponsor the models. Methods of urban runoff modeling range from simple to complex and include constant concentration or unit load estimates, preliminary screening procedures, statistical methods, the universal soil loss equation, rating curve or regression approaches, and hydrologic and pollutant buildup-washoff models such as the storm water management model (SWMM).

In addition to the magnitude of a pollution load, the location of a pollution source with respect to the affected resource, the mode of transport to the resource, and degradation of the pollution should be considered. For example, sources with a clear path to a waterway, such as pipes, ditches, and gullies, are more likely to cause adverse effects in a receiving water than similar sources that must travel through natural filters, such as forested or grassy areas, before entering a surface water body. Changes in loads, from the initial source discharge to the point where they affect the receptor, occur because of such factors as travel time, dilution, soil infiltration, and decay. Fate and transport of pollutants can be modeled using hydrologic and pollution buildup-washoff models (such as SWMM) which attempt to account for these factors deterministically. Since simpler methods (i.e., unit load or statistical) can only empirically estimate these factors, the level of uncertainty and error is higher. The level of uncertainty is high even with the deterministic models, though. Site-specific data are thus important to validate any model used.

Resource assessments. Resource assessments address the impact of pollutant sources on the resource of interest by taking the results of the pollutant source assessments and determining the effect of these pollution sources on water resources. Water resources can include water quality as well as aquatic life, sediment, and other characteristics of the water body. Methods to perform resource assessments can change from evaluation of water quality data and comparison with criteria, to mathematical modeling of receiving waters.

Urban runoff problems can be identified by evaluating available and newly collected data. Evaluation of available data is conducted with numerous tools, including spreadsheets, database management systems, geographic information systems, statistical analysis, and mathematical models. The data are compared to an acceptable resource criteria to determine the existence and severity of problems.

A useful measure of the condition of a specific water resource is to compare its water quality, sediment, or biological data with state water quality standards or EPA water quality criteria. State water quality standards define the quality of water that supports a particular designated use. EPA publishes water quality criteria that consist of scientific information regarding the concentrations of specific chemicals in water that protect species against adverse acute (short-term) effects on sensitive aquatic organisms, chronic (long-term) effects on aquatic organisms, and effects on human health from drinking water and eating fish. These criteria, often based on results of toxicity testing of sensitive species, are intended to be protective of all species. The Clean Water Act requires EPA to publish and periodically update these criteria.

The Safe Drinking Water Act of 1974, established to protect public drinking water supplies, requires EPA to publish maximum contaminant level goals (MCLGs), which are nonenforceable levels at which there are not known or anticipated health effects, and maximum contaminant levels (MCLs), which are enforceable levels, based on best technology, treatment techniques, and other factors including cost. EPA is also developing criteria for sediment similar to those for water quality for certain organic compounds.

States have surface water standards that classify surface water bodies into use categories, establish in-stream levels necessary to support these uses, and define policies regarding the protection and enhancement of these water resources. In addition, many states have groundwater standards that designate uses for various groundwaters, and water quality levels necessary to sustain these uses and protect groundwater quality.

Ecological effects can be assessed by examining the biological community structure. Specific parameters to consider include the relative abundance of pollution-tolerant and pollution-sensitive species as well as common indices such as Shannon-Weiner diversity, Simpson's dominance, and evenness. Various types of biological criteria or indices are available from the literature and can be used for comparative purposes.

Receiving water models are used to assess existing conditions and to simulate future conditions of a water resource under various pollution prevention and control scenarios. The models can also be used to assess the impact of alternative watershed control methods. These models receive input from runoff model results, field-measured parameters, and values of parameters found in the literature. The level of complexity of the receiving water model chosen should parallel that of the model used to assess urban runoff flows and loads. Some commonly used receiving water models include the Enhanced Stream Water Quality Model (QUAL2E), the Water Quality Analysis Simulation Program (WASP4), and the Exposure Analysis Modeling System II (EXAMSII). In addition, HSPF has a receiving water model component.

Institutional assessments. Assessment of the institutional constraints of a program provides watershed managers with perspective concerning the nontechnical issues affecting the program. The institutional issues of a program are assessed by evaluating the program's potential and limitations and by reviewing the requirements of involved agencies and the public. One major institutional issue that must be addressed for a watershed program is the determination of the responsibilities of each involved party, especially for programs involving multiple agencies; issues related to the control of the program (e.g., enforcement, maintenance, permitting, and funding) can affect the program's emphasis and the selection of its corrective measures. Another institutional issue involves the limitations of available technology. Implementability of controls can also be considered, particularly in areas involving limited access to private properties. The potential for eliminating or reducing a watershed pollution problem or improving affected water resources can be considered. Questions and concerns of the public might prove to be influential during the decision-making processes. Applicable regulations could force the sequencing of corrective measures so that those addressing compliance with the regulations are implemented first.

Goals and objectives assessments. The relative importance of a watershed problem can be assessed by comparing it to the program's resource- and/or technology-based goals and the objectives of the program's sponsor. For example, one water resource goal might be to "Provide improvements to water quality in areas where the most people will benefit." Comparison of the pollution problems to such a goal provides a perspective on which problems to tackle to achieve the goal. Comparing the pollution problems to the program's goals and objectives helps identify and focus on problems that are compatible with these goals. The assessments conducted on pollutant sources, water resources, and institutional aspects provide input to these determinations.

Selection of control practices

Urban watershed pollution problems present significant challenges because of the intermittent nature of rainfall and runoff, the number of pollutant source types, and the variable nature of the source loadings.

Since the expense of constructing facilities to collect and treat urban runoff is often prohibitive, the emphasis of urban watershed pollution control should be on developing a least-cost approach which includes nonstructural controls and low-cost structural controls to the extent possible. Nonstructural controls include regulatory controls that prevent pollution problems by controlling land development and land use. They also include source controls that reduce pollutant buildup or lessen its availability for washoff during rainfall. Low-cost structural controls include the use of facilities that encourage uptake of pollutants by vegetation, settling, or filtering.

All sources, both point and nonpoint, in a watershed need to be addressed. For urban areas, such sources often include urban runoff as well as overflows from separate systems [sanitary sewer overflows (SSOs)] or from combined sewer overflows (CSOs). Table 9.4 lists major categories and specific practices used to control urban runoff and sewer overflows. Various textbooks and manuals address these and other control practices in detail.

Selection of controls. Control selection may be accomplished in a variety of ways. A common approach is to (1) compile a list of the possible control options, (2) screen this list to eliminate those that are inappropriate based on a set of initial criteria, (3) develop more detailed control alternatives using the list of feasible controls, and (4) evaluate these alternatives to select the preferred watershed control plan. Sample initial screening criteria can vary based on whether the controls are nonstructural or structural, and examples are given in Table 9.5.

Alternatives are developed using the controls still under consideration after the screening process. The alternatives can include various combinations of controls. Source control and regulatory controls are often implemented for entire regions or jurisdictions. Structural controls can be directed at specific pollutant sources or implemented across geographic areas, required for new development in currently undeveloped areas or for retrofit in already developed areas. To address fully the urban pollution problems in a watershed, controls from all these categories are often required.

Several methods for developing alternatives are commonly used. One starts with known urban runoff problems and known pollutant reductions desired. Various combinations of controls are then developed as alternatives to address these specific problems. Another involves developing a range of possible control levels for evaluation. This method can be used to compare a range of pollutant reduction or resource use improvement levels, e.g., low, medium, and high, and the relative costs and benefits of achieving them. A third method involves applying a spe-

Table 9.4 Urban Runoff Pollution Controls

Stormwater controls	Sewer overflow controls (CSO, SSO)
Regulatory controls	**Source controls**
Land-use regulations	Water conservation programs
Comprehensive runoff control regulations	Pretreatment programs
Land acquisition	**Collection system controls**
Source controls	Sewer separation
Cross-connection identification and removal	Infiltration control
Proper construction activities	Inflow control
Street sweeping	Regulator and system maintenance
Catch basin cleaning	In-system modifications
Industrial and commercial runoff control	Sewer flushing
Solid waste management	Sewer rehabilitation
Animal waste removal	Sewer system relief
Toxic and hazardous pollution prevention	**Storage**
Reduced fertilizer, pesticide, and herbicide use	In-line storage
Reduced roadway sanding and salting	Off-line storage
Detention facilities	Flow balance method
Extended detention dry ponds	**Physical treatment**
Wet ponds	Bar racks and screens
Construction wetlands	Swirl concentrators and vortex solids separators
Infiltration facilities	Dissolved air flotation
Infiltration basins	Fine screens and microstrainers
Infiltration trenches and dry wells	Filtration
Porous pavement	**Chemical precipitation**
Vegetative practices	**Biological treatment**
Grassed swales	**Disinfection**
Filter strips	Chlorine treatment
Filtration practices	UV radiation
Filtration basins	
Sand filters	
Other	
Water quality inlets	

Table 9.5 Sample Screening Criteria

Nonstructural	Structural
Pollutants controlled	Pollutant removal rates
Legal authority	Watershed area controlled
Public acceptance	Land requirements
Agency acceptance	Soil and groundwater characteristics
Institutional feasibility	Public acceptance
Technical feasibility	Agency acceptance
Cost	Reliability
	Cost
	Environmental impacts

cific control throughout a project area. This method allows comparison of costs of specific controls implemented systemwide. Any one or a combination of these methods can be used to develop a watershed control plan.

After the alternatives have been developed, they are compared using a decision process that evaluates the relative merits of each plan. Because of the complexity of urban runoff control problems, a number of factors must be considered in assessing alternative plans. These alternatives are represented in Figure 9.4 as inputs to a decision process, and

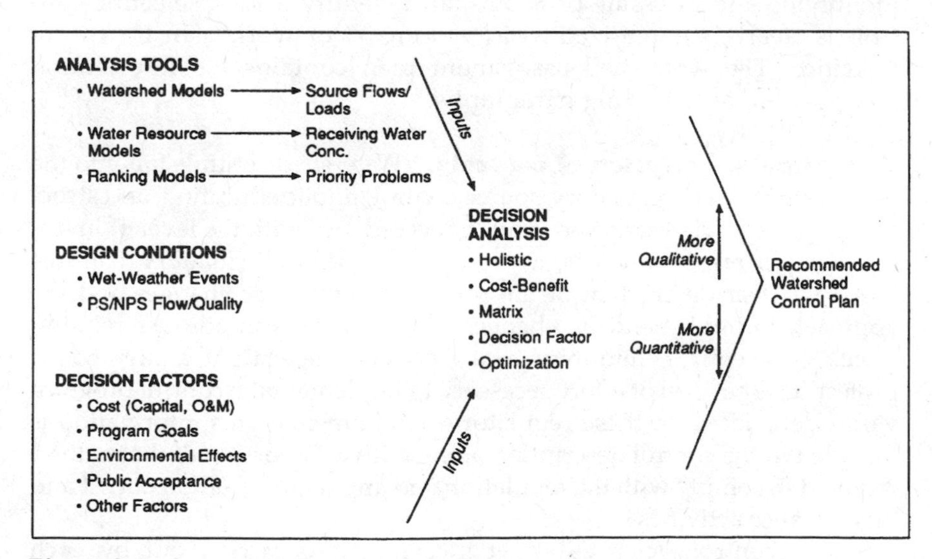

Figure 9.4 Conceptual diagram of watershed control selection method.

they include analysis tools (i.e., models), design conditions, and decision factors. The analysis tools are those used to assess and rank the existing pollution problems. The design conditions are the conditions under which the alternatives are compared. The decision factors are the criteria used to compare the alternatives. All these inputs are then used to evaluate the alternatives by using one or more decision analysis methods.

Assessing of alternatives involves an integration and comparison of a variety of factors, both quantitative and qualitative. This is an example of multiattribute decision making, and it can be performed with various decision analysis methods. Several decision analysis methods are available, including holistic, cost-benefit ratios, matrix comparisons, decision factor analysis, and optimization. These range, in order, from the most qualitative to the most quantitative. Matrix comparisons and decision factor analysis approaches are most commonly used. Due to the large number of qualitative factors involved, approaches such as formal optimization analysis are not feasible.

Implementation of the watershed program

The final step in the watershed planning process is to develop an implementation plan which sets forth the recommended control program. The information obtained through the earlier efforts of developing a watershed inventory, collecting and analyzing additional data, identifying and assessing problems, and identifying and selecting controls is clearly summarized as a "roadmap" or work plan for future activities. The watershed assessment plan contains the information described in the following paragraphs.

Conceptual description of controls. Watershed controls fall into the major categories of regulatory, source control, public education, and structural. Conceptual descriptions vary in accordance with the level of investigation performed and the category of control. Regulatory controls require a clear description of the proposed regulatory changes and the approach to implement the changes. The controls can address requirements for an entire community or can focus on a specific area targeted for protection. The level of effort necessary to implement the control program varies depending on these regulatory requirements. This information is included in the control description along with a discussion of the method required to comply with the regulation and any required enforcement and maintenance activities.

Source controls such as street sweeping are carried out by each responsible public entity. A plan needs to be prepared that details the frequency of conducting each practice, the locations at which the practice takes place, a schedule of activities, the required staffing, and the cost. Initial program start-up costs could cover training of staff and purchasing of equipment. In addition, source controls typically include ongoing operational costs—labor for public works and maintenance staff efforts. A record system needs to be designed to track activities and pertinent data (e.g., pounds of debris removed and areas swept).

Structural controls require engineering design and construction. Information needed includes a description, pictures, diagrams or concept sketches, design information and assumptions, as well as pertinent conceptual details. The details should indicate known site conditions such as existing structures; topography; and other site-specific information such as soil conditions, utility locations, and wetlands. Also included should be a general plan of the watershed showing locations of the recommended controls and the pollution sources they are designed to address. Final detailed design plans and specifications for each structural control are developed after the plan is approved. For each control, a cost estimate is also developed.

Schedule of activities. Because of the complex nature of watershed planning, implementing all the recommended controls in a short time generally is not possible. Implicit in developing an implementation schedule is the need to set priorities. An order of implementation (or phasing) is determined, taking into account important issues such as expected benefits and costs. These considerations are incorporated into a schedule with start and finish dates for major tasks and milestones. The schedule also includes interim dates for reporting results.

Depending on the program size, the schedule could be shown by means of a simple bar chart or a more complex critical-path method (CPM) system using project scheduling and management computer software. The type of schedule selected depends on the level of program complexity—the number of tasks and subtasks (activities) required, the number of involved entities, the length of time over which the program will extend, and the available program management resources.

Responsibilities for implementation. The individuals and entities responsible for implementing each aspect of the program must be identified. A well-defined institutional framework for the watershed control program is often lacking, and much of the effort for implementing plans must come from local and regional governments. To develop a plan, local officials must coordinate and initiate activities and motivate others in the community or other agencies to get involved. Obtaining firm commitments from these agencies prior to program implementation is important to the final success of the program.

Monitoring plan. A monitoring program should be conducted during and after program implementation to assist in determining the effectiveness of the watershed controls in achieving water resource goals. Monitoring during program implementation includes data collection to measure the overall program effects on water resources and to determine the effectiveness of controls. Existing water resource conditions determined during the planning process provide a good understanding of water resource quality before program implementation. A monitoring plan to assess water resource conditions during and after program implementation allows the level of resulting improvements to be assessed.

Summary of regulatory requirements. Regulatory issues that need to be addressed include both the implementation of regulatory controls and the application for regulatory approvals and permits needed to implement nonstructural and structural controls. The control program could involve the modification or strengthening of existing regulations, such as zoning, site plan review, subdivision, or wetlands protection, or the development of new regulations. In addition, appropriate regulatory approvals and permits must be obtained before implementation and construction of controls that could alter wetlands, waterways, or water quality, even if the control results in environmental benefit. These requirements are summarized as part of the control program. Coordination with appropriate agencies is advisable prior to application for the necessary approvals and permits. Agencies from which permits will be required should be contacted early in the planning process to determine requirements.

Public involvement. Support and involvement of the general public, including homeowners and businesses, are considered crucial to plan implementation and its ultimate success. While public involvement should be an integral part of the planning process, a public involvement program should be developed as part of the watershed program implementation. Components of public involvement programs can be wide-ranging, involving one or more of the following:

- Program meetings and presentations to provide information and updates
- Program materials such as newsletters, fact sheets, brochures, and posters
- School education programs such as special classes and tours
- Homeowner education programs on individual control of urban runoff-related pollution
- Consumer education programs on appropriate product purchasing and handling

- Business education programs
- Media campaigns including radio, newspaper, and television
- Coordinating and coalition building with local watershed or activist groups to support the program

The numerous other possibilities include setting up a program hotline, sponsoring special events, and conducting surveys. A task force can be set up to coordinate and help focus these activities.

Funding sources and mechanisms. A watershed program budget typically includes funds from a combination of sources. The actual funding sources utilized depend on many factors, including

- The sustainability of the funds
- The ease with which the funds can be obtained
- The administrative requirements of the funding option
- The correlation between the funding option and the problem
- The typical use made of the funding

The construction of a structural control, e.g., typically requires one-time, short-term funding that can be obtained through a grant or cost-share program. The development of a monitoring or maintenance program, however, typically requires continuing funding.

Regional, state, and federal storm water and nonpoint source (NPS) funding programs are usually intended for small-scale projects to collect data and demonstrate control methods. Larger-scale programs, therefore, have to be financed primarily through local mechanisms, such as general funds, long-term borrowing, pro rata share fees, storm water utilities, and special assessment districts.

Summary. The final recommended plan should be summarized in a document which can be used by responsible officials and agencies in plan implementation. By developing a thorough and accessible final implementation document and periodic reports, the plan will have a greater chance of success. While completion of the watershed plan signifies the end of the planning process, it is only the first step in the overall program. Plan implementation will likely be a long-term effort, and the planning is by no means over at this stage. As implementation and further monitoring occur, the plan might need to be updated, refined, and modified. When this occurs, the planning process may be reentered at any point. The program needs to be reevaluated and updated throughout implementation.

10
A Watershed Approach to Agricultural Nonpoint Source Pollution Abatement

Fred J. Brenner
Biology Department
Grove City College
Grove City, Pennsylvania

Elaine K. Brenner
Department of Chemistry, Chemical Engineering and Environmental Science
New Jersey Institute of Technology
Newark, New Jersey

Introduction

For over two decades, studies have documented that agriculture is a major contributor to the degradation of surface and groundwater systems in the United States and elsewhere in the world (Clark et al., 1985; Diebel et al., 1992; Hill, 1987; Brenner and Mondok, 1995). In the British Isles, e.g., the number of agriculture-related pollution discharges increased 475 percent in the 10-year period from 1974 to 1984; and of the total of 13,427 reported pollution incidents during this period, 74 percent were attributed to livestock production. In addition, 75 percent of the nitrate contamination of the drinking water in agricultural districts was traced to agricultural discharges (Payne, 1986). Likewise in the United States, Clark et al. (1985) estimated that nonpoint source pollutants, primarily from agriculture, account for 73 percent of the biochemical oxygen demand (BOD), 83 percent of the bacterial loads, and 92 percent of the suspended sediments in waterways. Moreover, the U.S. Environmental Protection Agency (EPA, 1983, 1987) estimated that 57 percent of the lakes, 64 percent of the rivers, and 19 percent of the estuaries are adversely impacted by the discharges from agricultural lands. Macharis (1985) indicated that croplands generate a major portion of nutrient loads in the Chesapeake Bay drainage basin, with agriculture by far being the major contributor to nonpoint sources in the watershed.

It has been demonstrated that overland flows from agricultural lands transport bacteria into receiving streams (Weber et al., 1989), and this is intensified by the disturbance of soil and vegetation (Blackie and Newson, 1986), thereby adversely impacting aquatic systems (Simmons and Li, 1980). In addition, manure disposal on croplands is also a principal source of nutrient and bacteriological contamination of freshwater ecosystems (Brenner and Mondok, 1995; Brenner et al., 1995). Hill (1987) reported that nitrogen concentrations in the runoff from agricultural land in the state of North Carolina were 4 to 5 times greater than they were from a forested watershed.

Section 208 of the federal Water Pollution Control Act (P.L. 92-500), as amended by the U.S. Congress in the Clean Water Act of 1977, recognizes that nonpoint sources are major contributors to sediment, bacteria, and nutrient loads of freshwater ecosystems. Recognizing the impact of agriculture on aquatic ecosystems, the Environmental Protection Agency is encouraging the implementation of watershed management programs to abate the adverse impacts of agricultural discharges on aquatic ecosystems. Agriculture is considered a major industry in England and Wales, as well as in large segments of the United States and other countries of the world. It is apparent, therefore, that pollution from agricultural enterprises should be monitored and regulated as stringently as that from the manufacturing segments of our economy. And since watersheds are defined geographical units, it is logical to approach the abatement of agricultural pollution on a watershed basis. By the use of case studies, this chapter addresses a watershed approach to the abatement of agricultural discharges on aquatic ecosystems.

Assessment of Agricultural Pollution Impacts

The potential of nonpoint pollution impacts from agriculture in a watershed may be assessed according to the rating system developed by the Pennsylvania Department of Environmental Protection. This model takes into account the relative contributions from surface water (20 percent), groundwater (15 percent), livestock contributions (25 percent), and the degree of management practiced by farmers within the watershed (Pennsylvania Department of Environmental Resources, 1990; Brenner and Mondok, 1995). The agricultural intensity within a watershed can be determined from aerial photographs, soil maps using a dot grid matrix, geographic information systems, or other methods of calculating land use. Additional information on agricultural intensity may be obtained from existing farm plans, livestock reports, or field observations. These factors are then integrated into a watershed nonpoint source pollution rating factor according to the formula

$$\mathrm{RF} = \frac{20\ \mathrm{WD}}{\mathrm{WD}_{\max}} + \frac{25\ \mathrm{AN}}{\mathrm{AN}_{\max}} + \frac{15\ \mathrm{GD}}{\mathrm{GD}_{\max}} + \frac{40\ \mathrm{MF}}{\mathrm{MF}_{\max}}$$

where RF = rating factor, AN = animal nutrient, WD = watershed delivery, and MF = management factor.

The maximum values are used to "normalize" the results to ensure that no one value unduly influences the rating factor. Each variable factor is expressed as a relative value in order to compare watersheds. Each component of the watershed rating factor is calculated as follows:

$$\mathrm{WD} = \frac{20\ \mathrm{SD}}{\mathrm{SD}_{\max}} + \frac{\mathrm{RCI}}{\mathrm{RCI}_{\max}} + \frac{\mathrm{HEL}}{\mathrm{HEL}_{\max}}$$

where *stream density* SD is the cumulative length of all perennial streams identified on 7.5-in USGS topographic maps within the watershed boundary divided by the watershed area; *row crop intensity* RCI is determined by dividing the watershed acreage in row crops by the watershed area; *highly erodible land* HEL is a measure of the potential erodibility of soils in the watershed, which is dependent on slope length and steepness (LS), rainfall intensity *R*, and the inherent erodibility of the soil type *K*.

$$\text{HEL index} = \frac{ST}{RK}$$

where T = soil loss tolerance value. Measured in metric tons per hectare per year, T is the maximum amount of soil that can be lost due to erosion and still sustain long-term productivity. Each soil type has been assigned a T value by the U.S. Natural Resources Conservation Service.

If LS exceeds the HEL index for a given soil type, the land is considered to be highly erodible. The LS values for the slopes of each soil type identified in the county soil survey are determined by selecting 40-ha sample blocks on soil maps within the watershed and then calculating the area of cropland within each soil mapping unit. If mapping units account for at least 80 percent of the cropland, the average LS value is divided by the HEL index for that mapping unit and is multiplied by the percentage of the area represented by that mapping unit. The summation of these values is HEL for the watershed.

Animal nutrient factor

$$AN = \frac{\Sigma(\text{animal units} \times \text{nutrient factor})}{\text{cropland/hayland acreage}}$$

This factor is considered to be both a surface water and a groundwater potential pollution factor. It is a means of estimating the pollution potential of livestock operations by considering the number and types of animals in a watershed, the amount of waste generated, and the amount of land available for spreading manure. The various animal types are equated by assigning the following nutrient factors: dairy cow, 1; beef cow, 1; horse, 0.7; hog, 1.4; sheep, 1.1; poultry, 2.7; and veal, 0.6. One animal unit is equated to 400 kg live weight. Animal waste production for each animal type is multiplied by the nutrient (nitrogen and phosphorus) in the manure (obtained from state agronomy guide), and the resultant values are standardized by using a base value of 1.0 of dairy cattle. Nutrient values for young animals should be adjusted according to the average weight of each age group. The number of the different types of livestock can be obtained from crop and livestock reports and/or the state extension service.

Groundwater delivery factor

$$GD = 15\left(\frac{AG}{AG_{max}} + \frac{SLP}{SLP_{max}}\right)$$

where AG = aquifer geology and SLP = soil leaching potential.

Aquifer geology is calculated from the geology associated with each soil mapping unit in 40-ha sample blocks and then is assigned the following value for rock porosity: carbonate, 10; glacial sand and gravel, 8; glacial till deposits, 4; noncarbonate rock formation, 2.

Soil leaching potential is an estimate of the ability of the soil to absorb or retain nutrients, which influences the ability of these pollutants to enter the groundwater. Based on soil characteristics, the Natural Resource Service has assigned SLP values from high (10) to moderate (7) to low (4) to very low (1). The appropriate value is assigned to the soil types within 40-ha sample blocks, summed, and the sum is multiplied by the percentage of the soil type within the watershed.

Management factor

$$\text{MF} = 40 \text{ weighed value}$$

Management can reduce the pollution potential inherent in the watershed soils and animal factors. It is a combination of conservation practices and animal and nutrient management. This factor is best obtained from site visits and farmer interviews, which makes it the most subjective factor in the overall rating system. Each of four subfactors is assigned a value of 10 for little or no management, 2 to 9 for some management, and 1 for all conservation practices in place. The MF value is the summation of the following four subfactors divided by the number of farms evaluated within the watershed.

The criteria for assigning the values are as follows:

Conservation practices

- Little or no management: No conservation plan, few implemented practices, observed evidence of major stream problems
- Full management: Fully implemented conservation plan, extensive use of conservation tillage, cover cropping, no visual evidence of stream problems.

Animal management

- Little or no management: Livestock have unlimited access to streams, evidence of stream banks being broken down by animals, overgrazed pastures, barnyards barren of vegetation, extensive manure buildup in buildings.

ment: Limited access to streams, rotational grazing sys- nted, vegetated riparian zones, concrete or impervious l where animals are concentrated.

Nutrient management

- Little or no management: Little or no routine soil testing, no manure nutrient analyses, no consideration of where or how much manure is spread, no nutrient management plan, soil test recommendations not followed.
- Full management: Implementation of nutrient management plan, manure storage facility in place, implementation of soil test recommendations, manure analyzed for nutrient content and incorporated in fertilizer application.

This model was used by Brenner and Mondok (1995) to assess the agricultural impacts within the Shenango River watershed in western Pennsylvania. In their study, 104 water samples were obtained from streams within the watershed, and there was a significant correlation between the overall rating factor and fecal coliforms, total phosphate, and nitrate concentrations in receiving streams. An analysis of the individual components of the rating factor revealed that the watershed delivery factor, animal nutrient factor, and management factor were significantly correlated with fecal coliforms and total phosphate in receiving streams, with the groundwater delivery factor being correlated with nitrate concentrations. Based on this study, it appears that overland flows from agricultural lands are major factors in deteriorating water quality in aquatic ecosystems within the watershed. However, nitrate apparently enters stream systems via the groundwater-surface water interface in the stream channel.

Agricultural Impacts on Water Quality

According to the National Research Council (1992), agriculture is a principal source of nutrient, sediment, and bacteriological contamination of aquatic ecosystems in the United States. Within Pennsylvania alone, agriculture is responsible for degrading approximately 700 mi of streams and rivers, and of these 20 percent are located in the Ohio River watershed. Fontaine (1994) estimated that agriculture contributes over 200 metric tons (t) of nutrients annually to aquatic ecosystems within the United States. In the Shenango River watershed in northwest Pennsylvania, Brenner and Mondok (1995) estimated that the total manure production from livestock was approximately 435,715 t/yr of which 68 percent was contributed by dairy cattle. The interaction between watershed size and livestock concentration has a significant impact on water quality in receiving streams. Brenner and Brenner (1995) found that fecal coliforms, phosphate, and suspended solids were inversely correlated with the size of the watershed. All these pollutants are transported by overland flows to receiving streams, and thus their concentrations are affected by rainfall events (Brenner et al., 1996). Brenner and Brenner (1995) reported that over an 8-year period, nutrient concentrations in eight streams in the Coolspring Creek watershed in northwest Pennsylvania were significantly correlated with rainfall events 5 and 14 days prior to sampling. Fecal coliforms, however, were correlated with amount of rainfall 5 and 30 days prior to sampling.

In a series of studies, Brenner (1996a, b) and Brenner et al. (1991) reported that land use and soil characteristics significantly affect water quality within the Otter Creek watershed in northwestern Pennsylvania. Otter Creek is a 12,400-ha watershed situated within the Glaciated Allegheny Plateau in Mercer County, Pennsylvania. The soils within the watershed are derived from glacial till and are generally poorly drained with seasonally perched water tables. The dominant soils include hydric Frenchtown and Ravenna, which has hydric Frenchtown inclusions along with hydric Wayland soils occurring in the floodplains. Frenchtown and Ravenna soils are characterized by a low permeable fragipan at depths of 25 to 40 cm, with a water table within 15 cm of the surface for at least 6 months per year. Wayland soils are poorly drained with a silt loam in the upper 45 cm underlined by a gravely loam. Although classified as hydric, the underlying sand and gravel of Wayland soils have a greater permeability than the clay-rich Frenchtown soil. The entire watershed is agricultural with pastures and row crops being the principal land use. All private homes within the watershed have on-lot (septic) sewage systems of varying operational efficiencies.

Samples were obtained from 10 stations situated through the watershed from June 1, 1979, until May 30, 1994, and analyzed for fecal coliforms and seven chemical parameters (pH, carbonate alkalinity, calcium, magnesium, phosphate, ammonia-N, and conductivity). Based on a Bonferroni's multiple comparison analysis (Neter and Wasserman, 1974), all water quality parameters—except pH, conductivity, and NH_3-N—varied significantly among the different sampling stations with significant seasonal impacts on water quality parameters (Table 10.1). However, there was not a significant sample station seasonal effect for

Table 10.1 ANOVA of the Station and Seasonal Effects on Water Quality within the Otter Creek Watershed

	DF	Mean square	*F* value	*P* > *F*
Station				
pH	9	0.170	1.04	0.4114
$CaCO_3$	9	0.179	5.19	0.0001
Ca	9	0.230	7.55	0.0001
Mg	9	0.252	2.49	0.0024
Conductivity	9	0.127	1.20	0.2751
PO_4	9	0.468	7.00	0.0001
NH_3-N	9	0.266	0.52	0.9122
Fecal coliform	9	12.044	23.37	0.0001
Season				
pH	3	2.672	16.27	0.0001
$CaCO_3$	3	2.661	77.27	0.0001
Ca	3	1.272	45.69	0.0001
Mg	3	1.633	15.88	0.0020
Conductivity	3	2.592	2.59	0.0001
PO_4	3	0.563	9.63	0.0001
NH_3-N	3	9.139	0.52	0.9122
Fecal coliform	3	8.093	15.71	0.0001

any of the different water quality parameters ($P > 0.919$). The increase in calcium, magnesium, phosphate, and fecal coliform concentrations during the summer months (June, July, August) and autumn (September, October, November) (Table 10.2) was probably due to the application of lime in excess of what the soil could assimilate. However, elevated water temperatures during these months were also probably a factor in maintaining elevated fecal coliforms during the summer months.

Throughout the watershed, $CaCO_3$, calcium, magnesium, and conductivity, along with phosphate and fecal coliform concentrations, were significantly higher at stations that lacked riparian buffer zones or wetlands (Table 10.3). In headwater areas lacking riparian buffers, fecal coliform and phosphate concentrations averaged 1542/100 mL and 1.35 mg/L, respectively, but coliform concentrations were reduced 41 percent and phosphate 23 percent after the stream flowed through 6.3 km

Table 10.2 Mean Seasonal Changes in Water Quality Parameters within the Otter Creek Watershed (N = number of observations per season)

Parameter	Winter	Spring	Summer	Autumn
pH	7.0	7.2	7.1	7.1
$CaCO_3$, mg/L	19.4	24.9	35.6	30.8
Ca, mg/L	48.7	48.3	63.2	65.0
Mg, mg/L	51.2	43.4	46.5	56.0
Conductivity, μ mohms	143.2	155.3	140.4	103.6
PO_4, mg/L	1.28	1.69	2.09	1.41
NH_3-N, mg/L	0.40	0.65	0.21	0.73
Fecal coliform, most probable number (MPN)/100 mL	941	895	1389	1064

Table 10.3 Comparison of Water Quality in Stations with and without Riparian and Wetland Buffer Zones

Parameter	Headwaters, no buffers	Riparian buffers	Wetland area	Downstream, no buffers
No. stations	4	3	1	2
pH	7.1	7.1	7.0	7.1
$CaCO_3$, mg/L	37.4	25.8	24.9	39.4
Ca, mg/L	69.4	47.5	47.8	51.7
Mg, mg/L	54.1	47.4	50.8	42.9
Conductivity, μ mohms	153.2	109.4	110.3	119.2
PO_4, mg/L	1.35	1.03	1.13	1.20
NH_3-N, mg/L	0.66	0.82	0.78	0.95
Fecal coliform/100 mL	1542	910	523	1009

of forested riparian buffer zone. Fecal coliform concentrations were further reduced by 42.5 percent after the streams flowed through 88-ha channel and riparian wetland system, but fecal coliform and phosphate concentrations increased as the streams flowed through pastures where cattle had direct access to the stream (Table 10.2). Likewise in a previous study, Brenner et al. (1991) reported that fecal coliform and phosphate

[illegible].4 Comparison of Mean Fecal Coliform and Nutrient Concentrations in Hydric and Nonhydric Soils

Soils	*N*	Fecal coliform/ 100 mL	PO_4, mg/L	NH_3-N, mg/L
Hydric	496	596 ± 10.4	198 ± 0.06	0.25 ± 0.01
Nonhydric	372	492 ± 18.6	1.61 ± 0.02	0.13 ± 0.02

SOURCE: Brenner et al., 1995.

concentrations were reduced in streams where at least 50 percent of the riparian zones were intact within 30 m of the stream channel.

In a 62-month study of a 1200-ha watershed, Brenner et al. (1996) found that fecal coliform and nutrient concentrations were significantly higher in samples collected from lysimeters in hydric than in nonhydric soils being used as cattle pasture ($P < 0.05$) (Table 10.4). Likewise, fecal coliforms per 100 mL averaged 1302 ± 84.4 at those stations receiving agricultural drainage compared to 541 ± 38.3 at nonagricultural stations. Brenner (1996b) and Brenner et al. (1996) reported that hydric soils accumulate water from the surrounding soils, and hence fecal coliforms, nutrients, and possibly other contaminants could be transported laterally along the fragipan, acting as a conduit into the receiving stream (Fig. 10.1).

Best Management Practices

Black (1996) defined watershed management as the planned manipulation of one or more factors of a natural or artificial drainage so as to effect a desired change in or maintain a desired condition of the water resource. Conventional best management practices generally include strip cropping, grass buffer strips, contour plowing, grass waterways, conservation tillage, and manure storage among other practices designed to reduce soil erosion while maintaining agricultural production. Brenner et al. (1990) reported that the effectiveness of these practices varied seasonally and in some instances increased nutrient concentrations in receiving streams. Brenner (1996a, b) reported that the installation of 608 ha of best management practices in the 12,400-ha Otter Creek watershed in Pennsylvania consisting of primarily vegetation management along with 19.4 km of drain tile had minimal, if any, impact on water quality within the watershed. Adelman (1996) indicated that although nutrient and pesticide loading to surface waters was lower with conservation tillage, the leaching of these chemicals to groundwater was always greater in conservation

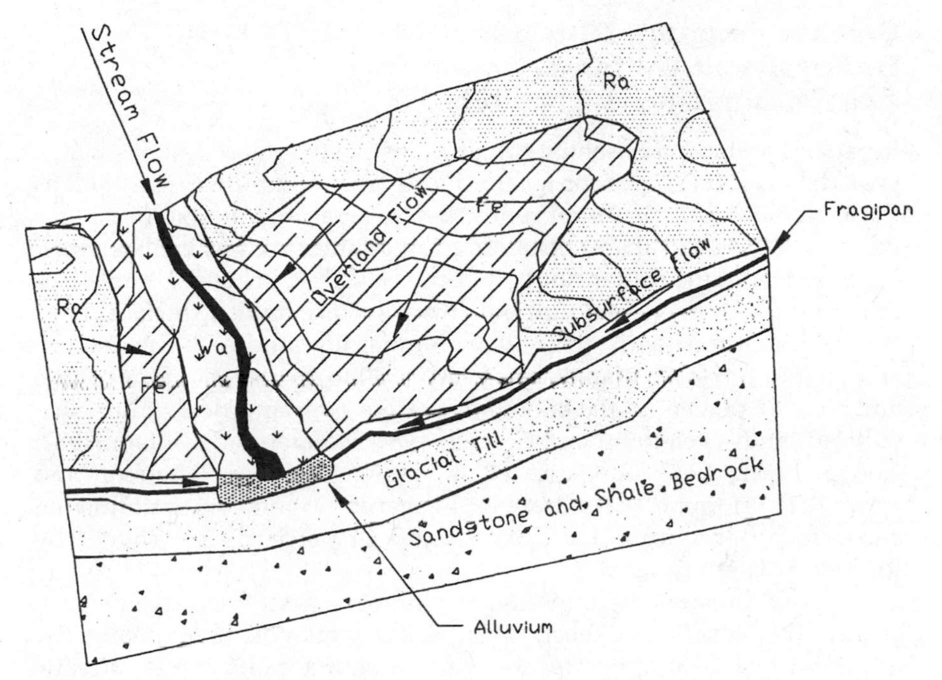

Figure 10.1 Model of the possible pathway of bacteria and nutrient migration from nonhydric Ravenna soils via hydric Frenchtown and Wayland soils into receiving streams. (*From Brenner, 1996b; Brenner et al., 1996.*)

tillage fields. Moreover, conservation tillage may also require more applications of herbicides than conventional tillage, which may have an adverse impact on aquatic ecosystems (Brenner et al., 1995). Quentin (1996) found that the concentrations of pesticides in the New York City watershed system was always below the maximum contamination level (MCL) established by the Environmental Protection Agency through the Safe Drinking Water Act (SDWA). But many of these chemicals have been reported to have long-term impacts on aquatic ecosystems as well as human health (Brenner et al., 1995).

Various state and federal agriculture agencies have been cost-sharing the implementation of best management practices for over five decades, and yet agriculture remains the major factor in the deterioration of aquatic ecosystems in the United States. In the past, there has not been an assessment of problem areas within watersheds prior to the application of best management practices nor an evaluation of which practices were necessary to improve water quality within the targeted watershed. Hence often after the installation of best management practices, there was no identified improvement in water quality.

Stream Fencing, Riparian Buffers, and Wetlands as Best Management Practices

Riparian habitats are essential components of stream ecosystems which provide a carbon source for macroinvertebrate communities as well as the abatement of agricultural runoff. These riparian buffers provide for the physical dispersal and filtration of overland flows as well as nutrient uptake, denitrification, and other biological and chemical processes that remove sediment and nutrients from surface and subsurface flows from agricultural lands (Fig. 10.2). Numerous studies have demonstrated that forest riparian buffers remove 80 to 90 percent of the nitrate and 50 to 75 percent of the phosphorus from overland flows from agricultural lands (Peterjohn and Correll, 1984; Cooper and Gilliam, 1987; Cooper et al., 1987; Lowrance et al., 1985). Likewise, Peterjohn and Correll (1984) found that nitrate and organic carbon concentrations in subsurface flows are reduced by 95 and 64 percent, respectively, by forested riparian buffers.

As stated previously in this chapter, numerous studies have demonstrated the beneficial aspects of riparian wetlands in reducing the adverse impacts of agricultural nonpoint source pollution on aquatic ecosystems (Brenner, 1996a, b; Brenner et al., 1991, 1996). Brenner (1996a, b) reported that the exclusion of cattle from a 6-ha riparian wetland adjacent to a stream resulted in a 53 percent reduction in fecal coliforms over 3 years ($\overline{X}$ of 1302/100 mL to an $\overline{X}$ of 689/100 mL) along with a 48 percent reduction in phosphate concentrations ($\overline{X}$ of 1.88 mg/L to an $\overline{X}$ of 0.97 mg/L). In a previous study, Brenner et al. (1996) found that the exclusion of cattle from a stream and the restoration of a riparian wetland resulted in a 30 percent reduction in fecal coliforms and a 60 percent reduction in phosphate concentrations over a 13-month period, and enhanced wildlife habitat within the watershed.

As stated earlier in this chapter, within the Otter Creek watershed, carbonate alkalinity, calcium, magnesium, and conductivity along with phosphate and fecal coliform concentrations were significantly higher at stations that lacked riparian buffer zones or wetlands (Table 10.3). In headwater areas lacking riparian buffers, fecal coliform and phosphate concentrations averaged 1542/100 mL and 1.35 mg/L, respectively, but coliform concentrations were reduced 41 percent and phosphate 23 percent after the stream flowed through a 6.3-km forested riparian buffer zone. Fecal coliform concentrations were further reduced by 42.5 percent after the stream flowed through 88-ha channel and riparian wetland system, but both fecal coliforms and phosphate concentrations increased downstream from the wetland system. The ability of wetlands

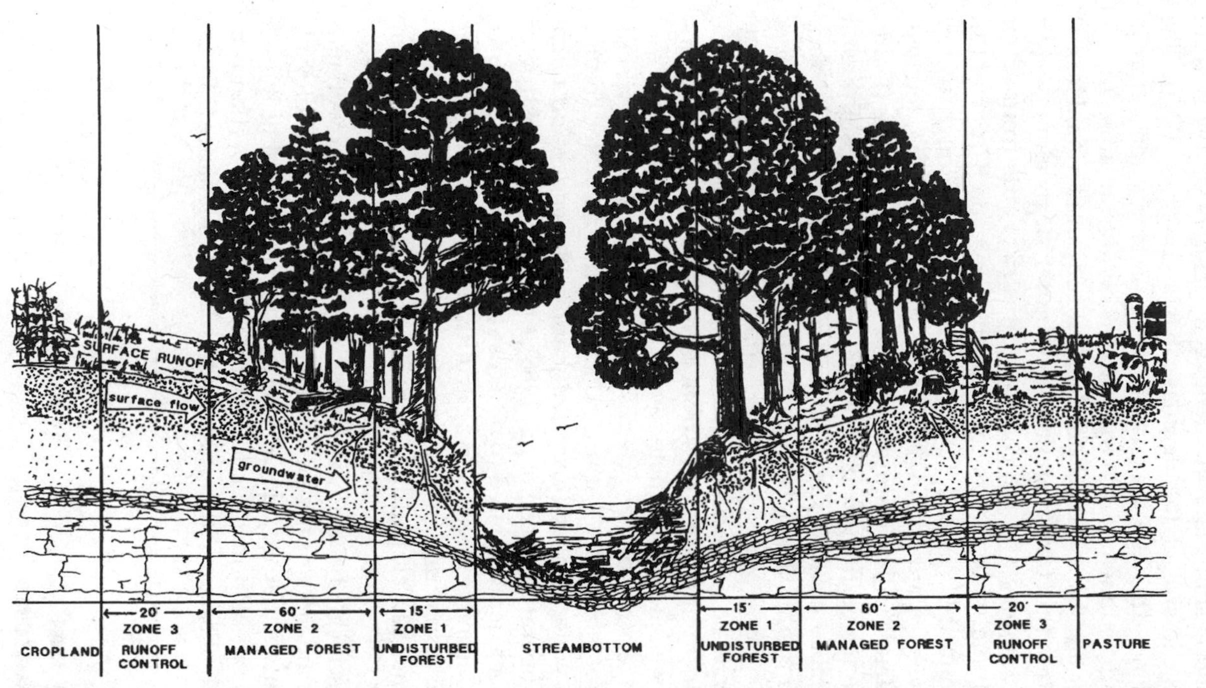

Figure 10.2 A diagrammatic summary of the role of riparian buffers in the abatement of nonpoint source pollution. In zone 3, the concentrated flows are converted to dispersed flows by water bars or spreaders to facilitated infiltration; grazing may be permitted under controlled conditions. The function of zone 2 is to allow for filtration, deposition, nutrient uptake, anaerobic denitrification, and other biological processes that remove sediment and nutrients from surface and subsurface flows. Periodic harvesting should be undertaken to enhance nutrient uptake through vigorous tree growth. The mature forest in zone 1 provides detritus, maintains water temperature, and provides other aspects necessary for a productive stream ecosystem. To protect stream ecosystems, harvesting and grazing should not be permitted in zone 1.

to reduce suspended solids and nutrients is dependent on the ratio of wetland area to the size of the drainage basin (Brenner, 1995). The higher the wetland–drainage area ratio, the greater the reduction in the nutrient and sediment discharges to receiving streams. In the Kissimmee River and Lake Okeechobee watersheds in Florida, Havens et al. (1996) reported that the installation of buffer strips along waterways, lagoons to trap solids, and fencing to restrict cattle from water courses reduced total phosphorus concentrations 63 and 87 percent in the lower Kissimmee River and Taylor Creek/Nubbin Slough watersheds, respectively. Throughout the Kissimmee River watershed, 11,000-ha of wetlands will be restored over the next two decades (Dahm et al., 1995). The restoration of these wetlands will have a significant impact on the abatement of agricultural nonpoint pollution, thus improving water quality throughout the watershed.

In addition to natural wetland systems, constructed wetland systems have been shown to be beneficial as passive treatment systems for agricultural wastes including discharges from feedlots and milking parlors. As with natural wetlands, constructed wetlands removed sediment and nutrients via vegetative uptake, deposition and adsorption by sediments, and microbial activity within the sediments and water column (Hammer, 1993) while fecal coliform reduction is dependent on retention time and temperature (Reed et al., 1988). Overall, these constructed wetlands have been demonstrated to be a cost-effective method of nutrient reduction in receiving streams in agricultural watersheds (Hammer, 1989, 1993). The actual design and construction criteria for wetland treatment systems for agricultural wastes are discussed elsewhere (Hammer, 1989, 1993; Rodgers and Dunn, 1993) and are not repeated here; but as Mitsch (1993) stated, we must tie our rivers and wetlands back together in an ecologically sustainable way. Successful ecological engineering requires that we take advantage of our increasing knowledge of ecology and its principles to construct and restore wetlands as part of a natural landscape with minimum human maintenance (Mitsch, 1993).

Conclusions and Recommendations

Since wetlands are definable units, it is only logical that we address the abatement of agricultural nonpoint source pollution on a watershed basis. The first goal in the restoration of agricultural areas is to identify the major problem areas and then apply the appropriate best management practices to address the problems. On the basis of available data, stream fencing and the restoration and enhancement of riparian buffers are a cost-effective method of reducing the adverse impacts of agricultural drainages on receiving streams. In addition, the retirement and enrollment of marginal agricultural croplands and pastures into the Conservation Reserve Program (CRP) and wetland restoration will further reduce the nutrient loading to receiving streams within the watershed. Overall, the application of these best management practices on a watershed basis appears to be most effective in reducing the adverse impact of agricultural nonpoint source pollution on aquatic ecosystems.

Literature Cited

Adelman, D. D. 1996. Surface residue and soil type effects on water quality. Pp. 425–434. In *Watershed Restoration and Management: Physical, Chemical and Biological Consideration.* J. J. McDonnell, D. J. Leopold, J. B. Stribling, and L. R. Neville (eds.). American Water Resources Association, Herndon, VA.

Black, P. E. 1996. Watershed management. Pp. 245–251. In *Watershed Restoration and Management: Physical, Chemical and Biological Considerations.* J. J. McDonnell, J. B. Stribling, L. R. Neville, and D. J. Leopold (eds.). American Water Resources Association. Herndon, VA.

Blackie, J. R., and M. D. Newson. 1986. The effects of forestry on the quantity and quality of runoff in upland Britain. In *Effects of Land Use on Fresh Waters: Agriculture, Forestry, Mineral Exploitation, Wastewater.* J. F. Solbe and L. G. Solbe (eds.). American Public Health Association. Washington.

Brenner, E. K. 1995. Effectiveness of sediment control structures for reducing nonpoint source pollution parameters entering a rural watershed in Mercer County, Pennsylvania. Master's Thesis. New Jersey Institute of Technology, Newark.

Brenner, E. K., and F. J. Brenner. 1995. Impact of precipitation and drainage area size on water quality in stream systems supplying a rural residential lake. *J. Penn. Acad. Science* **69:**88–92.

Brenner, F. J. 1996a. Impact of changing land use on water quality in an agricultural watershed. Pp. 189–194. In *Proceedings Conserv. 96: Responsible Water Stewardship.* American Water Works Association. Denver, CO.

Brenner, F. J. 1996b. Watershed restoration through changing agricultural practices. Pp. 397–404. In *Watershed Restoration Management.* J. J. McDonnell, J. B. Stribling, L. R. Neville, and D. J. Leopold (eds.). American Water Resources Association. Herndon, VA.

Brenner, F. J., R. P. Steiner, J. J. Mondok, and R. J. McDonald, Jr. 1990. Nonpoint-source pollution: A model watershed approach to improve water quality. Pp. 483–498. In *Water Resources in Pennsylvania: Availability, Quality and Management.* Pennsylvania Academy of Science, Easton, PA.

Brenner, F. J., J. J. Mondok, and R. J. McDonald, Jr. 1991. Impact of riparian areas and land use on four nonpoint source pollution parameters in Pennsylvania. *J. Penn. Acad. Science* **65:**65–70.

Brenner, F. J., and J. J. Mondok. 1995. Nonpoint source pollution potential in an agricultural watershed in northwestern Pennsylvania. *Water Resources Bull.* **31**:1101–1112.

Brenner, F. J., S. D. Yoder, and R. J. Blair. 1995. Impact of nonpoint source contaminants on ecosystems and human health. Pp. 87–99. In S. K. Majumdar, E. W. Miller, and F. J. Brenner (eds.). *Environmental Contaminants, Ecosystems and Human Health.* Pennsylvania Academy of Science, Easton, PA.

Brenner, F. J., R. P. Steiner, and J. J. Mondok. 1996. Groundwater-surface water interaction in an agricultural watershed. *J. Penn. Acad. Science.*

Cooper, J. R., and J. W. Gilliam. 1987. Phosphorus redistribution from cultivated fields into riparian areas. *Soil Science Soc. Amer. J.* **21**:733–739.

Cooper, J. R., J. W. Gilliam, R. B. Daniels, and W. P. Robarge. 1987. Riparian areas as filters for agricultural sediment. *Soil Science Soc. Amer. J.* **51**:417–420.

Clark. E. H., H. H. Haverkamp, and W. Chapman. 1985. *Eroding Soil: The Off Farm Impacts.* The Conservation Foundation. Washington.

Commonwealth of Pennsylvania. 1937. Water quality standards. Pp. 1–93, 130. Title 25, Subpart C. Article 11, Chapter 93. P.L. 1987, as amended. Harrisburg, PA.

Dahm, C. N., K. W. Cummins, H. M. Valett, and R. L. Coleman. 1995. An ecosystem view of the restoration of the Kissimmee River. *Restoration Ecology* **3**:225–238.

Diebel, P. L., D. B. Taylor, S. S. Batie, and C. D. Hearwole. 1992. Low-input agriculture as ground water protection strategy. *Water Resources Bull.* **28**:755–762.

Environmental Protection Agency. 1983. *Chesapeake Bay: A Framework for Action.* Chesapeake Bay Program. Chesapeake Bay Liaison Office. Annapolis, MD.

Environmental Protection Agency. 1987. *Nonpoint Source Guidance.* Nonpoint Source Branch. Office of Water Regulations and Standards. Washington.

Fontaine, A. E. 1994. The Everglades: Past, Present and Future. *Third Symposium on Biogeochemistry of Wetlands.* (Abstract). Orlando, Fl.

Hammer, D. A. 1989. Constructed wetlands for treatment of agricultural waste and urban stormwater. Pp. 333–348. In *Wetlands Ecology and Conservation: Emphasis in Pennsylvania.* S. K. Majumdar, R. P. Brooks, F. J. Brenner, and R. W. Tiner (eds.). Pennsylvania Academy of Science, Easton, PA.

Hammer, D. A. 1993. Designing constructed wetland systems to treat agricultural nonpoint pollution. Pp. 71–111. *Created and Natural Wetlands for Controlling Nonpoint Source Pollution.* R. K. Olsen (ed.). Environmental Protection Agency. Washington.

Havens, K. E., E. G. Flaig, R. T. James, S. Lostal, and D. Muszick. 1996. Results of a program to control phosphorus discharges from dairy operations in south-central Florida, USA. *Environmental Management* **21**:585–593.

Hill, C. L. 1987. Monitoring the effects of land management practices on water quality in Gulford County, North Carolina. Pp. 137–148. In *Monitoring, Modeling and Mediating Water Quality.* S. J. Nix and P. E. Black (eds.). American Water Resources Association. Herndon, VA.

Lowrance, R., R. Leonard, and J. Seridan. 1985. Managing riparian ecosystems to control nonpoint pollution. *J. Soil and Water Conservation,* Jan./Feb., pp. 87–91.

Macharis, J. 1985. Chesapeake Bay non-point source pollution. In *Perspectives on Non-Source Pollution.* Environmental Protection Agency. EPA 44015-85001.

Mitsch, W. 1993. Landscape design and the role of created, restored, and natural riparian wetlands in controlling nonpoint source pollution. In *Created and Natural Wetlands for Controlling Nonpoint Pollution.* R. K. Olsen (ed.). Environmental Protection Agency. Washington.

National Research Council. 1992. *Restoration of Aquatic Ecosystems: Science, Technology and Public Policy.* National Academy Press. Washington.

Neter, J., and W. Wasserman. 1974. *Applied Linear Statistical Models.* Richard D. Irwin Inc. Homewood, IL.

Payne, M. 1986. Agricultural pollution—The farmers' view. In *Effects of Land Use on Fresh Waters: Agriculture, Forestry, Mineral Exploitation, Urbanization.* Ellis Norwood Publishers. Chichester, England.

Pennsylvania Department of Environmental Resources. 1990. Criteria to develop ranking of subwatersheds. In 205 (j) Watershed Evaluations.

Peterjohn, W. T., and D. L. Correll. 1984. Nutrient dynamics in an agricultural watershed: Observations on the role of a riparian forest. *Ecology* **65:**1466–1475.

Quentin, D. H. 1996. Pesticide concentrations within streams of four New York City reservoir drainage basins. Pp. 153–163. In *Watershed Restoration Management: Physical, Chemical, and Biological Considerations.* New York City Water Supply Studies. J. J. McDonnell, D. J. Leopold, J. B. Stribling, and L. R. Neville (eds.). American Water Resources Association. Herndon, VA.

Reed, S. C., E. J. Middlebrooks, and R. W. Crites. 1995. *Natural Systems for Waste Management and Treatment.* McGraw-Hill. New York.

Rodgers, J. H., Jr., and A. Dunn. 1993. Developing design guidelines for constructed wetlands to remove pesticides from agricultural runoff. Pp. 113–130. In *Created and Natural Wetlands for Controlling Nonpoint Source Pollution.* R. K. Olsen (ed.). Environmental Protection Agency. Washington.

Simmons, D. S., and R. M. Li. 1980. Modeling of sediment nonpoint source pollution from watersheds. Pp. 341–373. In *Environmental Impact of Nonpoint Pollution.* M. R. Overcash and J. M. Davidson (eds.). Ann Arbor Science. Ann Arbor, MI.

Weber, J. B., P. J. Shea, and H. J. Streak. 1989. An evaluation of nonpoint sources of pesticide pollution in runoff. Pp. 69–98. In *Environmental Impact of Nonpoint Pollution.* M. R. Overcash and J. M. Davidson (eds.). Ann Arbor Science. Ann Arbor, MI.

11

The Case for River Basin Management

The Ohio River Valley Water Sanitation Commission (ORSANCO) as a Case Example

Alan H. Vicory, Jr.
P.E., DEE, Executive Director, ORSANCO

Peter A. Tennant
P.E., Technical Programs Manager, ORSANCO

To effectively develop and manage water resources is to achieve a balance and accommodation among the legitimate uses placed on it, while at the same time minimizing interference with the natural hydrological cycle and the disturbance of wildlife. Legitimate uses can include, but not be limited to, irrigation, industrial processing, raw supply for drinking water after treatment, navigation, and hydropower. Some uses are related substantially to quantity concerns (e.g., navigation) while others involve quality and quantity objectives (e.g., drinking water).

Competing uses can often come into conflict and become acute at times. For example, the use of a river for industrial processing or transportation can conflict with its simultaneous use as a drinking water supply when spills occur. Managing water resources for these particular uses requires good interactive dialogue among constituencies, such that the needs, capabilities, and concerns can be understood and respected.

Use conflicts may also be geographically based within a watershed, as activities in the upper reaches of a basin may have unforeseen impacts in the lower reaches. Impoundment of water in one area for a certain use or uses may cause shortages in other areas. Therefore, it is important also to understand the availability, movement, and fluctuations in water conditions and uses from a whole-basin context.

Watershed, or basinwide, scale approaches to development and management of water have proved highly successful and are gaining more prominence worldwide due to their inherent effectiveness in budgeting and balancing uses. There are a wide variety of specific institutional approaches that have been created, each accommodating the local hydrological and political landscape.

In the United Kingdom, river authorities have been established with broad powers to manage both the quality and quantity of rivers. Further east, in Germany, national legislation has established self-administration water associations for the management of the Emscher and Ruhr river basins. In France, six water agencies corresponding to the country's six large hydrographic regions have been created and have been effective in developing and executing policy decisions to enhance water resources.

In the United States, various watershed management approaches are in place to address small streams and major river and lake basins. Given the arrangement of state political jurisdictions, agreements among the states have been a necessity in certain instances to permit the watershed to be managed in other than a political-jurisdiction context. Examples of watershed management organizations in the United States are the Great Lakes Commission, the Ohio River Valley Water Sanitation Commission (ORSANCO), and, on a more local scale, the Anacostia River Restoration efforts in Washington, D.C.

Watershed management organizations on the international scale have been created for the Rhine River in Europe and Great Lakes (International Joint Commission) in the United States and Canada to address cross-border pollution problems. Through these agencies, aggressive programs have been created and agreed to by the countries sharing the watershed, with the input of water users, delineating the actions each will take to ensure collectively that water quality and quantity needs are met.

Significantly, at present, organizational efforts are underway among 11 nations to formulate a common approach to evaluate and resolve the water quality problems of the Danube River. Initial activities of the Danube River program have been funded by various international development agencies and currently include surveys to characterize water quality problems and administrative actions to establish a permanent management organization to direct the ongoing program.

Whether international or local in scope, all the examples cited contain a common thread, which is unification of the political jurisdiction sharing a watershed and involvement of the constituencies placing demands on the resource in the common development and management effort.

While there are a wide variety of institutional approaches, an examination of existing institutions suggests a general pattern of organization, illustrated in Fig. 11.1. The diagram suggests that an effective watershed management organization consists of a formally established legal authority or governing body retaining the authority for making key decisions and setting policy and programs. Members of the governing body should include high-level representatives of the political jurisdictions comprising the watershed. Also, it is appropriate to include on the management body individuals representing agencies of government whose missions involve water quality and quality management.

Finally, representation of different use perspectives is important. A properly represented governing body ensures that decisions will receive the input of relevant perspectives and that the implementation of decisions will be coordinated along the lines of interest.

The governing authority can greatly benefit from the services of a technical advisory board. Typically, this group consists of individuals with technical expertise in specific scientific disciplines such as environmental chemistry, biology, hydrology, limnology, and environmental engineering. Effective advice and input from such an advisory group helps ensure that the decisions made by the governing body are technically sound.

Critically important to the overall management process are the input and assistance of those constituencies placing demands on the resources. An opportunity for these constituencies to provide input to the governing body ensures decisions that are responsive to and coordinated among the needs and concerns of those affected.

A professional staff, who can be either employed by the watershed organization or assigned on loan from outside organizations or agencies, is a necessary component of the structure in order to provide day-to-day administration and implementation of the programs and initiatives established by the governing body.

Financing to operate the watershed management organization can take several forms depending on the organization's authorities and mission. Assessments of users for withdrawals or for the treatment of water discharges are commonly employed. Should the watershed management organization serve the function of coordinating programs among political jurisdictions, then assessments to the political jurisdictions may be most appropriate. It should again be emphasized that the appro-

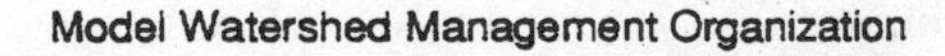

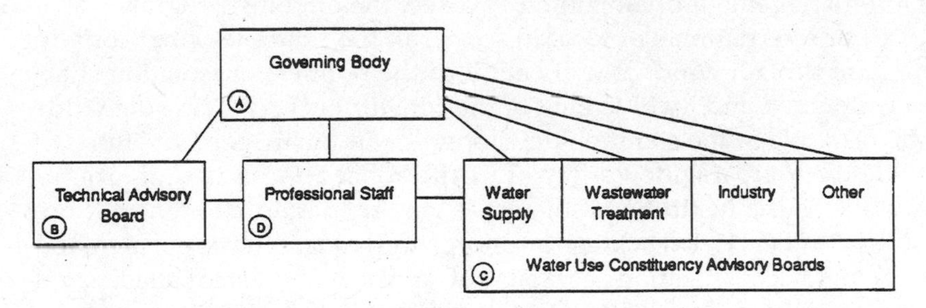

A) Legal authority comprising representatives of the political jurisdictions sharing the watershed (nations, states, provinces or cities) and users constituencies.

B) Technical Advisory Board of experts to address technical issues and report findings and recommendations to Governing Authority (A). Technical Board works closely with organization professional staff (D).

C) Groups of individuals employed in and representing views of users (e.g., industry, agriculture, water supply, wastewater treatment). Provides input and recommendations to governing body regarding program needs.

D) Provides professional staffing for day-to-day management of the basinwide management organization and implementation of authorities, policies and programs.

Figure 11.1 Model watershed management organization.

priate institutional arrangement to achieve water resources development and management must be based on a consideration of the specific mission to be undertaken and local political, economic, social, and environmental conditions.

Case History: The Ohio River Valley Water Sanitation Commission

The Ohio River Valley Water Sanitation Commission, an interstate water pollution control agency with the aim of improving water quality in the Ohio River basin, provides a case study in the implementation of river basin management concepts.

The Ohio River Valley, shown in Fig. 11.2, is situated in the northeast United States. The abundance of flowing streams, combined with major coal deposits, made the valley attractive for human settlement and industrial development beginning in the 19th century.

In order to address abatement of interstate water pollution, the governors of eight states—Illinois, Indiana, Kentucky, New York, Ohio,

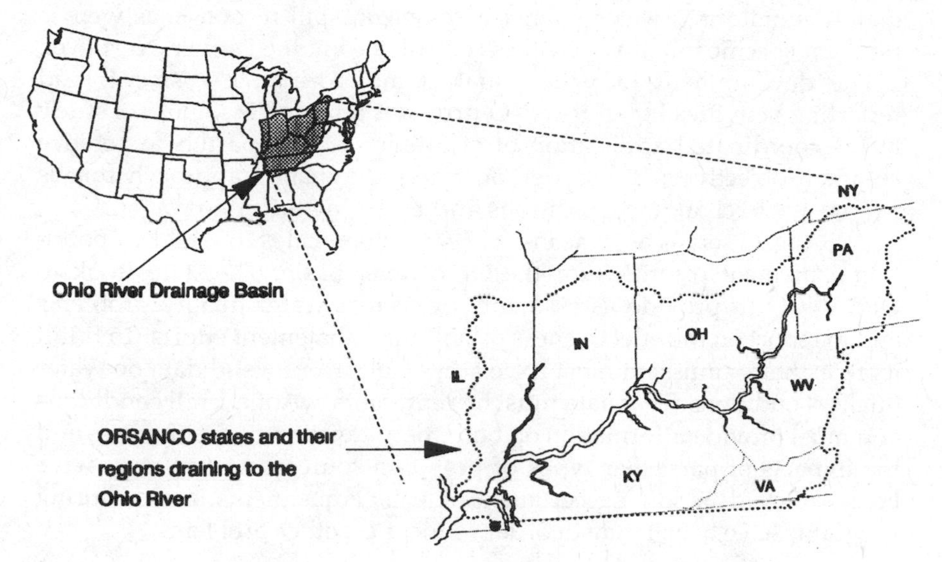

Figure 11.2 Ohio River drainage basin and ORSANCO states.

Pennsylvania, Virginia, and West Virginia—entered into the Ohio River Valley Water Sanitation Compact. The compact consists of a series of articles within which a mutual pledge of interstate cooperation is stated, the boundaries of the agreement are defined, and the mechanism for accomplishing the general goals, i.e., the Ohio River Valley Water Sanitation Commission, is established.

The commission consists of three representatives from each state as appointed by the governor. Specific supplemental state legislation typically provides that one of the three commissioners must be the administrator of the state environmental protection agency. To ensure representation of federal interests, three positions are reserved for appointment by the President.

The commission convenes three times annually to conduct business and establish policies and programs. A full-time staff is maintained to support administrative operations and carry out approved programs. ORSANCO's structure is similar, as shown in Fig. 11.1. Among those represented on its technical committee are the state environmental agencies, the U.S. Environmental Protection Agency, the U.S. Army Corps of Engineers, the U.S. Geological Survey, and the U.S. Coast Guard. Outside advisory committees represent publicly owned wastewater treatment plants, potable water suppliers, the power and chemical industries, and the general public.

Programs conducted by ORSANCO include development, adoption, and enforcement of effluent discharge standards for the Ohio River, water quality monitoring, water quality assessment, spill response, as well as problem-specific initiatives such as control of combined sewer overflows.

The development of water quality monitoring, assessment, and reporting activities by ORSANCO provides a more detailed case study in the coordinated application of efforts by several parties to achieve common objectives. It also demonstrates a dynamic approach that is responsive to changing conditions and emerging technology.

ORSANCO serves as a means for its member states to establish pollution abatement priorities for the Ohio River basin. The states look to ORSANCO to provide assessments of current water quality problems and to report on the effectiveness of pollution abatement efforts. To fulfill its role, the commission must have access to current, valid data on water quality conditions. That data must be representative of overall conditions and must provide information on both the attainment of desired uses and the impacts of particular types of pollution sources. As successes have been achieved in solving certain problems, adjustments in monitoring programs have been made in order to focus on other problems.

The Water Users Network

It was recognized at the inception of ORSANCO's activities that a comprehensive water quality monitoring network would be necessary in order to document existing conditions, define specific problems, and provide a means of demonstrating the effects of remedial actions. Implementation of such a network was beyond the means of the commission; it was therefore necessary to find a different way to build a monitoring system.

ORSANCO turned to the municipalities, utilities, and industries that used the Ohio River as a source of water for either potable or process purposes. To determine the treatment needs, each of the "water users" needed to perform tests on the river water at the intake. A number of tests, such as temperature, pH, hardness, and turbidity, were performed by a number of the users. By establishing minimal requirements for data collection (test method, frequency, etc.), ORSANCO was able to initiate a monitoring network.

The Water Users Network provided a reliable source of basic water quality data from a number of sites along the Ohio River and its tributaries at a minimal cost. Results were compiled monthly and were used to prepare annual reports on water quality conditions. There were, however, several limitations to this network:

- Since most participants were drinking water utilities, monitoring locations were generally upstream of major sources of pollution; therefore, results did not reflect worst-case conditions in the river.

- Parameters monitored were those of interest to water supply and industrial uses; parameters of concern to other uses, such as aquatic life protection, were not monitored.
- Because the data were provided on a voluntary basis, ORSANCO could not request extensive effort to provide additional analyses or increased quality assurance checks.

As demands for wider parametric coverage and increased quality assurance grew, the commission placed more emphasis on other data sources. The Water Users Network, however, remains a continuing source of data with a considerable historic period of record. It also provided the foundation for future cooperative monitoring efforts by ORSANCO and Ohio River water users.

Manual sampling program

In 1974, the member states of ORSANCO were faced with increased water quality monitoring requirements under the 1972 amendments to the federal Clean Water Act as well as revised state laws. It was recognized that, rather than states undertaking their own monitoring efforts on the Ohio River, it would be more efficient to agree on a single monitoring system to be carried out by ORSANCO. A work group of personnel from state and federal agency monitoring programs met and developed a strategy for routine ambient monitoring of the Ohio River and the lower reaches of its major tributaries.

The resulting program consisted of monthly sampling at 36 sites—22 on the Ohio River and 14 on tributaries. The Ohio River sites took into account locations of major wastewater discharges, urban areas, and state line crossings. The tributary sites were chosen to characterize conditions at the point of confluence with the Ohio. Parameters measured included physical (such as suspended solids), nutrients, metals, and organic chemicals (cyanide and phenols). Taken as a whole, the sampling network was designed to allow characterization of water quality conditions in the Ohio River to allow identification of problem areas and to provide indications of locations of major pollution sources.

With the delegation to ORSANCO of responsibility for routine monitoring of Ohio River water quality conditions, the member states found it logical to also delegate responsibility for preparing the annual assessment of water quality conditions, as required under section 305(b) of the federal Clean Water Act. Since Article 8 of the Ohio River Valley Water Sanitation Compact requires the commission to report to the states on the pollution problems of the compact district, ORSANCO was already performing a similar assessment that could be readily adjusted to meet the requirements of section 305(b). That report, now prepared on a biennial basis, provides a basis for evaluating pollution abatement priorities and the effectiveness of ongoing programs.

The monthly sampling data are ideal for use in trend assessment. In 1989, ORSANCO undertook the application of the seasonal Kendall test to the data from the manual sampling program. The results indicated that conditions were improving (i.e., decreasing concentrations) at most locations for parameters influenced by point source discharges. No trends were evident for parameters such as hardness and dissolved solids, which are affected more by natural factors.

Electronic monitors

In ORSANCO's early years, one of the most serious water quality problems in the Ohio River and several of its major tributaries was depressed levels of dissolved oxygen due to untreated, or inadequately treated, discharges of sewage and industrial wastes. Low levels of dissolved oxygen (DO) made sections of the river uninhabitable for desirable fish species and, at times, created malodorous conditions. Because DO concentrations were known to vary from day to day, and even from hour to hour, it was felt that more frequent monitoring than that provided by the Water Users Network was needed.

ORSANCO worked with an electronic instrument company to develop a monitor which would be able to automatically collect samples and analyze them for DO and other water quality parameters. The instrument was introduced in 1960. It was capable of providing continuous measurement of temperature, pH, conductivity, dissolved oxygen, and chloride. A data logger was added to record hourly readings. The commission worked with a computer company to develop a means of interrogating monitors at remote sites from a central location. These developments eventually led to the operation of a network of electronic monitors encompassing 21 sites—14 on the Ohio River and 7 on lower reaches of major tributaries. Data were received daily at commission headquarters on a routine basis; if necessary, monitors could be interrogated at any time.

Data from the electronic monitors were used in several instances to allow "real-time" water quality management. The primary application involved enhancement of dissolved oxygen levels. It was known that certain configurations of gates at the navigation dams would introduce greater quantities of oxygen to the water passing over the dams. Such configurations, however, were also believed to cause increased river bank erosion. The Corps of Engineers, therefore, wanted to operate the dams to maximize aeration only when the additional oxygen was most needed. Installation of electronic monitors on the upstream sides of navigation dams, and provision of the data to the Corps, allowed the gate settings to be adjusted to provide maximum aeration when dissolved oxygen levels threatened to fall below stream criteria, and to return to normal operation when conditions improved.

A second application involved reservoirs on the Allegheny River. The Kiskiminetas River, a tributary to the Allegheny, was degraded by acid mine drainage; pH values often fell below 5.0. To maintain pH levels above the 6.0 required minimum on the Allegheny, the Corps would release higher-pH water from certain of its reservoirs to offset the impact of the Kiskiminetas. Installation of electronic monitors at the reservoir outlets, on the Kiskiminetas, and on the Allegheny below the inputs allowed efficient operation of the dams and flow releases. This operation virtually eliminated historic acidity problems in the Allegheny River and upper Ohio River.

In the late 1960s, ORSANCO undertook the development of a river water quality model. The unique aspect of the ORSANCO model was to be its ability to forecast water quality conditions. The model was designed to take data from the electronic monitors, together with flow forecasts from the National Weather Service, and predict conditions for succeeding days. It was originally envisioned that this capability would then allow regulatory agencies to direct wastewater treatment facilities to cut back their discharges during critical periods. This arrangement did not prove practical, however, as treatment facilities did not have storage facilities to hold back some of their effluent, nor could they increase the treatment provided during critical flow conditions.

By the 1980s, most major dischargers were providing adequate treatment of their wastes. Stream criteria for pH, chloride, and dissolved solids (conductivity readings were correlated to dissolved solids) were met virtually 100 percent of the time. Occasional violations of temperature criteria occurred, but were due to natural conditions. While some violations of DO criteria were still recorded by the monitors, the violations were restricted to a few locations during the months of July through October. At the same time, concern over the effects of toxic substances in the river was increasing. With the growing need to increase monitoring for toxics, it became increasingly difficult to justify 24 h/day, 365 days/yr operation of the electronic monitors.

At the time when the commission was considering the termination of the electronic monitor system, development of hydroelectric power generating facilities was being pursued at Ohio River navigation dams. When the cost of fossil fuels was high, hydropower could provide a cost-competitive alternative. The potential generating capacity was not great—development of facilities at all Ohio River dams would yield the equivalent of one large coal-fired unit—but hydropower could provide a reliable source of power. ORSANCO's concern, however, was that if river flow were diverted through a hydropower facility, it would not pass over the dam and therefore would not receive the input of oxygen that normally occurs at the dams. This would be especially harmful at dams with the capability to provide increased aeration.

In response to the concerns raised by ORSANCO and others, the Federal Energy Regulatory Commission (FERC) included provisions in the licenses it issued for hydropower facilities on the Ohio River that required monitoring of dissolved oxygen, provision of the data to ORSANCO, and capability to mitigate low-DO conditions, by either mechanical aeration or curtailing operations.

With the knowledge that DO measurements would be available from hydropower facilities as they came into operation, and with agreements with the Corps of Engineers to operate temporary monitors at critical locations, the commission opted to discontinue its operation of the electronic monitor system in 1986. ORSANCO worked with the hydropower operators and the Corps to develop the capability to interrogate monitors at hydro facilities and Corps dams. A procedure has developed whereby the monitors are interrogated by ORSANCO on a weekly basis from May through October. If critical conditions appear to be forthcoming, interrogations are increased to daily. When low-DO levels are observed, ORSANCO notifies the operator of the monitor—either the Corps or a hydro operator—and requests confirmation of the low levels. If the levels are confirmed, mitigative measures are undertaken as available—adjustment of flow over the dam, curtailment of hydro operations, or injection of air by the hydro facility.

Spill detection

In the 1970s, a number of unreported spills of organic chemicals caused problems for Ohio River water supply utilities. It was learned that while a number of industries on the Ohio River and its tributaries were permitted to discharge various organic chemicals, no routine monitoring for those chemicals was taking place on the receiving waters.

With over 25 years of operation of the Water Users Network as background, ORSANCO and several Ohio River water utilities conceived of a cooperative monitoring system for organic chemicals. The system would consist of monitoring sites at key locations—downstream of potential spill sites and upstream of water supply intakes—on the Ohio River and certain of its tributaries. Each monitoring site would be provided with a gas chromatograph; all sites would utilize the same detector and common analytical procedures in order to ensure consistent detection capabilities. ORSANCO would provide quality assurance, coordination, and communications for the system.

The ORSANCO organic detection system (ODS) began operation at seven locations in 1978. Subsequent additions have brought the system to its current size of 14 locations. Basic operation of the system involves collection and analysis of a daily sample at each location (some participants analyze additional samples each day). If an unusual detection is found, the analyst notifies ORSANCO. Additional sampling and/or submittal of duplicate samples to a commercial laboratory may be initiated to confirm the initial detection. Upon confirmation, an alert is sounded, with notification of state and federal emergency response agencies as well as downstream water utilities.

While some utilities in the system operate their gas chromatographs to obtain low levels of quantification, the primary objective is to provide wide-range detection capabilities. The system utilized flame ionization detectors (FIDs) which are relatively durable and capable of detecting approximately 60 volatile organic chemicals. At present 22 chemicals are quantified in the routine operation of the system.

Automatic process samplers have been installed at several sites in the system. These units allow repeated sampling and analyses with minimal effort by the operator. This has been an important consideration for small utilities in the system, where laboratories are normally staffed for just one shift per day. Participants in the system would like to see additional development of this capability, including the installation of alarms that would sound automatically when an unusual detection is made.

Operation of the ODS generates a considerable amount of rather complex data. Voice transmittal of the results during spill events often resulted in confusion; it was felt that electronic transmittal would be preferable. ORSANCO therefore developed an electronic bulletin board in order to disseminate reports of spills, results of spill monitoring, and other pertinent information. The bulletin board is designed to allow posting of information from any site; in this way, ODS participants can post results as they become available. In instances where spills affect several water utilities, those affected first can post reports of their experience—what treatment modifications they took and how effective they were.

In January 1988, a tank collapsed at an oil terminal on the Monongahela River near Pittsburgh, resulting in a spill of over half a million gallons of diesel fuel. At the time of the spill, the river was in an extreme low-flow, low-temperature condition. The movement of the spill did not conform to time-of-travel predictions, and the concentration in the water column did not appear to diminish over time. It was necessary to mount an on-river monitoring effort. Working with personnel from the state agencies and the Corps of Engineers, ORSANCO tried several monitoring instruments. The instrument which proved successful was a fluorometer. When equipped with the proper combination of lamps and filters, the fluorometer would provide instantaneous readings of the level of diesel fuel in the water. Once this capability was established, daily profiles of the spill were taken by boat, and the results were posted on the electronic bulletin board.

The extreme cold weather during the spill made manual collection of samples difficult. Personnel from the Corps of Engineers developed a better alternative. By placing an intake line beneath the hull of the boat with the opening facing toward the bow, they were able to use the forward motion of the boat to draw water through the intake to the fluorometer. Thus they were able to take measurements while the boat was underway, greatly reducing the time it took to produce a profile of the spill as well as reducing the discomfort of the field crew. After the spill, the arrangement was modified to allow "monitoring on the fly" of other water quality parameters such as pH, temperature, conductivity, and dissolved oxygen. The system can also be used to collect samples for laboratory analysis. It has been used to track several spills, and to define the impacts of intermittent discharges such as combined sewer overflows.

Contact recreation protection

Water quality criteria for protection of water contact recreation usually prescribe a monthly geometric mean for bacteria (fecal coliform or *E coli*) based on at least five samples per month. Given the variability of the bacteria tests, such criteria are more valid statistically than single-number maximums; their use is entirely suitable in stable environments where actual bacteria densities would not be expected to vary greatly. In situations where rapid changes in bacteria levels are likely, the inherent time delay of the test, as well as the need for a month of samples to assess compliance with the criteria, makes it nearly impossible to provide timely notification of the public regarding the advisability of water contact recreation.

In the mid 1980s, the only bacteria data collected by ORSANCO on a 5 times per month frequency were those provided by the Water Users Network. Following an assessment of the available data that was undertaken to evaluate the support of contact recreational use, it was determined that additional monitoring locations needed to be established downstream of major cities. ORSANCO arranged for the collection and analysis of at least five samples per month from May through October (the period designated for contact recreation in the commission's pollution control standards) at locations downstream of the six largest cities along the river.

Results from the monitoring indicated that stream criteria for contact recreation were exceeded during periods of rainfall at all sites downstream of urban areas. Conversely, the criteria were generally met in dry-weather periods. It could be seen that the situation was subject to rapid change and that providing current information on compliance with stream criteria would be difficult.

ORSANCO identified local (county and city) health departments as the agencies charged with the responsibility for informing the public about the advisability of contact recreation. Arrangements were then made to provide the results of ORSANCO's bacteria samples to the health departments in the areas where the samples were collected. The data were conveyed by mail and by the ORSANCO electronic bulletin board. Some of the health departments chose to issue blanket advisories against contact recreation in certain portions of the Ohio River; others have sought to provide more timely information. The Cincinnati Health Department, e.g., supplemented ORSANCO data with its own sampling and developed an index to express the results. The index, similar to that used for informing the public of air quality conditions, was published weekly in the local newspapers. The Allegheny County Health Department (Pittsburgh area) based its advisories on the occurrence of combined sewer overflows. Frustrated in its attempts to alert the public through the local news media, the health department established its own warning system by providing facilities along the river with large orange flags that are flown during periods when combined sewer overflows would cause bacteria levels to exceed recreational criteria.

Data dissemination

The emphasis in water quality management has shifted from a few large, concentrated sources of pollution to numerous small, dispersed sources. This has brought about the involvement of an increasing number of entities in assessing monitoring data, thereby requiring more widespread dissemination of such data. ORSANCO has always sought to make the results of its water quality monitoring available to all who wish to receive them. The involvement of more parties has greatly increased the demand for the results, and has caused ORSANCO to continually incorporate new technology to increase its information dissemination capabilities.

In 1968, ORSANCO initiated monthly publication of *Quality Monitor.* The initial publication included results from the Water Users Network and Electronic Monitor Network. As other monitoring systems were developed, the format of *Quality Monitor* was adjusted. It contained listings of all sample results from the monthly manual sampling program as well as summaries of results from the other monitoring programs. *Quality Monitor* remained the primary means of disseminating results from commission monitoring programs until 1985. At that time, with the impending termination of the electronic monitoring system, it was decided to decrease the frequency of publication to quarterly.

In the meantime, the commission had initiated the issuance of *Quality Updates* to alert state and federal agencies, as well as affected water users, of impending water quality problems such as disruption of wastewater treatment facilities or periods of low flow and high temperatures. To maintain the timely dissemination of water quality information, it was decided to issue *Quality Updates* monthly during the warm-weather months (May through October). The monthly updates include information on river flow, dissolved oxygen, and bacteria levels, and are distributed to state and federal agency personnel as well as certain interested parties.

The ORSANCO electronic bulletin board provides a means for rapid dissemination of water quality data. While the bulletin board was originally developed for spill information, expansion of its capabilities allowed posting of other information such as dissolved oxygen and bacteria data. As utilization of personal computers with modems has spread, the bulletin board has become increasingly important to the commission's data dissemination efforts.

A new component of the bulletin board is the Biological Management Information System, a data storage and retrieval system available to all interested parties. This system provides an on-line, user-searchable database containing Ohio River fish and macroinvertebrate population data. The data in the system are contributed by a variety of researchers, including government agencies, private industry, and universities, in addition to ORSANCO itself. The system provides a means for widespread sharing of information on Ohio River aquatic life. Future enhancements of the system will include provisions for habitat data as well as additional groups of organisms.

The latest development in ORSANCO's efforts to make its information widely available is the establishment of a Home Page on the Internet. It appears to be the consensus of environmental professionals that this is the preferred means for making data available to all interested parties. ORSANCO will therefore utilize the Home Page to make its data available, as well as to provide information on the commission, the Ohio River, and the results of commission studies. This home page can be accessed at *http://www.orsanco.org*.

12

Twelve Years of Successfully Managing the Muddy River

Irene V. Gillis
Chair, Restore Olmsted's Waterway (ROW)
Boston, Massachusetts

It is one of the smallest rivers in the world and artificial. The Muddy River heads at Jamaica Pond, flows through the Bay Bay Fens to the Charles River, and forms a 3½-mi border between the city of Boston and the town of Brookline, Massachusetts. But it has special significance.

First, this watershed was designed in the late 19th century by Frederick Law Olmsted, known as the "father of U.S. landscape architecture." It runs through a landscaped linear park called the "Emerald Necklace."

Bordering the necklace is an incredible piece of real estate. Approximately 300,000 urban dwellers live in close proximity to this river, and closeby are three of Boston's major industries. Such tourist attractions as Boston's Museum of Fine Arts, the Isabella Stuart Gardner Museum, and Sears Chapel are on its banks. One of the world's most

For a complimentary copy of the ROW "Citizen's Guide to the Muddy River," please write ROW, 163 Kent Street, Brookline, MA 02146.

renowned medical centers, Harvard's Longwood medical area, gets its name through its proximity to the river, the Longwood area. Every year over a million patients come here from all parts of the world. And many of Boston's ever-growing educational institutions are in close proximity, including Northeastern University, New England Conservatory of Music, Boston University, Emmanuel, Wheelock College, Simmons College, and many other educational institutions with a combined student body of more than 250,000.

How is it, then, with thousands of well-educated people in close proximity, that the Muddy River is so neglected? That is a good question since there are probably more Ph.D.s and M.D.s per square foot close to this river than on any strip of land in the world.

The Muddy River is a microcosm of polluted rivers. As one wit put it, it should have been named De-nial.

When Frederick Law Olmsted designed the Muddy River, he felt he was cleaning up the longstanding badly polluted Muddy River marshes. These marshes had an interesting history. In colonial days when Boston Common became a crowded place for pasturing cows, the early settlers drove the cattle through the marshes up to a summer pasture area now known as Brookline and back again to the Boston Common pasture in the fall. In fact, when this once-rural area, now Brookline, broke away from Boston in 1705, it was incorporated as the Town of Muddy River. When Olmsted constructed his "chain of pleasant water," he wrote that he expected the two municipalities would protect it forever. He did not anticipate the 100 years of "buck passing" that followed or the proliferation of the automobile.

As a microcosm, the Muddy River with its badly ensnarled problems invokes the thought: What must it be like to try to clean up the Mississippi? The Muddy River passes through just two municipalities—not 6 states, 14 counties, and 30 towns and cities. But somehow responsibility for this pollution becomes entangled among 40 different bureaucracies, to name a few, the Boston Water and Sewer Commission, the Brookline Department of Public Works, two different conservation commissions, the state's Department of Environmental Protection (DEP), and the Department of Environmental Management (DEM). On the federal level, there is the U.S. Environmental Protection Agency (EPA) and the U.S. Department of the Interior Fish and Wildlife Services—just a few of the agencies one encounters in trying to clean up this river.

Even to document its obvious pollution has been a hassle. Two citizens set out to do something about this river nearly 12 years ago, namely, Edward Shoucair, an urban planner with a master's degree from Massachusetts Institute of Technology (MIT), and Irene Gillis, a retired journalist and publicist. A graduate of the University of Rhode

Island, she was raised on a 200-acre Rhode Island tree farm—a land of trout-filled, clear babbling brooks. Shoucair had worked in North Carolina near another Olmsted park.

While working as a reporter for the *TAB* newspapers, Irene became freshly concerned about the environment as a result of covering stories on Boston's drinking water supply, the Quabbin Reservoir, and air pollution and noise pollution problems. She decided to volunteer one day per week to work in the environment. Her free services were welcomed by the New England office of Clean Water Action. The director, Polly Hoppin, undertook to educate Irene on grassroots environmentalism, including participation in legislative hearings and visits to local groups throughout Massachusetts.

One day Hoppin (now Polly Hoppin, Ph.D.) declared, "You need a backyard project. What do you know about the Muddy River?" Gillis's answer was, "Absolutely nothing," although she lived in Brookline. She set out to learn.

First, she fell upon a report that had been done for the Metropolitan District Commission (MDC) by the engineering firm C. E. Maguire. It documented fecal coliform readings up and down the Muddy River within the hundreds of thousands of parts per millimeter. Totally dismayed, Gillis called Clean Water Action friend Rick Bird, who had a master's degree in environmental science (and is now an M.D.).

"Rick," she asked, "what do these readings mean?"

"Let's put it this way, Irene," he said. "They close beaches at 250 ppm."

An avid local newspaper reader, Gillis was dismayed that she had not seen anything about this startling report in the papers. So she decided to make inquiries at a Brookline Board of Selectmen's hearing.

The selectmen brought forth the head of the town's Department of Public Works, William Griffiths. Griffiths declared that night that he did not know where the sewage could be coming from. "All I know is we're not putting a teaspoon of anything in the River," he said. "It must be coming from Boston."

Gillis went home discouraged. Her husband, a former military officer, asked what was wrong. When she told him, he said, "A good general never admits defeat. Whatever happens, he declares victory." The next day, she decided "to declare victory," using whatever ammunition she had. She called the newspapers. But she realized she needed troops. Soon afterward, the New England Rivers Association held a conference in Fitchburg, Massachusetts, and she decided to go, in the hopes of learning what other groups were doing about such problems.

She took a train to Fitchburg and bummed a ride back to Brookline with a young man named Ed Shoucair. He was fascinated when she described the problems of the Muddy River and particularly intrigued by its connection to Frederick Law Olmsted.

Meeting regularly in Gillis's living room, they gathered their friends and neighbors to form the coalition to Restore Olmsted's Waterway (ROW), now 300 members strong with a history of progress.

In addition to hot coffee and warm brownies, Gillis supplied a knowledge of journalism and public relations, and Shoucair, the technological knowledge learned in urban planning courses at MIT. Neither was afraid to speak out, and they agreed they must focus attention on the river's pollution.

At that time, Shoucair lived in Jamaica Plain, Boston, near Jamaica Pond, the headwaters of the Muddy River. It is a beautiful freshwater body, a geological kettle hole, fed by underground springs. Eight species of freshwater fish live in Jamaica Pond. It is stocked regularly by the state and is a source of joy for many fishers.

But as these fish try to swim downstream, they become "the canaries in the coal mine." There are perhaps only four species that survive in the big pond down river created by Olmsted, called Leverett Pond. And farther down, as you head toward the Charles River, only carp survive. Where was this fish-fatal pollution coming from? ROW decided to find out.

With the help of Brookline's state representative, John Businger, Shoucair and Gillis and other ROW members organized a lobbying effort at the State House to get money for a fresh engineering study of the river.

In the end, the legislature approved a grant of $150,000 to undertake research on the sources of the sewage. The firm of Metcalf & Eddy was hired for the task. Metcalf & Eddy went "up the pipes" that dumped into the river and revealed that most of the sewage was coming from illegal cross-connections. Over time, many buildings had connected their sewer pipes to the storm drains that led to the river.

Metcalf & Eddy identified five main problems in the river: the illegal cross-connections; leaking underground storage tanks; street runoff (nonpoint source pollution); hydrology, the sluggishness of the river, depriving it of dissolved oxygen; and the overgrowth of reeds, called *Phragmites*.

But this was 1989, near the end of Governor Michael Dukakis's administration, and the state was almost bankrupt. There was no money to implement Metcalf & Eddy's recommendations: including about $45 million worth of dredging.

So ROW turned to the federal government. With the help of Senators Edward Kennedy and John Kerry and Representatives Barney Frank and Joseph Moakley, ROW petitioned Congress to send the U.S. Army

Corps of Engineers (ACE) to the area. In a dramatic announcement on the banks of the Muddy River at Longwood in August 1992, Senator Kennedy announced that Congress had appropriated $250,000 to begin the Corps' work.

The next announcement was a sad blow. ACE could not go by anyone else's study. The Corps can base its work only on its own studies. And so began 5 years of ACE studies.

Five years and $750,000 later, the Army Corps of Engineers is still studying the Muddy River, but has not done one shovelful of work. Its $250,000 Reconnaissance Study was essentially a regurgitation of the work already done by Metcalf & Eddy. Once this was complete, ACE asked for $500,000 more for a "feasibility study"—a matching grant with $250,000 from the Commonwealth of Massachusetts.

This latter study has yet to be made public. Drafts recommend a pumping proposal to cost about $4.5 million which would send water from the Charles River several miles upstream to create a cascade to restore dissolved oxygen, to aid the fish and wildlife.

There is constant disagreement as to how much dredging is needed and who can afford it. Who should pay how much—the federal government, the state, the local communities? And what will happen to these badly polluted sediments? They will need to be disposed of as hazardous waste, at considerable expense.

The floods of September 18 and October 20, 1996, added a new dimension to the problems of the Muddy River. When the Army Corps of Engineers first came to town, it announced that the Muddy River did not have a flooding problem; despite citizens' testimony to the contrary. "If there is flooding, it does not meet our cost-benefit analysis," ACE engineers said. Translation: "There's not enough damage done by flooding to justify the cost of correctional measures."

Well, the fall rains of 1996 changed that viewpoint. The tunnel of the Massachusetts Bay Transportation Authority (MBTA) Green Line trains alone experienced $40 million in damages from the floodwaters of the Muddy River on October 20, 1996. Northeastern University, Wentworth Institute, and the Art School of the Museum of Fine Arts altogether experienced another $10 million in flood damages.

In Brookline, 300 homes reported flood damage to apartments, basements, backyards, and submerged automobiles. People living near the Longwood area of Muddy River were unable to get out of their front doors and had to crawl through second-story windows to escape. Even Brookline's Water Department, located close to the river, had to be evacuated. And $100 million in flood damages is probably a conservative estimate. The water itself was not nice, fresh rainwater or street runoff. Most of it was laden with sewage from broken sewer caps and pipes. Citizens were advised to spray flooded areas with disinfectants. All the equipment in a day-care center near the river had to be destroyed, for example.

This recent reality forced the Army Corps of Engineers to rethink its flooding analysis and further delayed its report by many months. Subsequently, the ACE has been working with the Federal Emergency Management Agency (FEMA) to design a flood prevention program. It is focusing mostly on the drainage systems near the old Sears Roebuck building in the Fenway area.

The water of this section of the Muddy River comes not from Jamaica Pond but from a tributary stream called Stoney Brook. It includes a combined sewer overflow (CSO) system and gatehouse that has long been a source of local controversy, and is located right behind Boston's Museum of Fine Arts.

CSOs were declared illegal around 1984 when Boston's Massachusetts Water Resources Authority (MWRA) began its court-mandated cleanup of Boston Harbor. The Muddy River flows into the Charles River which in turn dumps into Boston Harbor. The point where the Muddy River joins the Charles River has long been identified as the most polluted spot on the Charles.

The ROW coalition counts as its most effective accomplishment the intervention of the Environmental Protection Agency. Shoucair presented the EPA with the documentation from Metcalf & Eddy and other reports. In 1994, the EPA issued an order to Boston and Brookline that they must specifically identify the buildings whose sewage pipes are illegally connected to storm drains. The EPA advised the municipalities they would be fined $25,000 per day for violating the Clean Water Act of 1972. No one may pollute the waters of the United States.

Town officials of Brookline have finally hired engineering firms to identify these sewers dumping into the Longwood, Tannery Brook, and Village Brook, main tributaries to the river. Thus far, hundreds of homes and apartment houses, even an old Brookline fire station, have been identified as culprits, dumping many, many thousands of gallons of totally untreated human sewage every day into this small urban stream. The concentration, for so small a stream, is horrendous. It is like adding two cups of sugar to one quart of lemonade.

How could this have happened? Total denial. The foul smell formerly always present was attributed to mud, dog feces, and bird droppings. The murky stream looked like dirty laundry water. The toilet paper hanging on the branches of trees near the river after heavy rains was called nothing but dried oak leaves.

ROW observers now feel that the color of the water is much improved but the polluted sediments remain a problem. Also the huge stands of reeds, *Phragmites,* have been fed daily banquets of nitrogen and phosphorus from the sewage and now are eutrophying the stream in many places. No one argues that the huge reed *Phragmite* is out of control on the Muddy River. They often obstruct the view from one river bank to the other. Because of *Phragmites,* the year 1994 went down in the history of the ROW coalition as the battle of Rodeo.

Phragmites are a problem in all parts of the world and have been for centuries. These tall (growing up to 20 ft) reeds are the reeds in which Moses was placed—in the bull rushes.

On the Muddy River they have not been controlled by regular mowing by either municipality, Boston or Brookline. When the Boston Parks Department (BPD) became anxious to landscape the river banks, the *Phragmites* stood in the way of their plans. The quick and cheap solution was chemical spraying.

Citizens immediately protested, led by those opposed to toxic spraying in general and those likely to be personally affected, the 350 Fenway Victory Gardeners. When the BPD referred the matter to the Boston Conservation Department for a decision, ROW and the gardeners organized a separate group on this issue called CAPS (Citizens Against Pesticide Spraying). The three hearings that followed were an intensive course in political science.

The BPD organized its political supporters and so-called experts to testify on the harmlessness and effectiveness of Rodeo, a chemical also known as glyposate; CAPS did likewise. The debate went three separate rounds—the third and final Rodeo hearing at Boston City Hall went from 6:30 p.m. to 1 a.m., the longest hearing in the history of the Boston Conservation Commission. ROW and CAPS thoroughly researched the issue. Rodeo, they declared, was, as the BPD declared, "no more harmful than table salt."

ROW's research disclosed that although the proponents declared Rodeo was permitted by the EPA, the Environmental Protection Agency had performed no tests on it. The EPA simply accepted the test results of the manufacturer, Monsanto Chemical Co., which of course wanted to sell lots of it.

During the first two rounds, it seemed that the citizens who testified had a chance. By the third round, the mayor of Boston appeared at the doorway, presumably to see what was going on, but was not announced or introduced. After dozens of speakers on both sides of the issue, the commission, an appointed board, voted to approve this pesticide spraying.

One surprise piece was presented by ROW board member Dr. Fred Youngs, a chemist who asked what surfactant or bonding agent would be used along with Rodeo and what chemicals would result in the river water after the spraying? Basic chemistry, no? The river is already polluted with PCBs (polychlorinated biphenyls), lead, mercury, and heavy metals, according to scientific research findings. Of course, the Rodeo proponents could not answer these questions.

The Fenway Victory Gardeners, whose plots in this urban area date back to World War II, were particularly adamant because a previous chemical spraying by another chemical circa 1990 had made their entire vegetable crop inedible.

In the end, after a parade of dozens of speakers on both sides of the issue, the commission voted to allow the spraying. Because of the politics of it all, this was no surprise to ROW and CAPS. They had prepared for a possible defeat and had lined up an attorney to appeal the decision.

Attorney Ken Kimmell filed the appeal before the Massachusetts Executive Office of Environmental Affairs (EOEA). After studying both sides, Executive Secretary of the EOEA Trudy Cone found for the citizenry, and the BPD was forbidden to use Rodeo.

Lawyers, of course, are expensive, and ROW and CAPS worked together to pay the legal bills. Raising this money was a lesson in organizing. Rodeo opponents learned that planning fun events like a Sunday brunch in a Back Bay restaurant is a good way to raise money.

The BPD was allowed by the EOEA to experiment with other methods of eradicating *Phragmites* such as covering them with black plastic. But results eventually were declared "inconclusive."

The *Phragmites* continue to thrive with little or no mowing. There is hope that extensive dredging of the sediments may be the answer. The *Phragmites'* rugged roots, called rhyzomes, are firmly stuck in the mud, and dredging is a sure-fire solution to the problem. It is remarkable that the sewer drains are so closely related to the dense stands of *Phragmites*. Or is it? The phosphorus and nitrogen in the sewage are a daily banquet for them.

The question remains: Why has a well-educated urban population—the citizens of the Athens of America—tolerated these health-threatening conditions for so long?

As time goes on and we become aware that many old urban areas are polluting U.S. waters with sewage, possible answers include the concentration in the 1970s and 1980s on chemical pollution that followed the chemical revolution, which began in the 1940s. Environmentalists were preoccupied then with stories of chemical dumping from factories and chemical plants. If sewage was a problem, waterborne diseases could all be treated with modern antibiotics. Not to worry!

This perspective still prevails among many people.

Despite constant revelations through the media, the ROW coalition has had difficulty getting the public to relate high levels of fecal coliform to waterborne diseases such as hepatitis A and B, Salmonella, cholera, and diphtheria. ROW has been trying desperately to raise the concerns of medical professionals in Harvard's Longwood medical area.

One step was an all-day symposium staged on November 16, 1996, at the Harvard School of Public Health. That day, two ROW scientists, Harry Boehrs of the Boston University Medical Center and Dr. Fred Youngs, Director of the Citizen's Environmental Laboratory in Cambridge, announced the findings of their first round of pathogen testing—testing the water and sediments of the Muddy River for specific diseases. To the surprise of everyone but the ROW zealots, extremely high levels of a skin disease gangrene were found in testing done by a certified laboratory. This very expensive testing was paid for by a grant from the Kellogg Foundation.

Finally the message of a public health threat is getting through. In today's world, common sense no longer prevails. Citizens must document their concerns with flawless scientific data to gain any credibility, but skeptics still remain.

Why? The ROW leaders feel one answer is that sewer work is not fashionable. Sewers are out of sight, out of mind. Landscaping is far more rewarding, or so many think. And money for beautification is therefore far easier to obtain.

Often these are known as "open-space" projects and translate to land for parks, playgrounds, baseball, football, and soccer fields. A downside of open space from the watershed perspective is that it often results in such destructive measures as filling in wetlands, cutting forests, and covering fields with astroturf. As Sam Bass Warner at MIT accurately details, "We are constantly destroying nature's sponge."

This has been true along the Muddy River over a period of 100 years. Olmsted, a master landscaper, planted hundreds of oaks, maples, beeches, and other trees to try to give the "Emerald Necklace" the feeling of the New England woods, and he succeeded in that.

What he did not envision was the burgeoning urban population and its dependence upon the automobile. High-rise apartments, offices, and hospital buildings rose along the riverway and with them asphalt-covered roadways and parking lots, each destroying more and more of nature's sponge.

Now with changing global weather patterns, the threat of more flooding is imminent. This must be undone, but how?

Real estate interests have long felt that fine school systems, parks, and playgrounds will attract families and raise property values. But they have largely ignored the fact that municipalities need good infrastructures and money must be spent on the invisible. Development must be curbed. Our attention turns again to the warnings of Sam Bass Warner and the writings of Jane Holtz Kay and her new book, *Asphalt Nation.* We must question the wisdom of asphalt bike paths along rivers, of widened highways, and huge parking lots when we consider wetland protection.

A Bit More on Organizing

The job of organizing citizen activists seems to be getting much harder—perhaps for the same reasons that getting out the vote is harder: cynicism, two working couples, couch potatoes addicted to nighttime TV shows.

ROW has broadened its base through working with nearby schools and colleges. The Winsor School, a private girls' school right on the Muddy River, has "adopted" it and uses it as a living laboratory for science courses. ROW members are frequently invited to speak there.

At one point, the Winsor School girls organized a dance with several private boys' schools and raised $3500—better than we adults have ever done! ROW has held two of its annual meetings at Winsor, and it is the knowledge that these young girls have of the Muddy River and their caring that give us hope.

A project well worth the effort was ROW's "Citizen Guide to the Muddy River." Produced under the leadership of Shoucair, it is a beautiful piece of work. Professional photographer Liz Linder was hired to illustrate both the negative and the positive aspects of the park and the Muddy River.

The text written by Shoucair has saved ROW many thousands of hours of "reinventing the wheel," telling people over and over about its Olmsted history and its well-documented problems.

The Massachusetts Department of Fisheries and Wildlife paid for the guide's printing costs. Following rave reviews from as far away as Quebec City, Canada, and Liverpool, England, the guide has just been updated, reprinted, this time with funding from a Kellogg Foundation grant. All in all, ROW considers the printing of this "flyer" its most successful effort to date.

Over the years, the citizen-based ROW coalition has learned many, many hard lessons in organizing. It has operated almost exclusively with volunteer help. Its 15-member board of directors does all the work. This has both a plus and a minus side: ROW leaders have made very little effort to procure grants, fearing that accepting grant money sacrifices independence. Grants also lock a group into commitments to projects, and grantors, they say, seek out innovative ideas…thus diverting attention from the primary mission.

The ROW board has learned it needs to keep focused on water quality, documenting the pollution and its sources and lobbying for solutions. ROW has made what might be described as mistakes. It once held a large "RiverFest" on the banks of the Muddy River. It was a tremendous amount of work; lining up popular bands, dancers, face painters, etc., and then it was victimized by the New England weather. Being rained out and having to reschedule ran up a large debt that had to be paid. Yes, such events bring people to the park and the river, but are they worth it? Do they detract from the original mission? Some ROW leaders think so. Others disagree.

ROW has learned the value of credible research. You must always be ready to scientifically document what you have to say, or you will get no respect from the media. The bureaucracies will also shoot holes in everything you say. Quotes from lawyers and doctors (Ph.D.s.) are always necessary to support the perfectly correct commonsense observations of people.

A good leader learns to share the responsibilities, to give credit where credit is due, to operate by consensus of the board, and to take the raps sure to come from rival groups, media, politicians, and bureaucrats.

The bureaucrats are the most discouraging—people paid to do what citizens are trying to accomplish for free. Paid workers who do not really know or care, they do not live there.

Sometimes ROW cynics quip, "It's not the river that smells, it's the politics." We are taken back to the thought: What must it be like to try to clean up bigger rivers?

Trying to clean up a river is Political Science 901, and it should not be. If only one agency held the responsibility instead of 40, it would be so much easier and rivers might become much much cleaner.

13
Watershed Management in Russia and the Former Soviet Union

John G. Aronson
President, AATA International, Inc.
Fort Collins, Colorado

Introduction

Russia and what is now the Former Soviet Union (FSU)—including Kazakstan, Turkmenistan, Uzbekistan, Azerbaijan, Kyrgyztan, Ukraine, Belarus, Estonia, Georgia, and others—have been undergoing dramatic social, economic, and political change since the collapse of the Soviet Union. Russia and the FSU represent about one-ninth (17 million km^2) of the total land mass of the planet Earth and stretch across the northern hemisphere for at least 11 time zones. As could be expected, Russia and the FSU are home to some of the world's greatest rivers and their watersheds. Some of these rivers are known to the western world, as they have received the attention of historians, writers, scientists, and others. The Volga River and Lake Baikal are arguably the most recognizable Russian river and lake, respectively. However, many other huge rivers and their watersheds exist in Russia and the FSU and are little known to the western world. These include the Lena, Yenisey, Don, Amur, Ob, Yana, Neva, Tom, Dniepr, Kolyma, Pechora, Kuban, Terek, Omolon, and many others.

This chapter was developed on the basis of the author's experience working in Russia, Kazakstan, and the FSU as an environmental consultant to Russian/FSU international joint-venture mining, port development, and oil and gas projects. Interviewing networks of local contacts, standard literature research, and modern Internet data and literature resource acquisition were performed. The primary goals were to provide background information about some of the representative large watersheds in Russia and the Former Soviet Union, to document the legacy of environmental problems that plague some of these watersheds, and to outline some of the case histories of modern watershed management activities that have been initiated in Russia and the FSU. The author wishes to convince the reader that application of integrated watershed management programs throughout Russia and the FSU must be pursued with increased vigor and funding to counteract the years of neglect and abuse that occurred in many of the world's greatest watersheds. This will require a tremendous infusion of interdisciplinary, "ecological paradigm" education, thinking, planning, technology transfer, action, and funding.

Watersheds of Russia and the FSU

Russia is home to four of the world's top 20 rivers as ranked by discharge at the mouth, including the Yenisey (6), Lena (9), Ob (13), and Amur (15). For comparison, the Mississippi is ranked 7th, the Nile is 33d, the Danube is 21st, and the Rhine is 38th. Other Russian and FSU rivers in the top 100 rivers of the world based upon discharge include the Pechora (25), Kolyma (27), Dvina (30), Neva (35), Pyasina (36), Indigirka (40), Dniepr (41), and Yana (49). With respect to total length, Russia is home to three of the top 10 rivers of the world, Yenisey (5540 km, 4th), Ob-Irtysh (5410 km, 7th), and Amur (4444 km, 10th). A summary of major Russian and FSU rivers and their watersheds is provided in Table 13.1. Russia also possesses the longest continuous coastline of any country in the world (32,180 km) along the Arctic and Pacific oceans, and has significant coastline along the Black Sea and Caspian Sea.

The watersheds of Russia and the FSU can be divided into three broad geographic areas (see Fig. 13.1): European Russia (west of the Ural Mountains), Siberia (Urals to the Lena River), and the Russian Far East (east of the Lena to the Pacific, including the Kamchatka Peninsula). European Russia is dominated by relatively low-relief areas, with a preponderance of lakes, swamps, and bogs. The western Siberian lowlands also contain

Table 13.1 Summary of Selected Russian/FSU Watershed Characteristics

Watershed system	Drainage area ($km^2 \times 10^3$)	Average discharge ($km^3 \times 10^3$)	World rank (by discharge)	Length (km)	World rank (by length)
Aldan	N/A	N/A	N/A	2273	58
Amur	1844	12.0	15	4444	10
Angara	N/A	N/A	N/A	1779	89
Don	224	1.20	N/A	1870	82
Dniepr	502	1.67	41	2200	61
Kolyma	645	3.79	27	2513	49
Lena	2424	15.49	9	4400	11
Neva	282	2.61	35	74	>100
Northern Dvina	360	3.51	30	1302	>100
Ob-Irtysh	2975	15.0	13	5410	7
Pechora	326	4.08	25	1809	85
Volga	1380	8.0	N/A	3530	23
Yana	246	0.99	49	1492	>100
Yenisey	2590	17.39	6	5540	5

poorly drained swamps. The central Siberian platform has more incised rivers with increased relief (500 to 700 m). The eastern Siberian uplands can range up to 3200 m. The Russian Far East boasts some of the most pristine rivers of the world. The Volga River is the most important inland waterway in Russia, with over half the country's total river traffic. A series of seven major dams were constructed on the Volga, along with canals to the Don River and Baltic Waterway, to enhance navigation. Moscow is connected to the Volga system through the Moscow Canal. Rivers of the central Siberian area flow to the Arctic Ocean and, as such, are of limited usefulness for large-scale transport. The Amur River is the principal east-west river for transportation in the Russian Far East. With such a huge territory, there is virtually every type of northern watershed imaginable, from arid to wet, from small to huge, from steep to flat, from pristine to completely devastated.

The Soviet era brought on incredible development and exploitation of the empire's water resources, including substantial hydropower programs on the Volga, Don, Yenisey, Angara, and Ob-Irtysh rivers, accounting for up to 13 percent of total annual electrical production. Huge irrigation projects were also developed, which resulted in some of

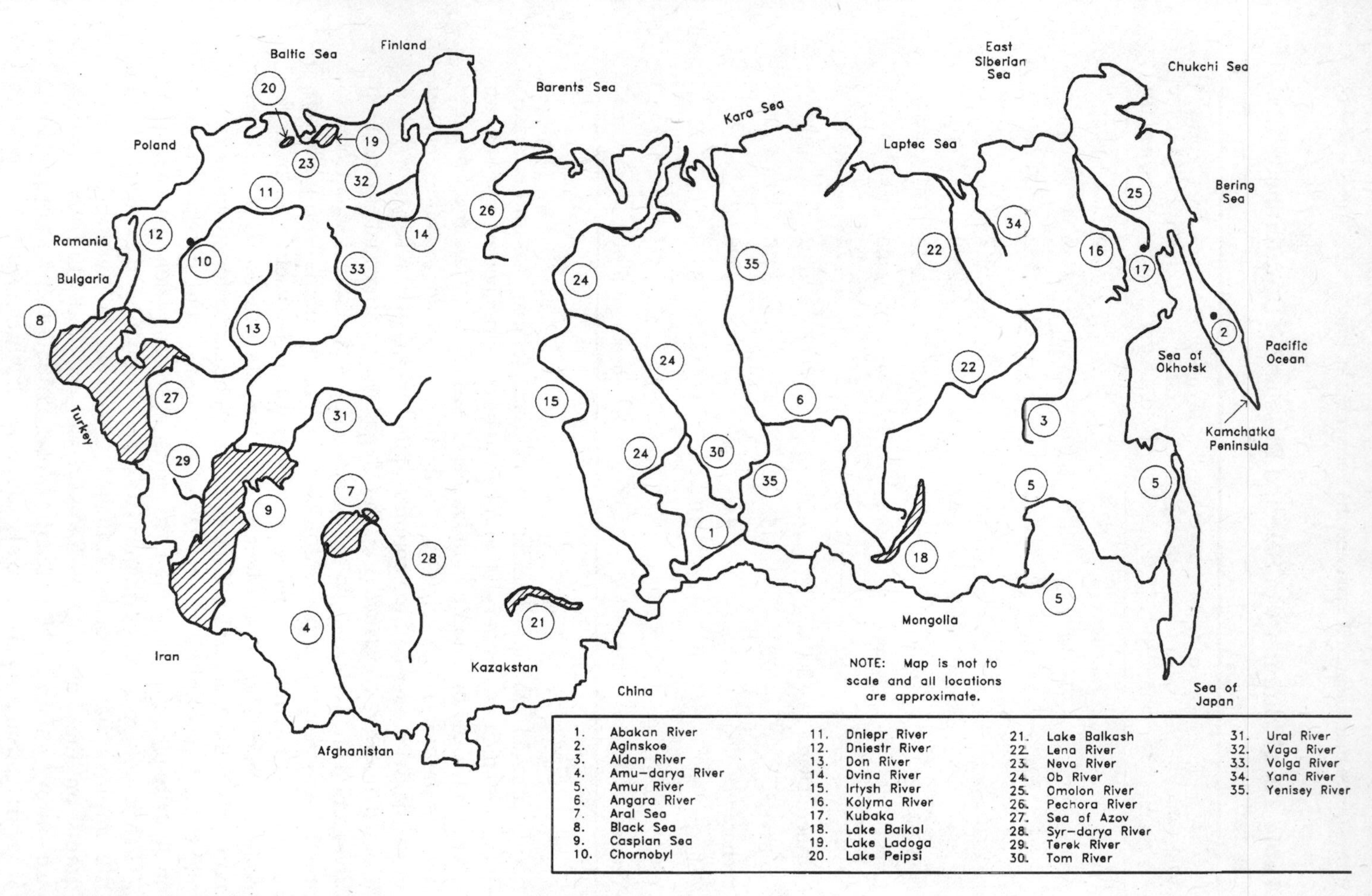

Figure 13.1 Generalized map of important Russian and FSU watersheds.

the worst ecological impacts ever recorded from such agriculturalization. Accidental and planned nuclear releases have been documented to have impacted large watersheds. Agricultural, industrial, and domestic water supplies in many areas are polluted with various metals, organics, pesticides, and other waterborne pollutants. Water quality maintenance and monitoring are major challenges. Technology for water and wastewater treatment is generally old and ineffective. Lack of electricity, modern equipment (such as modern pumps, aerators, mixers, digesters, electronic controls), replacement parts, or chemical treatment supplies further confounds effective treatment. Municipal, industrial, and agricultural effluents are mostly untreated. Water and wastewater management is largely a function of local governments, many of which are in or near bankruptcy.

According to sources within the Ministry of Nature Protection and the Environment, about one-half of the population of Russia uses water which does not meet national drinking water standards, and about one-third of the underground water supplies of the entire area are contaminated. Oil spills have been common to watersheds in many areas, and industrial pollution from metal processing plants has gone unchecked in many areas, resulting in what are now termed *ecological disaster zones*, where virtually all vegetation has been lost. Municipal and industrial wastewater discharged to rivers and streams is largely untreated. With the economic hardships that have occurred, necessary repairs, treatment chemicals, and basic operational electric demands have not been able to be met. The Soviet economy was previously run at the high cost of public health and the environment. Needless to say, there are many watersheds in need of attention. Details of the selected environmental problem areas related to specific watersheds are presented below.

Historical, Regulatory, and Sociological Perspectives on Watershed Management in Russia and the FSU

There has been little coordinated effort to develop watershed management infrastructure in Russia and the FSU. What organizational efforts have been implemented have been directed largely toward debating upstream and downstream claims of impact from one jurisdiction to another. Prior to 1967, watershed issues were the responsibility of state health departments. Some watershed committees have been developed over the years, with various oblasts, krais, and autonomous republics attempting to develop some cooperation. For example, the Volga basin is divided into upper, middle, lower, and delta watershed management divisions. Historically, there have existed separate committees for regulation of water use (Water Resources Committee) and protection of water resources (Water Resources Protection Committee). Prior to the collapse of the Soviet Union, these two committees made up the Ministry of Cultivation and Protection of Water Resources. Currently, there is a massive reorganization going on within the Russian government, and within the various sovereign republics, to reorganize the water resource management agencies. The principal environmental management agency at the oblast (regional) level is GOSKOM-PRIRODA, which literally translated means State Committee for Nature Protection and the Environment. The federal ministry level of this organization (PRIRODA) is also being reorganized at this time.

Within Russia and the FSU, rather extensive environmental regulations have been developed over the years. There exist detailed water quality standards for most of the chemical species that might be expected to occur as waterborne pollutants within Russian and FSU watersheds. Some of the standards are more stringent than those of the rest of the modern world. As an example, the standard for total suspended solids is 0.25 mg/L above background, a level which would be considered quite unachievable and practically unquantifiable in most situations. Unfortunately, water quality regulations are not generally enforced. When enforcement actions are taken, a complex set of equations is used to quantify damages. In most cases, the fines have not kept up with inflation over the years, and it is cheaper to pay the fines than pay for treatment or pollution prevention. In many cases, the most polluting industrial entities were state-owned enterprises, which simply ignored the fines or passed the costs along to the central government. Payments to workers in many sectors of society have been argued to be more important, and have taken priority over fines.

Special water use permits are required for significant water uses. Annual permission is granted by the basinal or territorial administrations which regulate the use and protection of waters following receipt of approvals from the following agencies: the state health department, the Fisheries Protection Agency, and the geological agency responsible for granting permission for groundwater use. This is one of the few areas where there is an indication of watershed-based organization.

Environmental scientists and engineers are generally well trained within what could be termed a "pillar society," whereby one advances within a relatively narrow field from lower echelons to higher levels of responsibility with time and experience. The concept of applied interdisciplinary environmental problem solving is foreign to most areas of the technical society. Normally, foresters did not talk with hydrologists, water quality experts did not work with civil engineers, zoologists did not communicate with toxicologists, etc. Thus, integrated watershed management requiring complex and intensive multidisciplinary work groups could not be implemented effectively in this type of pillar society. The modern "ecological paradigm" which focuses on an interdisciplinary approach to problem solving, and which is a key element of successful watershed management programs, was not able previously to be implemented with any significant level of success. Only a few forward-thinking research institutes have employed multidisciplinary approaches.

Bau (1995) supported a new paradigm for interdisciplinary work groups for integrating technical watershed management activities, and summarized well: "The change of paradigm in developing countries and eastern Europe has only begun. It is still limited to narrow scientific circles, and has no significant influence on the decision-making process. Centralized plans for water resources management and environmental protection actions still follow (with few exceptions) the traditional lines of monodisciplinary, large-scale thinking—an end-of-the-pipe approach instead of source control and local disposal, reaction instead of prevention, effluent dilution instead of selective concentration and reuse."

In February 1994, President Boris Yeltsin and Environment Minister Viktor Danilov-Danilyan issued a draft edict "On state strategy of the Russian Federation for protection of the environment and provision of stable development" which was created by various environmental specialists within the Ministry of Environmental Protection and Natural Resources. The primary interest was to create a Russian "Agenda 21" that could begin some movement toward sustainable development. The agenda included cleaning up freshwater resources and protecting natural rivers. More recently, a wide spectrum of international cooperative efforts have begun to have some effect on creating a more interdisciplinary approach to ecological problem solving. The reorganization of existing research institutes, agencies, and other government bureaucracies toward interdisciplinary structures has not occurred to any great extent, although some movement is occurring.

Watershed Problems of Russia and the FSU

The following discussions are provided to focus on some of the more notable watershed-related problems in Russia and the FSU and are by no means exhaustive. The level of detail is generally related to the level of environmental degradation and related research response. Obviously, there is a general paucity of data on the many thousands of smaller watershed problems occurring throughout the region due to lack of interest and/or funding.

Chornobyl

Chornobyl (the now-preferred Ukrainian spelling for Chernobyl) is one of the world's most well-known ecological catastrophes. At 1:23 a.m. local time, on April 26, 1986, the number 4 RMBK-type reactor of the V.I. Lenin Nuclear Power Plant located on the Pripyat River in north central Ukraine, just 25 km upstream of Chornobyl, exploded and released from 30 to 40 (with some estimates as high as 200) times the radioactivity of the atomic bombs dropped on Hiroshima and Nagasaki at the end of World War II. The Pripyat River is tributary to the Dniepr River, which flows to the Black Sea. The Kiev water supply reservoir is located immediately downstream of Chornobyl, upstream of Kiev.

Swedish scientists were the first to notice elevated fallout levels from the accident and alerted the world. Radioactive fallout contaminated large areas of the Ukraine, Belarus, and Europe. A wide spectrum of radionuclides were released to the atmosphere, including cesium 137; cesium 134; strontium 90; iodine 131; tellurium 132; barium 140; ruthenium 103; plutonium 238, 239, and 240; americium 241; cerium 144; curium 242, 243, and 244; antimony 125; and other miscellaneous species. The radionuclides of greatest interest, due to their large quantities and food chain mobility, included cesium and iodine. Iodine 131 has a relatively short half-life of 8 days, but cesium 134 and cesium 137 have half-lives of 2 and 30 years, respectively. The U.S. Department of Energy estimated that a total of 2,400,000 curies (Ci) of cesium was released (Goldman, 1987). Some of the plutonium species have half-lives ranging from 13 to 24,000 years. According to Freemantle (1996), about 85 percent of the total release consisted of radionuclides with half-lives of less than 1 month, 13 percent with half-lives of several months, 1 percent with half-lives of about 30 years, and 0.001 percent with half-lives of more than 50 years.

Wind and rainfall patterns (linked to wet deposition of the radioactive particles and gases emitted) were responsible for the heterogeneous distribution patterns of the fallout. Winds transported the radioactive debris and noble gases emitted mostly toward the northwest, where snow and rain interacted with the plume. The effects could be measured worldwide due to long-range transport. It has been estimated that the Chornobyl accident resulted in contamination of more than 200,000 km^2 of the earth's surface at levels exceeding 37,000 Bq/m^2 or 1 Ci/km^2 (3.7×10^{10} bequerels = 1 curie).

The worst impacts have been borne by the immediate area surrounding the plant, where the heavier elements deposited, including plutonium 239 and strontium 90, at levels exceeding 300 MBq/m^2. Belarus and Russia also had considerable wet and dry deposition which resulted in the contamination of large areas of forest and agricultural lands exceeding health standards. Over 400 settlements have been evacuated in Belarus, and about 2.2 million people have been exposed to prolonged impact of Chornobyl-derived radionuclides. A 30-km-radius exclusion zone was established around the plant from which an estimated 135,000 people were evacuated. (AATA staff have visited the evacuated city of Gomel and have participated in evaluation of the alternative cleanup methods for Belarus.)

Acute effects to coniferous forests in the areas with highest deposition were observed. An area known as the "Red Forest" near the site was completely killed. Concerns over redistribution of radionuclides from forest fires prompted remediation efforts in 1987, whereby dead trees were cut and hauled and the top soil layers removed. Wastes were buried in trenches and covered with sand. Recent investigations have found that ecosystem resiliency has been impressive. Elimination of the human impacts of forestry, land clearing, cultivation, and grazing has resulted in good regrowth of disturbed areas and enhanced recovery of terrestrial ecosystems. Long-lived radionuclides such as cesium 137 stay in soil, undergo uptake by plants and crops, and are transferred to animals, including meat and milk. Studies have shown that migration of cesium 137 has been generally limited to the upper 5 to 10 cm of the soil layer. Migration depends on a wide variety of factors including soil type, pH, rainfall, tillage, and organic content. Of course, this is the layer that is most susceptible to erosion and waterborne transport downstream through the watershed.

Aquatic ecosystem studies have shown that natural sedimentation processes have been effective in limiting impacts. Immediately following the accident, the radioactivity level in the Pripyat River was about 10 kBq/L, in the Uzh River about 5 kBq/L, and in the Dniepr River about 4 kBq/L. These levels were mostly the short half-life iodine 131. By 1989, the levels in the Kiev Reservoir had dropped to 0.4 Bq/L. Levels of cesium 137 in the downstream Kanev and Kremenchug reservoirs were at 0.2 and 0.05 Bq/L, respectively. Historical monitoring throughout the European Community has shown normal background levels to be on the order of 0.1 Bq/L. Fish samples have been analyzed from various locations around the site and the northern European area. Progressive decreases in concentrations were found from the source downstream to the Kiev, Kanecv, and Kremenchug reservoirs. It is interesting to note that no long-term direct effects on aquatic organisms in the plant's cooling water reservoir (area of maximum deposition) have been documented.

Groundwater contaminated beneath the plant and in the maximum deposition zone has the potential to reach the Pripyat River, the downstream water reservoir for Kiev, the Dniepr River, and the Black Sea. Vovk et al. (1994) indicated that strontium 90 is the radionuclide of greatest concern within the 30-km exclusion zone with respect to groundwater, and it could contaminate drinking water supplies in 10 to 100 years. According to OECD (1995), "The contamination of the water system has not posed a public health problem during the last decade; nevertheless, in view of the large quantities of radioactivity deposited in the catchment area of the system of water bodies in the contaminated regions around Chornobyl, there will continue to be for a long time a need for careful monitoring to ensure that washout from the catchment area will not contaminate drinking-water supplies."

Outside of Russia and the FSU, some Scandinavian lakes have required restrictions on the sale of fish with levels of radiation greater than local standards, incidents linked to Chornobyl. Pike and cod taken from the Baltic Sea in 1987 showed levels of cesium 137 of 15 to 30 Bq/kg, about 5 times the pre-Chornobyl levels (Ikaheimonen et al. 1988). Freshwater fish showed much higher levels of bioconcentration, with pike at 10,000 Bq/kg and perch at 16,000 Bq/kg (Saxen and Rantavaara 1987). Fishes from lakes in the Bryansk region, Russia, showed about 10 times more cesium 137 than those from rivers, with the highest levels being around 21,000 Bq/kg (Fleishman et al. 1992). In the Chornobyl cooling reservoir, cesium 137 levels reached 5,000,000 Bq/m^2 in the sediments and 420,000 Bq/kg fresh weight in pike muscle (Kryshev et al. 1995).

In a speech at the Vienna conference acknowledging the tenth anniversary of the Chornobyl disaster, United Nations Secretary Boutrous-Boutrous Ghali summarized the current situation, stating that "Even today, its health, social, economic, and environmental dimensions, both immediate and long term, remain to be defined." Clearly, integrated watershed management strategies over many different watersheds of the region with need to be employed with respect to the Chornobyl disaster for years to come. A wide spectrum of watershed management technical, political, and sociological issues will need to be addressed. Land-use planning, soil conservation, hydrologic conditions (especially floods), colloidal chemistry, bioconcentration, and overall sediment flux throughout the system are all important technical considerations as the radioactivity moves toward the Black Sea from the site, and throughout the other watersheds affected.

Aral Sea

Considered by many to be one of the world's worst ecological disasters, the Aral Sea has been severely impacted by the development of large-scale irrigation and application of pesticides for cotton production within

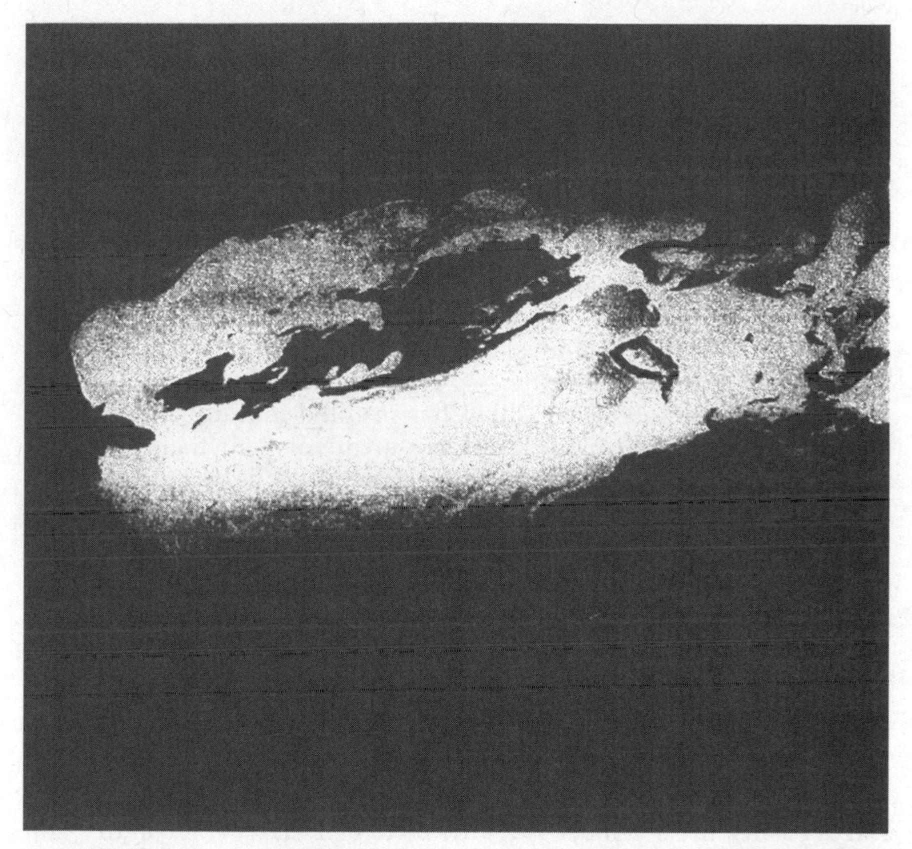

Figure 13.2 The Aral Sea. (*Photo taken by Shannon Lucid, from MIR Space Station, 1996.*)

the Amu Darya and Syr Darya watersheds. Diversions have resulted in massive desiccation of the Aral Sea to about one-half its historic size, once the world's fourth largest lake in area. Some coastal fishing villages are now more than 70 km from the new shoreline. Observations of the lake's shoreline from low earth orbit (see Fig. 13.2) using human-directed photography from the MIR Space Station and Space Shuttle have been dramatic. Natural evaporation coupled with increased diversions increased salinity from 10 percent in 1960 to 27 percent at the end of the 1980s. This caused eradication of 20 of the 24 local species of fishes and virtual elimination of the productive commercial fishery. The Aral Sea crisis includes pollution of drinking water supplies by waterborne pesticides and salts, duststorms which carry high salt loads, and loss of local species. Desertification is proceeding rapidly in the region, with massive amounts of airborne salt-laden dust being transported over 1000 km to the Pamir

Mountains, where spring snowmelt is accelerated. According to Dr. Oral Ataniyazova, Center for Human Reproduction and Family Planning, Bukus, Karakalpakstan, Uzbekistan, a regional health survey revealed that over the last 10 to 15 years, kidney and liver diseases have increased more than 30 times, arthritic disease has increased 60 times, and chronic bronchitis has increased 30 times.

The Aral Sea watersheds exist in one of the world's driest climatic zones, with rainfall of the region varying between 30 and 200 mm. According to Glantz et al. (1993), about 44.8 km^2 of the annual Amu Darya flow of 72 km^2 is diverted (62 percent), and almost all the Syr Darya annual flow of 36 km^2 is diverted principally for cotton production which was instituted by the Soviets in the late 1950s.

The breakup of the Soviet Union has resulted in multifaceted problems since now several distinct countries are involved in the watersheds of the Aral Sea, requiring new and more complicated interactions between and among Kyrgystan, Uzbekistan, Kazakstan, Turkmenistan, and Tajikistan. Bedford (1996) provided an excellent review of the incredible situation. The Aral Sea disaster was recognized as such during the Gorbachev *glasnost* period, but a series of resolutions fell short when the Soviet Union collapsed. The efforts of the World Bank, United Nations Environmental Program (UNEP), and United Nations Development Program (UNDP) working with the new republics resulted in a current institutional framework for dealing with the international crisis being created in 1992 and 1993. Two river basin commissions (BVO is the Russian acronym) were set up, one each for the Syr Darya and Amu Darya basins. In September 1995, the "Nukus Declaration" on the Aral Sea basin was adopted by the Nukus International Conference on Sustainable Development of the Countries of the Aral Sea Basin. The declaration reiterated support for a concerted multisectoral effort, world funding, and the Interstate Council for the Aral Sea (ICAS). In January 1996, the presidents of Turkmenistan and Uzbekistan signed an agreement on water management which allocates equal portions of the Amu Darya flow to each country at the level of 12.9 Mm^3/yr. The U.S. Agency for International Development (USAID) is instituting a number of projects to provide safe drinking water, improve sanitation, and increase efficiency of water use. The World Bank, in cooperation with the European Commission and the United Nations, is implementing pilot programs to implement water distribution, conservation, and allocation programs, with the ultimate goal of delivering more water to the Aral Sea for recharge.

Despite the immediate and serious need for cooperation using integrated watershed management approaches for conservation, protection, rehabilitation, and resource allocation, the BVOs have been largely ineffective so far. According to Bedford (1996), "The BVOs are incapable of genuine basin-wide management, and indeed are barely able to meet basic maintenance costs already in place on two large rivers." The existing cooperation among various central Asian republics has fostered an illusion of unity and has resulted in little progress in the areas of water conservation, or more equitable, environmentally responsible water resource allocation. Various technology transfer programs are now being advanced to deal with the Aral Sea situation, but much more coordination and funding are needed. Remote sensing, geographic information systems, and spatial analysis and modeling of desertification have been advanced in support of Aral Sea programs. Hardware such as canal liners, pumps, well-drilling rigs, drip irrigation, and evaporation covers have been advanced to assist nearly 35 million people potentially affected by the situation.

There are currently reported to be about 50 genuine projects addressing the Aral Sea disaster, but there has been little or no progress in dealing with conservation or replacement of the 33 to 36 km^2 of annual evaporative losses from the lake. The sustainability of regional agricultural production falls into question when both water quantity and water quality are considered. Without further, more intensive and decisive integrated watershed management action, it has been predicted that the Aral Sea will be virtually dry by the year 2020.

Lake Baikal

Lake Baikal, located in southern Siberia, is the world's oldest known lake, deepest lake, and largest freshwater source, containing about 20 percent of the world's freshwater. It is approximately 700 km long and 1637 m deep, and it contains an estimated 23,000 km^3 of water. Lake Baikal is revered in Russia, and has been called, among other things, the Sacred Sea, the Pearl of Siberia, and the Blue Eye of Siberia. Lake Baikal constitutes a unique ecosystem in the world. Of the roughly 2365 plant and animal species identified from the lake so far, more than 25 percent exist nowhere else on earth. A total of 336 rivers flow into Lake Baikal, but only one, the Angara, flows out. The watershed of Lake Baikal includes portions of Mongolia and China, both of which contribute water, natural geochemicals, nutrients, and various pollutants to the lake system.

The initial ecological movement in Russia was formed to prevent accelerated drainage of the lake by dynamiting the famous "black rock" at the Angara outlet, and to draw attention to the protection of the special ecological attributes of the lake. In the late 1960s, a paper plant was installed at Baikalsky city, and began discharging effluents to the lake.

This provided a heightened level of protest and a new focus for the budding ecological movement in Russia. Some limited success was gained when Lake Baikal protection legislation was passed in 1971 and 1977. Various nature reserves were set up around the lake, and deforestation of certain areas was forbidden.

The pulp mill continues to operate, employing approximately 3500 and polluting a reported 200 km^2 of the lake. Integrated watershed management initiatives have not been implemented on the Selenga River, the primary (about 50 percent) water supply to the lake, which has been reported to carry domestic and industrial wastes from upstream in Mongolia. More than 100 businesses operate along the shores of the lake, with very little in the way of pollution control. Further expansion of existing and hydropower dams on the Angara-Yenisey river systems has the potential to alter lake levels significantly. Nearby coal-burning power plants and other industrial operations contribute air pollutants that have the potential to enter the lake ecosystem.

Substantial international scientific cooperation and research are being conducted on Lake Baikal, in an attempt to gain important information on the ecology of the lake before significant changes occur. The United Nations Educational, Scientific, and Cultural Organization (UNESCO) has considered classification of Lake Baikal as a natural world treasure, which would heighten international protection status. Various nongovernment organizations (NGOs) have been organized to help increase information exchange, seek better public involvement, and attract international attention. Various entities have been increasing their levels of cooperation to deal with critical issues surrounding Lake Baikal, including enforcement of the Russian "green zone" land ordinances, delineation of protected forest areas, protection of threatened species, such as the Baikal seal (*Phoca sibirica*), and control of industrial, municipal, and agricultural pollution to the lake. The United Nations Economic Development Organization (UNEDO), through a team of selected specialists, studied the Baikalsk pulp and paper plant to determine the most appropriate course of action for the plant. The report evaluated the impacts of plant operations upon the environment, and provided recommendations to local government representatives for improvements and further modernization.

A land-use plan was drafted by a USAID-funded NGO in association with local representatives of the Irtutsk administration. Cooperation has also been established between the Irtutsk representatives and a Land Use Bureau in Germany. Recent exchange programs sponsored by various U.S. research organizations have focused on Lake Baikal related to local and regional institutional building, monitoring program design, and modeling. Substantial new research on the basic limnology, geology, and chemistry of Lake Baikal is being conducted through joint programs between western universities and government research agencies (such as the U.S. Geological Survey) and Russian scientific institutes.

Yeltsin and Danilov-Danilyan included Lake Baikal as a priority in their February 1994 draft edict "On State Strategy of the Russian Federation for Protection of the Environment and Provision of Stable Development." Effective integrated watershed management for Lake Baikal will require considerably more cooperation at all levels, including the local, regional, and federal government representatives, the landowners, the industrial operators, NGOs, the government of Mongolia, multilateral development institutions, scientific organizations, cooperating worldwide universities, and international funding agencies. Further refinement of the overall land-use plan is essential, and should include substantial integration of modern watershed management techniques.

Pechora (Komi) oil spill

The primary Pechora (Komi) oil spill occurred near the town of Usinsk, Komi Republic, just south of the arctic circle in northern Russia, on or about October 1, 1994. A major crude oil pipeline from Kharyaga to Usinsk had been leaking historically, with prior spills being contained behind a series of earthen dams. In September and October of 1994, heavy rains weakened the dam, and when the dam finally gave way, approximately 102,000 metric tons (t) of oil drained into the surrounding tundra, considered the third-largest oil spill in history. By comparison, the Valdez oil spill was about 34,000 at sea. The local drainage of the area is to the Kolva River, which is tributary to the Usa River. The Usa joins the Pechora River which flows to the Pechroskoye Sea and the Barents Sea. The area has been inventoried and found to be important for a wide variety of species, including reindeer, the beluga whale, Bewick's swan, eiders, ducks, guillemots, and other shorebirds, and a variety of typical arctic riverine fishes.

Substantial efforts were made to clean up the spill, with burning being the principal cleanup methodology. Containment berms were constructed, and oil was removed where possible. Following the spill, the World Bank and the European Bank for Reconstruction and Development (EBRD) provided funding of about U.S. $125 million to clean up the spill. A U.S.-Australian joint-venture company specializing in oil spill cleanup was hired, and progress was made in gaining control of the situation. According to World Bank officials tracking the cleanup, efforts in early 1995 were successful in keeping oil from reaching the Kolva-Pechora river system.

Mikhail Danilov, Russian State Duma Deputy from Arkhangelsk who chairs an intragovernment commission on the Komi spill, proposed that a large-scale, internationally supported environmental monitoring effort be mounted to include satellite imagery analysis and electronic communications in an attempt to gain early detection of oil spills, which are entirely commonplace within Russia and the FSU. Oil spills represent a significant existing and potential environmental threat to watersheds throughout Russia and the FSU. Modern spill prevention, control, and countermeasures (SPCC) programs are lacking within Russia and the FSU. Much work is needed to protect priority watersheds from acute and chronic impacts from petroleum leaks and spills. Much of the existing pipeline network throughout the region is in poor condition, with substantial potential for further failure. Early detection coupled with modern cleanup technology will be required as foreign investors expand their joint-venture activities with Russian and FSU oil and gas operations to protect watersheds in oil-producing areas.

Black Sea, Caspian Sea, and other watersheds

The Black Sea receives runoff from a number of huge watersheds representing roughly one-third of the land area of continental Europe. The major portions of 17 different countries, 13 capitals, and over 160 million people live in the watersheds tributary to the Black Sea, including the Dniepr, Don, Danube, and Dniestr. The Black Sea has a very long residence time and a large natural supply of nutrients, resulting in the world's largest anoxic basin. Agricultural, municipal, and industrial pollution has added to the deterioration of water quality in the Black Sea, and increasingly anoxic conditions. The bottom waters of the lake, mostly below 100 m, are toxic, containing high concentrations of hydrogen sulfide. Fisheries have declined significantly over the years. Cooperation on watershed management activities for the Black Sea started in earnest with the Bucharest Convention of 1994, which included protocols for control of land-based pollutants, dumping of waste, and coordinated response to accidental spills. Recently, the Global Environmental Facility (GEF) of the World Bank, UNDP, and UNEP have advanced the Black Sea Environmental Programme, with collateral funding by EU participants (PHARE-TACIS) and contributing countries. Integrated coastal zone management and watershed management activities are also being supported in Georgia, with significant cooperative efforts to upgrade old wastewater plants, to rehabilitate old waste dumps, to integrate multiagency efforts, and to prioritize allocation of resources for environmental improvements. Other NGO efforts are being coordinated for identification and protection of wetlands and other special watershed protection areas.

The Caspian Sea is the largest water basin in the world that is separated from the oceans with an area of about 380,000 km^2, containing an estimated 80,000 km^3 of water. The Caspian Sea is bordered by several countries, including Iran, and those of the newly independent states of the Former Soviet Union (Russia, Kazakstan, Azerbaijan, Turkmenistan). Changing water levels, petroleum pollution, untreated municipal and industrial wastewater, and agricultural nonpoint source pollution have altered water quality in the Caspian Sea. The principal watersheds of the Caspian Sea are the Volga and Ural. Since 1977, the water level has increased approximately 2 m; the factors affecting the overall water balance of the system are poorly understood and the subject of cooperative research. Increasing water levels have flooded existing historical petroleum facilities and thousands of square kilometers of coastline. Internal seiches of increasing magnitude have caused damage to many coastal areas. The Caspian Sea is the world-renowned center for caviar production, based upon the large natural sturgeon populations. Currently, a total of 13 species of sturgeon are considered threatened by water pollution, poaching, habitat alteration and destruction, and hydrologic modifications (dams and canals). New oil and gas production developments are planned for the near-shore and offshore Caspian. Large western joint-venture projects are advancing in the area, and new ecological challenges are anticipated. NATO held a workshop on the circum-Caspian region in 1996 (Glantz 1996) which focused international attention on sustainable development, biodiversity, regional cooperation, information exchange, education, research, "usable science," and progress within the Caspian Sea. Substantial progress is being made on developing cooperative efforts to deal with the Caspian Sea and its watersheds, including establishing lines of communication, conducting multidisciplinary ecological characterization, and promoting integration and prioritization of various world organizational efforts.

Other watersheds of note within Russia and the Former Soviet Union that exhibit ecological problems include the drainages of the nuclear-testing and production areas such as Semipalatinsk, Tomsk-7, and Chelabyinsk; the drainages of major metropolitan areas (Moscow, St. Petersburg, etc.); the watersheds of Murmansk and associated regional nickel smelting areas; the industrial drainages of the Tom River (Kuzbass area); and many of the smaller subwatersheds where uncontrolled municipal and industrial discharges have reduced the rivers, lakes, and streams to wastewater conduits. The extent of widespread small watershed problems is significant in most industrial and municipal areas, even in remote regions.

Modern Watershed-Related Programs and Technology Transfer

There are several important areas in which cooperation and interaction with the West are beginning to have positive influences on the environmental situation of Russia and the FSU, including the area of integrated watershed management. Environmental data collection, analysis, and information exchange are happening with increased frequency and intensity in many areas. There are myriad cooperative programs being advanced between Russian and FSU organizations and western universities, research institutes, government agencies, and nongovernment organizations. Organizations such as the International Association for the Promotion of Cooperation with Scientists from the Newly Independent States of the Former Soviet Union (INTAS) have created impressive communication networks, proposal solicitation and review, funding sources, project implementation, and reporting structures within many of the FSU countries.

Multilateral development banks are now funding projects with environmental provisos, requiring a level of environmental management and protection that meets recognized international standards. World Bank, International Finance Corporation (IFC), EBRD, Overseas Private Investment Corporation (OPIC), Export-Import Bank (EXIM), and other international financial institutions require that Russian/FSU joint ventures with the West have environmental audits, impact statements, reviews, and management programs. This has had a demonstrable positive effect on new joint-venture project areas, as western specialists and Russian and FSU specialists work together to achieve the requirements in an expeditious manner, so that projects can be approved and funded. Infusion of western capital with an effective environmental program component is what is needed to get many of the industrial problems turned around. The United Nations Environmental Program, Global Environmental Facility (GEF), United Nations Development Program, and the U.S. Agency for International Development have provided much needed funding and expertise, and a commitment to environmental integrity of new joint-venture projects. The new Global Water Partnership, a joint program involving the World Bank, UNDP, and the Swedish International Development Agency (SIDA), promises to direct from U.S. $30 to $40 billion over the next decade to achieve a more comprehensive approach toward water management. A discussion of selected and representative areas of progress for the development of integrated watershed management in Russia and the FSU is provided below. The list of programs is not exhaustive, and is presented in no particular order.

The Lake Peipsi/Chudskoye Lake project

Lake Peipsi (Chudskoye in Russian) occupies about two-thirds of the boundary region between Estonia and Russia, and with the nearby connected Lake Pskov, represents the fifth-largest lake (80 km^2) in Europe. The accepted international transboundary terminology recommended for the lake is *Lake Peipsi/Chudsko-Pskovskoe.* The breakup of the Soviet Union exacerbated transboundary issues of pollution, fishing rights, and ethnicity of inhabitants for this area. In 1993 an interdisciplinary project team was formed with the cooperation of Estonian, Russian, and U.S. researchers to develop strategies for dealing with the critical watershed problems facing Lake Peipsi/Chudsko-Pskovskoe, which include eutrophication, industrial pollution, deforestation, and overfishing. Nonprofit NGO registration was achieved for the Lake Peipsi Project (LPP) in 1994. The successes of the LPP are many, including establishing communication among key stakeholders, researchers, and the western world; organizing international and regional studies and conferences; establishing a database of environmental information and legislation for the region; drafting and supporting intergovernment agreements; and achieving the first environmental agreement between Estonia and the Russian Federation. The Finnish Ministry of Environment played a key role in keeping the Estonian and Russian sides working toward a common goal.

According to the LPP director, Gulnara Ishkuzina-Roll, the focus on the local level was the key element in these successes. Getting the various parties to communicate was greatly enhanced by modern e-mail and Internet resources. The Peipsi/Chudskoye Lake Project established a very active World Wide Web presence through its newsletter, *The Lake Peipsi Quarterly.* Cooperation and communication have been achieved with most of the interested Estonian and Russian parties, including researchers, government officials, the transboundary Environmental Information Agency, Dartmouth University, the international sponsors, PHARE-TACIS, UNDP-GRID, the Finnish Ministry of Environment, other NGOs, and the outside western world.

At an Environmental Governance Conference convened in Tartu, a series of mutually acceptable recommendations were formulated for management of the Lake Peipsi/Chudsko-Pskovsko watershed. Continued and increased communication between and among stakeholders was stressed. Coordination of monitoring efforts, sharing of resources and data, public information and outreach, environmental impact assessments, and recognition of transboundary conventions already established were included in the recommendations. The Lake Peipsi Project has started considering other subwatersheds, including the Leilup River basin, and Daugave and Zapadnaya Dvina river basins as natural extensions of their work. Cooperative ecological research efforts, further discussion and agreements, and continued opening of communications between the two sides will be required to achieve important watershed management goals.

One of the most important concepts to be instituted by the Lake Peipsi/Chudsko-Pskovskoe activities has been to initiate the prioritization of watershed activities according to modern integrated watershed management approaches. Since there is limited funding for any type of watershed activities in Russia and the FSU, it is vitally important that the greatest benefits be achieved with least-cost alternatives. Pollution prevention, land-use planning, enforcement of existing environmental laws, and recognition of the various nature reserves and parks ("green zones," zakazniks and zapovedniks) are among the top priorities on the basis of their relatively low cost. The Lake Peipsi/Chudsko-Pakovskoe program should serve as an excellent template for other such watershed management efforts in Russia and the FSU.

Ballerina, Kola Net, and other Internet resources for watershed management

Ballerina is an acronym for Baltic Sea Region On-Line Environmental Information Resources for Internet Access (*http://www.baltic-region.net*). The program is sponsored by UNEP-GRID Arendal and the Swedish Environmental Protection Agency. The Great Lakes Information Network (GLIN; *http://www.great-lakes.net/*) of the United States provided assistance and guidance for the project. Cooperative efforts to develop an interdisciplinary research communication vehicle for the Baltic Sea region have resulted in several useful hot-linked sites around the world.

A similar approach to the Ballerina project is being taken by Kola Net (*http://193.156.27.70*) and the Kola Science Center, a consortium of various technical institutes and organizations from the Kola peninsula area to cover the Barents Sea region. The charter of the Kola Net consortium is to create a cooperative Internet group for promoting significant application projects such as environmental monitoring, mapping, and management. This is another attempt to break down the "pillar society" approach to science and to establish real-world, real-time working relationships that can tackle multidisciplinary problems. It is the intention of Kola Net to provide free and open communication of information on all aspects of the Barents regional environment, including levels of chemicals, toxic gases, heavy metals, etc. in the regional terrestrial and aquatic ecosystems including lakes, forests, rivers, and seas. The group has already established itself as a popular Web site and lists most of its publications. Multidisciplinary approaches are encouraged, which can form the basis for watershed-based decision making.

Arctic Monitoring and Assessment Programme

The eight circumpolar countries of the world—Norway, Sweden, Finland, the United States, Russia, Greenland, Iceland, and Canada—adopted the Arctic Environmental Protection Strategy in 1991. The Arctic Monitoring and Assessment Programme (AMAP) is the international organization established to implement certain components of the strategy. AMAP objectives focus on measuring levels and assessing effects of pollutants within the arctic ecosystem. Transport of various pollutants, including radionuclides, from the northern-flowing Siberian rivers to the Arctic Ocean is an important consideration. There are many organizations which are cooperating on interdisciplinary monitoring, modeling, and assessment of the Russian arctic environment. The Australian Centre of the Asian Spatial Information and Analysis Network (ACASIAN), the Nansen International Environmental and Remote Sensing Center (NIERSC), and the Arctic Research Consortium of the United States are but a few of the many cooperating groups becoming linked around the world via the Internet. Integrated watershed management strategies and approaches are slowly finding their way into these types of regional programs.

NATO initiatives

The North American Treaty Organization, Committee on Challenges of Modern Society, North Atlantic Cooperation Council (NATO/CCMS/NACC) is responsible for coordination of investigations of defense-related cross-border environmental problems within the context of NATO's commitment to world peace. Norway and Germany proposed a study entitled "Cross-border environmental problems emanating from defense-related installation and activities" which was approved by CCMS in 1992. The purpose of the work was to identify, survey, and assess pollution from defense-related installations and activities with respect to the Barents and Kara seas, Baltic Sea, and Black Sea. One portion of the study was dedicated to nuclear, and one to chemical pollutants. A total of 23 countries participated in the study, and results were published in 1995 (NATO et al. 1995). Transport of radioisotopes by the Ob and Yenisey rivers, potential releases from sunken submarines, dumping of radioactive wastes near Novaya Zemlya, potential nuclear accidents, nuclear waste management, munition dumps, and decommissioning of the Russian nuclear fleet were assessed. The program identified various "hot spots" around the area for both aquatic and terrestrial ecosystem concerns, and selected risk assessment models were utilized to screen various selected sites. The study recognized the importance of modern integrated watershed management techniques and recommended the development of operational prediction models for the rivers Elbe, Odra, Wisla, Dniepr, Dniestr, and others of the area, as have been developed for the Rhine and Danube, so that warnings could be provided in case of spill events. Other modeling efforts were also recommended to provide reconnaissance, identification, recovery, transportation, intermediate storage, dismantling, final destruction, and management of obsolete chemical weapons destruction. The risk assessment-based approach applied to watersheds, coupled with modern information management and modeling technology, provides a model for industry-specific environmental management. NATO is an active organization for cooperative programs regarding environmental management, protection, and cleanup of the defense industry activities, including those areas affected by Russia and the rest of the FSU.

Cooperative watershed studies—Technology leapfrogging

Many excellent examples of "technology leapfrogging" are being used to define, study, evaluate, and report watershed management initiatives within Russia and the FSU. The relatively highly trained and technologically sophisticated research scientists and environmental engineers in Russia and the FSU are able to utilize advanced technology of the West such as modern computers, new geographic information system (GIS) software, remote sensing tools, modern field monitoring instrumentation, and the Internet.

A cooperative U.S.-Russian demonstration project conducted by the University of Minnesota and funded by the International Research and Exchange Board (IREX), entitled "GIS as an integrating tool for multidisciplinary environmental efforts in the headwaters of nations' rivers: The Mississippi and the Volga," is a good example of *technology leapfrogging,* whereby advanced western computer hardware and software capabilities, linked with an integrated watershed management approach, are introduced to Russia and the FSU to study common problems. Opening lines of communication, sharing approaches and study results, and creating a common computer-based framework for database management, analysis, and reporting were important components of this project. Interaction with appropriate Russian water management research organizations was established, including Moscow State University, Research Station of the State Hydrology Institute, Committee on Hydrology and Meteorology (Novgorod Region, NOVGIDROMET), Russian National Conference on Water Problems and Solutions (Russian Academy of Sciences), Forest Research Institute, and the Committee of Russian Federation on Geology and Subsurface Usage, among others.

Other areas of technology leapfrogging for integrated watershed management include advanced computer modeling, database management, remote sensing, GIS/remote sensing integration, and advanced Internet communications. The Baltic Drainage Basin Project of the University of Stockholm, Sweden (Sweitzer et al. 1996), stresses the interdisciplinary nature of natural and social sciences with focus on eutrophication. Point and nonpoint source nutrient evaluation using advanced modeling techniques is being integrated with economic valuation, uncertainty analysis, and institutional analysis to provide a basis for strategic prioritization of watershed management alternatives within the entire Baltic region. The group has created an on-line Baltic Sea drainage basin GIS, map, and statistical database which can be accessed throughout the region via the Internet. The basic level of Internet usefulness has been establishing open, effective, and timely communication of data, research results, proposals, assessments, and correspondence. The next logical higher level is for interactive analysis, modeling, data visualization, and advanced modeling, perhaps in real time. Integrated watershed management will benefit greatly from Internet resources applied in such a way.

Kubaka and other joint-venture resource development projects

New western and Russian and FSU joint venture natural resource development projects (mining, oil and gas, forestry, etc.) have been increasing over the past few years. Many of the mining sites that have been offered for tender have included areas with previous placer, open-pit, and underground mining. In general, environmental management at these sites has been quite rudimentary or nonexistent prior to the joint-venture activities. Modern joint-venture projects are held to both the existing Russian standards and internationally recognized standards, since outside financial institutions (World Bank, IFC, EBRD, OPIC, EXIM, etc.) are generally involved. The new joint-venture projects occur in many areas throughout Russia and the FSU, in many different types of watersheds. The Russian Far East (RFE) was a focal point for gold, silver, and other mining activities in the Soviet Union, supported in the early years by the forced labor camps of the Stalinist era. The biggest historical impact to watersheds was uncontrolled placer mining which increased significantly the sediment loads for many rivers, including some of the largest rivers in the RFE, resulting in significant decline in water quality and displacement or reduction of fisheries. These have included sites in the Amur Oblast, Magadan Oblast, and Yakutia, where a wide variety of watershed conditions exist. The Aldan River in Yakutia has been affected by placer mining activities for over 70 years, completely altering large areas of the watershed and causing a radical change in water quality and fisheries.

The first joint-venture gold mining project of the RFE anticipated to be commissioned early in 1997 is the Kubaka Gold Project, operated by the Omolon Gold Mining Company, a joint venture of Amax Gold USA (Denver, Colorado) and Russian partners. The site of the mine and mill is in a remote area of the RFE, just south of the arctic circle, approximately 600 km northeast of Magadan, Russia. (AATA provided environmental management and permitting support for this project since its early inception.) A team of Russian scientists comprised of experts from various research institutes in Magadan was organized to complete interdisciplinary studies for the project. A comprehensive Environmental Impact Analysis Report (OVOS) was prepared. The Eastern Scientific Research Institute for Gold and Precious Metals (VNII-1), the Institute for Biological Problems of the North (IBPN), the Hydrological and Meteorological Institute (GIDROMET), and other entities were organized into a research team. An integrated watershed management approach was taken for the project. The exploration and pilot phase of the project operated by the Russians prior to the joint venture was not subject to any environmental regulations. According to local interviews, sediment impacts were significant to the local Kubaka River, a class I fishery with arctic grayling, for the first 3 years of activity. When the western partners established the Kubaka project joint ventures, a series of environmental management actions were taken to improve sediment and erosion control; to establish on-site environmental monitoring and management; to establish an effective spill prevention, control, and countermeasures (SPCC) program; and to improve the environmental living conditions of the workers. A new sewage treatment plant was constructed to handle the anticipated 500+ construction work force, one of the few (if any) facilities of this kind at any mine site in the RFE. Technology transfer at the Kubaka project included training of Russian specialists with modern environmental monitoring equipment; installation of modern, solid-state hydrological and meteorological monitoring stations; creation of buffer zones around local streams; installation of check dams and sediment traps; erosion control planning; and implementation of a complete SPCC program with modern western materials and supplies (sorbents, booms, pads, etc.). Although the Russians were generally familiar with the theory of sediment and erosion control, they had not utilized these methods generally at mining sites. Prior to western involvement, the Russians did not have access to geotextiles for lining runoff channels, sediment fencing, filter fabric, or other such advanced materials for sediment control.

Inspections of the operations have been conducted by the local regulators, EBRD, OPIC, and others, and they were found to meet recognized international standards of performance for such facilities. A comprehensive environmental monitoring program documents upstream and downstream conditions and reports to the State Committee for Natural Protection and the Environment (GOSKOMPRIRODA). Other Russian-western joint-venture projects have operated in a similar fashion, increasing the level of environmental management at remote sites and implementing a set of standard operating procedures for attaining locally and internationally recognized standards.

In general, western joint-venture activities can be expected to have a demonstrable positive effect on watersheds compared to the operational situations which existed prior to western involvement. Environmental requirements of multilateral development banks have had a positive environmental influence on watersheds throughout the natural resource developments of Russia and the FSU, by requiring project sponsors to adhere to a combination of local and international environmental standards that might not otherwise be recognized.

Summary, Observations, and Recommendations

The vast area of Russia and the FSU has experienced some of the worst watershed impacts that humans can muster. Many of the watersheds of Russia and the rest of the FSU are in critical condition, with an extensive set of problems including irrigation overdrafting, lack of wastewater treatment, oil spills, poor nuclear waste management, chemical pollution, lack of erosion control, agricultural pollution (nutrients, pesticides), uncontrolled discharges, and lack of enforcement of environmental regulations. The breakup of the Soviet Union has caused increased uncertainty in the management of important watersheds and created a need for more international cooperation. There is no coherent, effective, interdisciplinary watershed management organizational structure in place which can deal with these types of problems. There is little or no money available to fund rudimentary environmental protection or watershed management activities in most areas.

The ecological paradigm shift to integrated watershed management as a component of sustainable development is just now finding its way into a small segment of the society. There are a wide variety of outside interests, cooperation, and funding being focused on improvement of the environmental situation in Russia and FSU. Government, NGO, university, and private groups of the west, Japan, and other countries are having a positive influence in achieving some degree of technology leapfrogging, especially as related to increased communications and information exchange. The Internet has become an important new tool for establishing and maintaining communications. Of particular value have been the multilateral development bank environmental standards and requirements, which established basic international environmental standards for a wide variety of funded projects. Improvements in the environmental conditions of watersheds where multilateral development bank funding is involved can be expected to occur, since the basic environmental management, spill and erosion control, and water quality protection are virtually nonexistent in most areas. Increased levels of funding are needed to address the many watershed problems. Decision support systems for prioritization of watershed-related programs need to be advanced so that the greatest environmental improvement can be achieved in the most cost-effective manner, as the various problem areas are simply too large to all be addressed effectively. The integrated watershed management approach can be a useful tool in prioritization, and in making better environmental decisions. Cooperation can be expected to continue to increase among a variety of interested watershed-related groups, as Internet resources and communications in general improve. Much more institutional building must be accomplished at the local, regional, national, and international levels. Increased communication, coordination, and sharing of resources are needed to make sound decisions.

Perhaps a level of common environmental understanding can be reached at the local, regional, and national levels based upon the watershed concept that will allow cooperation and the democratic process to proceed in these troubled lands.

References

Aarkrog, A. 1994. Source terms and inventories of anthropogenic radionuclides. Riso National Laboratory, Roskilde, Denmark.

Aarkrog, A., Y. Tsaturov, and G. G. Polikarpov. 1993. Sources of environmental radioactive contamination in the former USSR. Riso National Laboratory, Roskilde, Denmark.

Albrecht, J., P. Faller, D. Kurbjuweit, and W. Saller. 1994. Russia's Environmental Mess. Die Zeit. Nov. 11, 1994. Hamburg, Germany.

Aleksakhin, R. M. 1993. Radioecological lessons of Chernobyl. *Radiation Biology and Ecology* **33:** 3–80.

Battista, C. J. 1994. Chernobyl: GIS model aids nuclear disaster relief. *GIS World,* March 1994, pp. 32–35.

Bau, J. 1995. Cooperation among water research and development institutions of Europe. *Water International* **20:** 129–135.

Bedford, D. P. 1996. International water management in the Aral Sea Basin. *Water International* **21:** 63–69.

Belt, D. 1992. Russia's Lake Baikal: The world's great lake. *National Geographic* **181**(6): 2–39.

BISNIS. 1996. Water resources and water management in Turkmenistan. July 24, 1996.

Bjorgo, E. 1996. The Nansen International Environmental and Remoter Sensing Center, St. Petersburg, Russia. NIERSC.

Caritat, P. de, C. Reimann, M. Ayras, and V. A. Chekushin. 1995. Heavy metals content in stream water from eight catchments on the Kola Peninsula (NW Russia) and in adjacent areas of Finland and Norway. In R. D. Wilken, U. Forstner, and A. Knochel (eds.). Tenth International Conference on Heavy Metals in the Environment. *Proceedings* **1:** 192–195. Hamburg, Germany. September 18–22, 1995.

Center for Biological Monitoring. 1996. RADNET, nuclear information about source points of anthropogenic radioactivity. Section 10: Chernobyl fallout data, annotated bibliography. <http://home.acadia.net/cbm>

Chernousenko, V. M. 1991. *Chernobyl, Insight from the Inside.* Springer-Verlag. Berlin.

Dzepo et al. 1995. Hydrogeological effects of the principal radioactive waste burial site adjacent to the Chernobyl NPP. *Proceedings of an international conference, Sarcophagus Safety '94, Zeleny Mys, Chronobyl, Ukraine,* March 14–18, 1994. OECD/NEA, Paris.

European Commission. 1994. Environmental radioactivity in the European Community 1987–1988–1989–1990. European Union report 15699.

Feshbach, M. 1995. *Introduction to Ecological Disaster: Cleaning Up the Hidden Legacy of the Soviet Regime.* Twentieth Century Fund Press, New York.

Feshbach, M. 1995. *Environmental and Health Atlas of Russia.* Paims Publishing House, Moscow.

Feshbach, M., and A. Friendly. 1992. *Ecocide in the USSR: Health and Nature under Siege.* Basic Books, New York.

Fleishman, D. G., V. A. Nikiforov, A. A. Saulus, and V. T. Komov. 1990–1992. Cesium-137 in fish of some lakes and rivers of the Byransk Region and North-West Russia in 1990–1992. *Journal of Environmental Radioactivity* **24**(2): 145–158.

Freemantle, M. 1996. Ten years after Chernobyl consequences still emerging. *Chemical and Engineering News,* April 29.

Frost, E. G. 1996. *Central Asia Research and Remediation Exchange.* San Diego State University, San Diego, California.

Glantz, M. H. (ed.) 1996. Scientific, environmental, and political issues of the circum-Caspian region. *NATO Advanced Research Workshop Proceedings.* Kluwer Academic Publishers, Dordrecht, Holland.

Glantz, M. H., A. Z. Rubenstein, and I. Zonn. 1993. Tragedy in the Aral Sea Basin: Looking back to plan ahead? *Global Environmental Change* **3**(2): 174–198.

Goldman, M. 1987. Chernobyl: A radiobiological perspective. *Science* **238:** 622–623.

Halleraker, J. H., M. V. Kozlov, C. Reimann, E. Zvereva, and E. Berge. 1996. The state of ecosystems in the central Barents Region: A guide to published data. *Third International Barents Symposium: Environment in the Barents Region.* Geological Survey of Norway, Kirkenes, Norway. September 12–15, 1996.

Holmberg, J. 1996. The global water partnership—a new form of international cooperation. *Stockholm Water Front* 3:9.

Ikaheimonen, T. K., E. I. Ilus, and R. Saxen. 1988. Finnish studies of radioactivity in the Baltic Sea in 1987. Supplement 8 to Annual Report 1987, Number STUK-A74, Report Number STUK-A82. Finnish Centre for Radiation and Nuclear Safety. Helsinki, Finland.

Ilyin, L. A., and A. O. Pavlovskii. 1987. Radiological consequences of the Chernobyl accident in the Soviet Union and measures taken to mitigate their impact. *IAEA Bulletin* 4.

Izrael, Y. 1991. Radiation circumstances in the European part of the CIS and in the Ural Region in 1991. *Soviet Meteorology and Hydrology* **11:** 1–12.

Kryshev, V. A., et al. 1995. Radioactive contamination of aquatic ecosystems following the Chernobyl accident. *Journal of Environmental Radioactivity* **27**(3): 207–219.

Kuchlick, J. R., et al. 1994. Organochlorines in the water and biota of Lake Baikal, Siberia. *Environmental Science and Technology* **28**(1): 31–37.

Likhtarev, L. A., et al. 1989. Radioactive contamination of water ecosystems and sources of drinking water. Medical Aspects of the Chernobyl Accident, Technical Document 516, International Atomic Energy Agency, Vienna, Austria.

Lulla, K. P. (ed.). 1996. Earth observations and imaging: A journal of human-directed remote sensing from the space shuttle and international space station (MIR). NASA Space Shuttle Earth Observations Program, Johnson Space Center. <http://ersaf.jsc.nasa.gov/newsletter/mir/shannon.html>

Micklin, P. P. 1988. Desiccation of the Aral Sea: A water management disaster in the Soviet Union. *Science* **241**(4870): 1170–1176.

NATO/CCMS/NACC. 1995. Cross-border environmental problems emanating from defense-related installations and activities. Summary Final Report Phase I, 1993–1995. Report Number 206. Oslo, Norway.

Nawrocki, T., and C. Johnston. 1996. GIS as an integrating tool for multidisciplinary environmental efforts in the headwaters of nations' rivers: The Mississippi and the Volga. <http://gpl.nrri.umn.edu/abstract.html#volga>

OECD Nuclear Energy Agency. 1995. Chernobyl Ten Years On. Radiological and Health Impact. An assessment by the NEA Committee on Radiation Protection and Public Health. November.

Ozturk, M. 1995. An overview of the environmental issues of the Black Sea region. Paper presented at the NATO Advanced Research Workshop: Scientific, Environmental, and Political Issues of the circum-Caspian Region. Moscow, Russia. May 13–16, 1996.

Paluszkiewicz, T., L. F. Hibler, M. C. Richmond, D. J. Bradley, and S. A. Thomas. 1995. Modeling the potential radionuclide transport by the Ob and Yenisey Rivers to the Kara Sea. *Marine Pollution Bulletin.*

Pavoni, B., et al. 1993. Ecological analysis of point and nonpoint pollution on watersheds. INTAS-93-1918. International Association for the Promotion of Cooperation with Scientists from the Newly Independent States of the Former Soviet Union.

Pavoni, B., A. Voinov, and N. Zharova. 1995. Basin (watershed) approach as a methodological basis for regional decision making and management in the ex-USSR. Annual reports of the Russian Hydroproject Institute. Moscow.

Privalova, L. I., et al. 1994. An approach to detecting delayed effects of radioactive contamination on industrial-urban-area dwellers. *Environmental Health Perspectives* **102**(5): 544–545.

Saxen, R., and A. Rantavaara. 1987. Radioactivity of freshwater fish in Finland after the Chernobyl accident in 1986: Supplement 6 to Annual Report STUK-A55. Report No. STUK-A61. Finnish Centre for Radiation and Nuclear Safety. Helsinki, Finland.

Schaidler, C., M. K. Hobish, and H. A. Geller. 1996. Draft report on a Russian environment and health mapping project using the strategic environmental distributed active archive resource.

Stone, R. 1996. RUSSIA: Academy fights to maintain research in the "Wild East." *Science* **272**(5266): 1259.

Stone, R. 1996. Trans-Pacific alliance draws up ecology plan. *Science* **272**(5266): 1260.

Sweitzer, J., S. Langaas, and C. Folke. 1996. Land cover and population density in the Baltic Sea Drainage Basin: A GIS database. *Ambio* **25**(3): 191–198.

Voinov, A., V. Starostin, and A. Nikitin. 1993. The hierarchical database for ecological expertise of environmental impacts associated with hydroconstructions. *Annual Reports of the Russian Hydroproject Institute* **73:** 125–161.

Vovk, I. F., et al. 1994. Geological and hydrogeological features of the ChAES 30-kilometer zone and possibilities for the deep or shallow burial of radioactive wastes. *Proceedings of an International Conference, Sarcophagus Safety '94.* Zeleny Mys, Chronobyl, Ukraine. March 14–18, 1994. OECD/NEA, Paris.

World Bank/UNDP/UNEP. Aral Sea Basin Program—Phase I: Proceedings of the donor's meeting held in Paris on June 23–24, 1994, unpublished.

World Bank (Europe and Central Asia Region Country Department 3). 1995. Aral Basin unit, Aral Sea Basin Program Phase I: Progress Report No. 2. World Bank, Washington, DC.

14 Environmentally Sustainable Management and Use of Internationally Shared Freshwater Resources

David Smith (Resource Economist) and Walter Rast (Deputy Director)

Water Branch, United Nations Environment Programme, Nairobi, Kenya

"...All the water on Earth is all the water there is..."

Introduction

Freshwater is the most precious of the earth's natural resources, being absolutely necessary for survival of all forms of life on this planet (including humans). It also is a basic ingredient of socioeconomic development. Indeed, humans use freshwater for more diverse purposes than any other natural resource, including drinking, cooking, cleaning, growing crops, livestock, fish and other sources of food, recreation, aesthetics, transport of commercial goods, and hydroelectric power generation. Freshwater resources also are used to assimilate many unwanted by-products and wastes of industrial and other activities associated with economic development. Often overlooked is its critical function in maintaining the structure of our natural ecosystems.

Given its central role in human well-being and development, it is ironic that humans do not use this resource in an environmentally sustainable manner. Indeed, we often forget that the earth's freshwater resources are both *finite* and *irreplaceable*. Humans cannot "make" water. Nor are there any substitutes for its many diverse purposes; we cannot make steel with apple juice, or paper with milk. Indeed, even the apple juice and milk would not be possible without freshwater resources! Rather, we must use (and reuse) our existing resources. Unfortunately, our freshwater resources also are very *sensitive* to our actions.

Nor is the absolute quantity of water the only concern. Pollution can drastically reduce the uses to which humans can put freshwater without expensive and time-consuming pretreatment. Thus, virtually all socioeconomic development carries an associated price tag of some degree of environmental disturbance or depletion. The key, therefore, is to attempt to find the correct balance between human demands for water resources, on one hand, and the protection and conservation of the natural environment that supplies this critical resource, on the other hand. This concept is at the very core of the notion of sustainable development, whether considered on a global, regional, or river basin scale.

To this end, this chapter describes the integrated use of tools and techniques for enhancing the environmentally sustainable management and use of freshwater resources, particularly freshwater drainage basins. Howe (1995) suggested that water problems are problems of management, rather than problems of tools or techniques; i.e., environmentally sustainable management of freshwater resources is not so much a problem of the lack of appropriate tools, or guidance on how to use them, but rather a failure to use them in an integrated manner on a drainage basin or watershed scale. Accordingly, this chapter first provides a rationale for fostering the environmentally sustainable use of freshwater resources and then introduces a framework for integrated use of appropriate tools and techniques on a drainage basin scale.

The Problem

There is no doubt that the environmentally unsustainable use of freshwater is an increasing problem* (World Bank, 1996). This reality is leading to a future that is not sustainable on an economic, social, or environmental level. This concern focuses on both water quantity (supply) and water quality (pollution). Further, if this situation continues to be inadequately addressed, the human future can well lead to increasing social conflict, individual misery, and economic as well as environmental costs over the longer term. "Current water management practices and policies have resulted in stark and terrible failures. But the problems we witness today are only an indication of what may lie ahead" (Serageldin, 1995).

The Causes

Among the major reasons that humans use freshwater resources in an environmentally unsustainable manner are the following:

- Water as an issue is not high on most political agendas.
- Authority for different water uses and allocations is typically fragmented among a number of agencies and organizations.
- Available assessment tools and techniques for sustainable water management are applied in an incomplete and/or unintegrated manner.
- There is uncontrolled population growth and urbanization (in some areas, particularly developing countries) and associated increased water needs.
- Difficulty is experienced by many governments in trying to adequately consider myriad social, technical, legal, institutional, economic, and political factors affecting the management and use of freshwater resources in a comprehensive, holistic manner.

*The term *environmentally unsustainable* is used in this chapter to denote human use of water faster than it can be naturally replenished in rivers, lakes, and underground aquifers. It also means polluting water to the degree that its quality declines significantly—often to the extent that drinking it can result in serious illness or death.

Why Environmentally Sustainable Water Management and Use Should Have High Priority on National Agendas

The environment is the fundamental resource

Some suggest that one reason that freshwater resources are managed and used in an environmentally unsustainable manner is that there may be no other choice in a given situation. However, this view of the environment as a competing water use ignores the critical reality that the environment is a fundamental component of sustainable development. As a primary source of needed natural resources, it is the entity that makes development *sustainable* in the first place. It has been stated that "the environment [is] not just another consumptive user of water, but the water resource itself and...degrading the quantity and quality of water in rivers, lakes, wetlands and aquifers can inextricably alter the water resources system and its associated biota, affecting present and future generations" (World Bank, 1996).

Water resources are finite and irreplaceable

As noted above, there is a fixed volume of water on earth, and there is no way of increasing the total volume. We cannot "make" water, and there is no substitute for it for virtually all its purposes. Thus, unsustainable use of this resource can readily lead to the depletion of rivers, lakes, and groundwater aquifers, as well as pollution of the resource to the extent that it can no longer be used by humans without expensive and time-consuming pretreatment and/or transport.

Human population and associated water demands continue to increase

Given that the global water supply is finite at the same time as the world's population is continuing to increase rapidly in many regions, the amount of water available per capita is decreasing. Thus, humans must continue their cycle of water use, abuse, and reuse. If water is used in an environmentally sustainable manner (without reducing sustainable yields or unduly degrading its quality), it can be reused more readily. For example, water used for industrial purposes and treated properly prior to its discharge can be used again downstream for fishing or domestic water supply.

The economic costs of environmentally unsustainable water use can be very high

Another important reason for facilitating environmentally sustainable water use is that not doing so can result in enormous costs over the long term. The reality is that the significant environmental, economic, and social costs often are not adequately considered in the design and implementation of water projects. Some may argue that these costs are not significantly great in many cases; however, experience suggests that these costs are large enough to render many water projects and policies uneconomical over the long term. A large irrigation scheme in west Africa provided an example of this problem. Upon its completion, it was found the project's downstream "costs" exceeded the total estimated benefits. Unfortunately, not adequately considering all significant economic and environmental costs and benefits *before* its completion and implementation ensured the water project was deficient on both economic and environmental grounds (Barbier et al., 1991, 1993).

Long-term costs. The economic costs of environmentally unsustainable water use can vary greatly. Further, some may become apparent only over the long term. When a decreasing availability of water results in decreased levels of activities for those endeavors requiring substantial water resources, the typical costs can include lost production and underutilized investment. As a significant example, the environmentally unsustainable management and use of water resources for agricultural irrigation in the Aral Sea drainage basin decreased water levels and quality in this large inland lake to such an extent that the previously robust fishing industry completely collapsed. As a result, nearly all investment in this industry now lies idle. In addition, the inefficient use of irrigation water in this semiarid region has led to increased salinization and subsequent decreases in agricultural production (United Nations Environment Programme, 1993; United Nations Development Programme, 1995).

It also may be argued that the need for water in some cases or locations is so urgent that water withdrawal for irrigation or other purposes in excess of a drainage basin or aquifer's sustainable yield is justified. However, it also is important to consider the long-term costs of such expedient actions. Depending on the total quantity of available water and the magnitude of its withdrawal, eventually the water system will no longer supply the needed water—even sooner if the water is being polluted as well. The result can be a developed water infrastructure, as well as a variety of water users, both dependent on a water supply that is in fact no longer available.

Unfortunately, the demands of the present may be such that decision makers will decide to use available water resources now and will concern themselves about the economic, social, and environmental consequences later. Nevertheless, if the major consequences of unsustainable water use are identified *before* negative impacts occur, comprehensive water planning and incentive schemes can more readily be justified and implemented to achieve sustainable water use. This approach also provides a solid basis for environmentally sustainable economic and social development over the long term.

Costs of water subsidies. The long-term consequences of the costs imposed on society for supplying water at subsidized prices also contribute to unsustainable water use. An example is the supply of subsidized irrigation water for agricultural production. Such subsidies can appear to have benefits in the short term. Nevertheless, they impose economic costs on the sectors of society paying the subsidy, including the general public. In fact, the funds used for subsidizing water often can be used by society for other purposes, examples being to supply water for other human water uses.

Subsidies can also lead to inefficient water use, and in many cases water could be used elsewhere for more highly valued uses. In the United States, e.g., the federal government subsidizes the cost of providing water for irrigation in California. While the efficiency of irrigation in this region is greater than the global average, a result of the subsidy has been to enhance the growing of water-intensive rice crops in semiarid regions, at the same time as some large urban areas are experiencing water shortages. This approach also has resulted in environmental impacts in the region (Howe, 1995). Therefore, although water subsidies may be justified in some situations, it is important that decision makers calculate *all* the significant costs and impacts of providing subsidies for water resources, before deciding to apply them.

Costs of restricting other water uses. The use of water resources for one activity in such a way that it restricts other uses can impose economic costs on others, by significantly restricting the range of potential water uses. An example is the economic costs imposed by building a dam to generate hydroelectricity, which also restricts the spawning of fish and/or the annual inundation of fertile silt on low-lying farmland (Barbier et al., 1991,

1993; Aylward et al., 1995). Further, as mentioned above, excessive water use for agricultural irrigation can impose water use limitations or shortages in nearby cities.

Costs of water pollution. Pollution of water resources can impose significant social and health costs on water use. For example, agricultural production, although necessary for food production, is characterized by the generation and use of a wide range of water pollutants, including sewage, nutrients, and persistent organic compounds. These pollutants can result in human health problems, as well as restricted use of the water resource. Such impacts obviously impose their own socioeconomic costs. Fish catches can decrease, resulting in reduced income for fishers. People can contract dysentery when consuming contaminated water, imposing human sickness and treatment costs as well as lost economic production costs due to absentee workers.

Once a water body is significantly polluted and so is no longer suitable for many human water uses, restoration of the water quality to a usable standard can be prohibitively expensive or technically difficult. If water is used in a sustainable manner in the first place, such costs can be minimized, and uninterrupted use of the water resource can continue.

Costs for ecosystem damage. More difficult to calculate are the costs of damage to natural ecosystems that can result from unsustainable water use. This damage is caused by such factors as pollution, siltation, increased salinization, and depletion of water flows below critical levels. These factors can result in declining stocks of fish, plants, and other organisms essential to a balanced and properly functioning ecosystem. While impacts such as declining fish catches can be quantified, other environmental costs typically cannot easily be assigned a monetary value. This often results in their being dismissed from consideration in assessing potential environmental impacts of economic development. It is nevertheless critical that such potential impacts of water use be identified before they occur.

Costs of misallocation at the sector level. Where unsustainable use of water occurs due to sectorwide policy, such as subsidies for irrigation projects and irrigation water, a large-scale misallocation of resources can result. A much higher level of economic investment in irrigation and irrigation-based activities can be required than would have been the case if the full costs had been paid for irrigation projects and water. This situation represents a misallocation, in that funds invested in irrigation could have been used elsewhere in the economy for higher-value uses, examples of the latter being medical clinics, growing different types of food, and reforestation.

Another example of policy leading to misallocated resources and unsustainable water use is the situation in which water allocation between the urban and agricultural sectors is based on historical policy that has not been adjusted for changing circumstances. "Thus, in many areas, the much higher value of water in domestic and industrial uses, compared with agricultural uses, indicates a high economic cost of the existing allocation" (World Bank, 1993). As an example, some critics of water policy and projects claim that hundreds of millions of dollars have been wasted on large-scale irrigation schemes (Harrison, 1987).

Applying Tools and Techniques for Environmentally Sustainable Management to Watersheds

As previously noted, from a hydrologic perspective, the watershed is a fundamental management unit for freshwater resources. Further, there are a large number of tools and techniques available for the environmentally sustainable management of freshwater drainage basins. These include diagnostic studies, water quality assessments, hydrologic models, cost-benefit analysis (CBA), and environmental impact assessment (EIA). In fact, almost all these tools and techniques have been available for many years. However, they often are *not* applied in a coherent, integrated manner.

Integrated, environmentally sustainable water management requires more than simply carrying out environmental impact assessment procedures. It requires integration of policy formulation, project appraisal, water management institutions, and law across the breadth and depth of the decision-making process regarding use of freshwater resources (see Fig. 14.1).

The above obviously are not the only elements necessary for environmentally sustainable use of water resources. Further, the difficulties of implementation of this (or any) model or approach to ensure environmentally sustainable water use are fully acknowledged. However, the long-term economic, environmental, and social costs of failing to ensure sustainable water use are such that these difficulties must not be used by policymakers as a reason for rejecting this or similar approaches. An outline for such an approach is presented in the following sections.

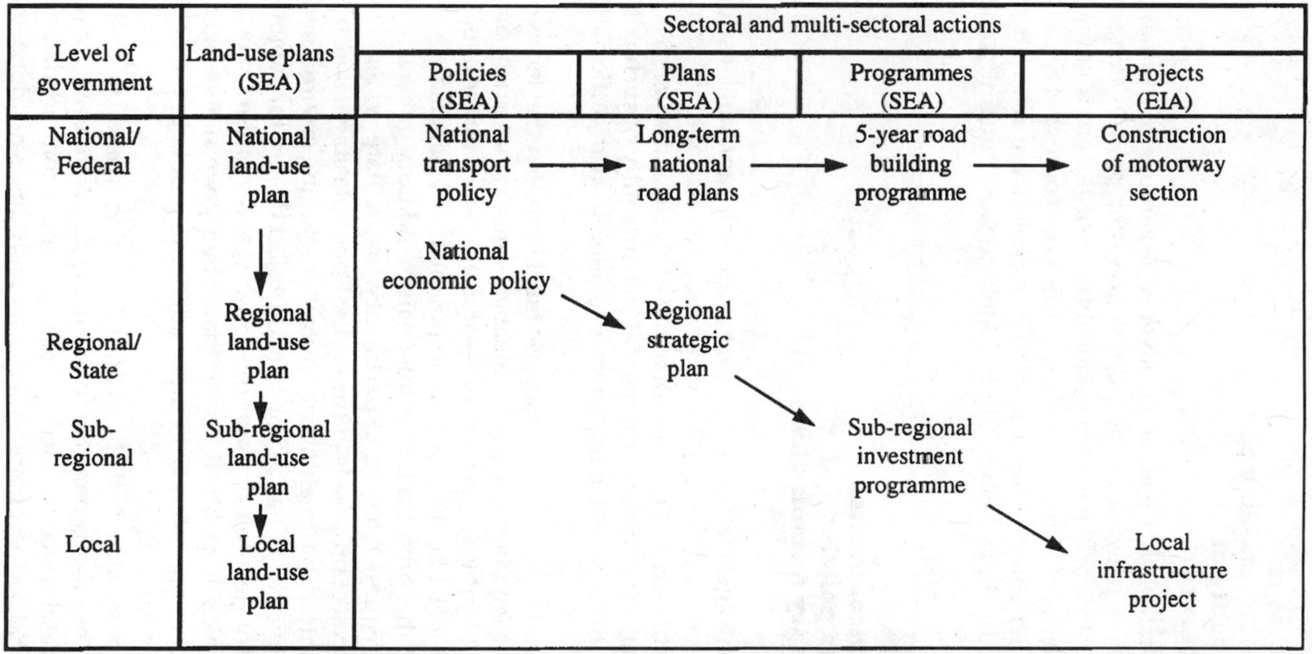

Level of government	Land-use plans (SEA)	Sectoral and multi-sectoral actions			
		Policies (SEA)	Plans (SEA)	Programmes (SEA)	Projects (EIA)
National/ Federal	National land-use plan	National transport policy →	Long-term national road plans →	5-year road building programme →	Construction of motorway section
	↓	National economic policy ↘			
Regional/ State	Regional land-use plan ↓		Regional strategic plan ↘		
Sub-regional	Sub-regional land-use plan ↓			Sub-regional investment programme ↘	
Local	Local land-use plan				Local infrastructure project

This is a simplified representation of what, in reality, could be a more complex set of relationships. In general those actions at the highest tier level (e.g. national policies) are likely to require the broadest and least detailed form of strategic environmental assessment.

Source: N. Lee and F Walsh, Strategic environmental assessment: an overview, *Project Appraisal, Vol. 7., No. 3*

Figure 14.1 Sequence of actions and assessments within a tiered planning and assessment system.

Diagnostic studies on a watershed scale to establish the quantity and quality of freshwater resources

A first step is to establish a watershed level of inventory of available freshwater resources relevant to the particular water policies or uses being considered. This will enhance establishment of the environmentally sustainable level of human water use within the context of readily available freshwater sources. Because both parameters are relevant to human water needs, the inventory should include water quantity and water quality.

Inclusion of environmental aspects of water policies and programs in policy formulation by governments

While government agencies typically formulate relevant policies for freshwater resources and design the water projects, national environmental agencies often do not have input at the early design stages. As a result, the implications of the environmental sustainability of such policies or projects often are not adequately considered at a sufficiently early stage.

Policy design in other economic sectors also must be assessed in terms of its implications for environmentally sustainable water management and use. Land-use policy provides an example—land-use policy that results in serious siltation of water bodies is not environmentally sustainable.

One solution to this problem is to implement mandatory government procedures that request relevant departments to circulate water-relevant policy and project proposals at the earliest possible stage. It also is advantageous if interdepartmental committees (including environmental agencies or ministries) finalize policy and project design for submission to government decision makers. In such cases, the environmental agency (often weaker in political power than other government agencies) should have an opportunity to make adequate inputs to the decision-making process.

An additional possibility is that a central screening agency (e.g., the President's office) may be responsible for ensuring that *all* relevant views have been adequately incorporated in government recommendations regarding water and socioeconomic development. If this is not done, the relevant policy documents could be returned to the lead agency or department for modification. It also may be necessary to develop a formal Cabinet manual, outlining the appropriate procedures.

Integrated economic, environmental, and social policy and project appraisals

Another need in ensuring sustainable development is for integrated economic and environmental appraisals of policy and project proposals. As previously noted, sound decision making requires an accurate, realistic analysis of the economic, environmental, and social costs and benefits of policy and project options. This is simply a reflection of the fact that society has multiple objectives and usually insufficient resources to meet *all* these objectives. Therefore, key decision makers have the difficult task of attempting to balance the often competing demands for water and other resources required by society. The most efficient balancing act requires multiple-objective planning, with decisions based on a *transparent, systematic,* and *integrated* assessment of all three factors—environmental, economic, and social. This assessment will not necessarily provide accurate monetary values of *all* costs and benefits, but it should work to ensure their identification and consideration. The weights to be attached to the different costs and benefits then becomes a political decision.

The elements of such an integrated assessment (Howe, 1995) include

- Cost-benefit analyses from the national perspective
- Cost-benefit analyses from the regional perspective
- Environmental impact analysis
- Social impact analysis (usually without attributing monetary values)

There are many excellent references on cost-benefit, environmental impact, and social impact analyses. The problem remains, however, that such analyses are often not conducted in an integrated manner. Thus, the major problem focuses on the integration of cost-benefit and impact analyses. While this discussion concentrates on integrating environment impact analysis and cost-benefit analyses (CBA), the same concerns are also generally applicable to social impact analysis.

Environmental impact assessment and strategic environmental assessment. The process of environment impact assessment (EIA) was designed to identify negative environmental impacts and provide recommendations for their amelioration. Nevertheless, despite decades of effort, hundreds of publications on how to carry out EIA, and many millions of dollars spent on a global scale, the environmental impacts of economic policies and projects still have not been adequately addressed in many cases. One reason is the lack of comprehensive integration of EIA and project appraisal and implementation (World Bank, 1996). As outlined below, a number of steps are necessary to carry out integrated economic and environmental policy and project appraisals.

Environmental impact assessment should be carried out at both the policy and the project level. At the policy level, it is often referred to as *strategic environmental assessment* (SEA); at the project level, environmental impact assessment. Figure 14.1 highlights the relationship between SEA and EIA.

Conduct environment impact assessment of project or policy development. Note that a comprehensive EIA may not be necessary in all cases. Nevertheless, identification of the key environmental implications of a project or policy in a transparent manner is fundamental in ensuring sustainable development.

The reality, however, is that EIAs of proposed economic development projects often are not carried out in an adequate manner, particularly in developing countries. This is due to such factors as lack of adequate facilities, data, and/or trained personnel to carry out a comprehensive EIA. However, this is only a partial explanation for this situation. In fact, "back of the envelope" scoping EIAs are feasible in many cases; certainly they can be better than no EIA at all. However, another reason for not carrying out EIAs in such situations is that environmental conservation and protection are issues that do not necessarily have a high profile on national development agendas.

This unfortunate reality can result in a vicious circle. On one hand, EIAs are not carried out because conservation and protection of the environment are not seen as being particularly important, particularly in comparison to desired socioeconomic development. On the other hand (and for this very reason), EIAs are most needed to highlight *all* major costs of environmentally unsustainable water use. Indeed, in such situations EIAs can help clarify the economic benefits of environmentally sustainable use of freshwater resources.

Conducting a scoping EIA during the project or policy formulation stage also can help policymakers to narrow project or policy options. By helping to steer project options toward more environmentally sustainable development activities, a reduction in efforts needed at a later date to assess environmental impacts often can be achieved.

Conduct EIA at early stage of project or policy development cycle. Another requirement for addressing integrated economic and environmental management is to carry out an EIA at a sufficiently early stage to actually influence policy choices. This is fundamental to ensuring an EIA is an integral part of the project or policy appraisal cycle. Unfortunately, a typical scenario is that a proposing government department often does not discuss the project (or project alternatives) with the government department responsible for environmental protection and conservation until key decisions have already been made. Thus, by the time the EIA is completed, it is too late to effectively reduce the environmental impacts of a project or policy.

Consideration of project and policy *options* also is important. While there may be opposition to consideration of such options, projects and policies themselves are only a means to an end—not an end in themselves. Thus, consideration of such options can sometimes deliver an equivalent end product at lower cost and reduced environmental impact. As one example, water authorities could reduce water shortages in large cities either by constructing a new dam and pipeline or by repairing leaking pipes. Thus, to minimize environmental damage and encourage sustainable use of freshwater resources requires assessment and consideration of the environmental impacts of *both* options *before* final decisions on project implementation are made.

Calculate economic costs and benefits of EIA recommendations. Another requirement for integrated economic and environmental policy making is calculation of the economic costs and benefits of options (including alternative policies or projects) to improve the environmental sustainability of water use. As an example, an EIA for a proposed irrigation project may result in a recommendation for less groundwater withdrawal and use. In the absence of adequate explanation, this action could result in protests from government agencies and/or the public. However, an EIA can also be used to highlight the economic and environmental costs of environmentally unsustainable water withdrawals from the groundwater aquifer (e.g., the aquifer may be seriously contaminated or depleted within 10 years, resulting in a significant decline in agricultural production). In such cases, it may persuade policymakers to reduce the excessive groundwater withdrawals. This is especially true if the EIA weighs the costs of investing in more efficient water supply and use technologies versus the costs of finding and developing new water sources to replace an aquifer.

Unfortunately, water analysts often do not consider these types of costs and benefits, mainly because they may not have the necessary background or skills to calculate such costs. Yet, as noted above, inclusion of all significant environmental costs and benefits in project and policy design requires that an EIA be an integral part of the project or policy cycle. Environmental assessment analysts should work closely with economists and engineers to calculate environmental costs and benefits and include them with other project costs and benefits.

The tools and techniques for calculating environmental costs and benefits certainly are imperfect (Hufschmidt et al., 1993). Nevertheless, the primary point is that major costs and benefits should at least be *identified* in a given situation. Detailed calculations of major costs and benefit figures can be a second-order priority. A point to emphasize is that if a cost-benefit analysis does not include environmental and social costs and benefits, it likely will not give an accurate picture of the impacts of the project or policy on the well-being of society.

An example of an approach for integrating environmental, economic, and social assessments is illustrated in Fig. 14.2. "A narrow benefit-cost analysis would include only those factors outlined in double lines…by extending the benefit-cost analysis, as indicated by the single-line boxes, the whole array of effects on the natural system, the receptors, and the economy are incorporated" (Hufschmidt et al., 1993).

Risk and sensitivity analyses

No cost-benefit analysis is perfect. This is because many project components have an element of risk, with known probabilities. Additionally, a number of project values are best guesses or estimates. Therefore, it is necessary to calculate the potential impacts of risk and errors in project values on the project's net benefits. The results of such risk and sensitivity analysis should be considered along with the best-estimate results of the CBA.

Economic incentives for environmentally sustainable management and use of water resources

In a given situation, a government's policy formulation process may be relatively comprehensive. Further, policy and project appraisals may integrate economic and environmental factors in a satisfactory manner. Nevertheless, human water use ultimately is driven by human water demand. It is crucial, therefore, that water users be provided with some incentives for practicing environmentally sustainable water use.

A fundamental aspect of the sustainability issue is water pricing. It often is argued that poor or impoverished people cannot afford increased water prices. Nevertheless, the long-term implications of inadequate charges for water are too costly to ignore. To this end, price systems that can address the situation for the poorest can be implemented in most cases. For example, progressive pricing structures that charge a small price for a basic level of water consumption can be implemented. Further, subsidies can be provided for the poorest of citizens, who typically already pay higher prices per cubic meter of water for

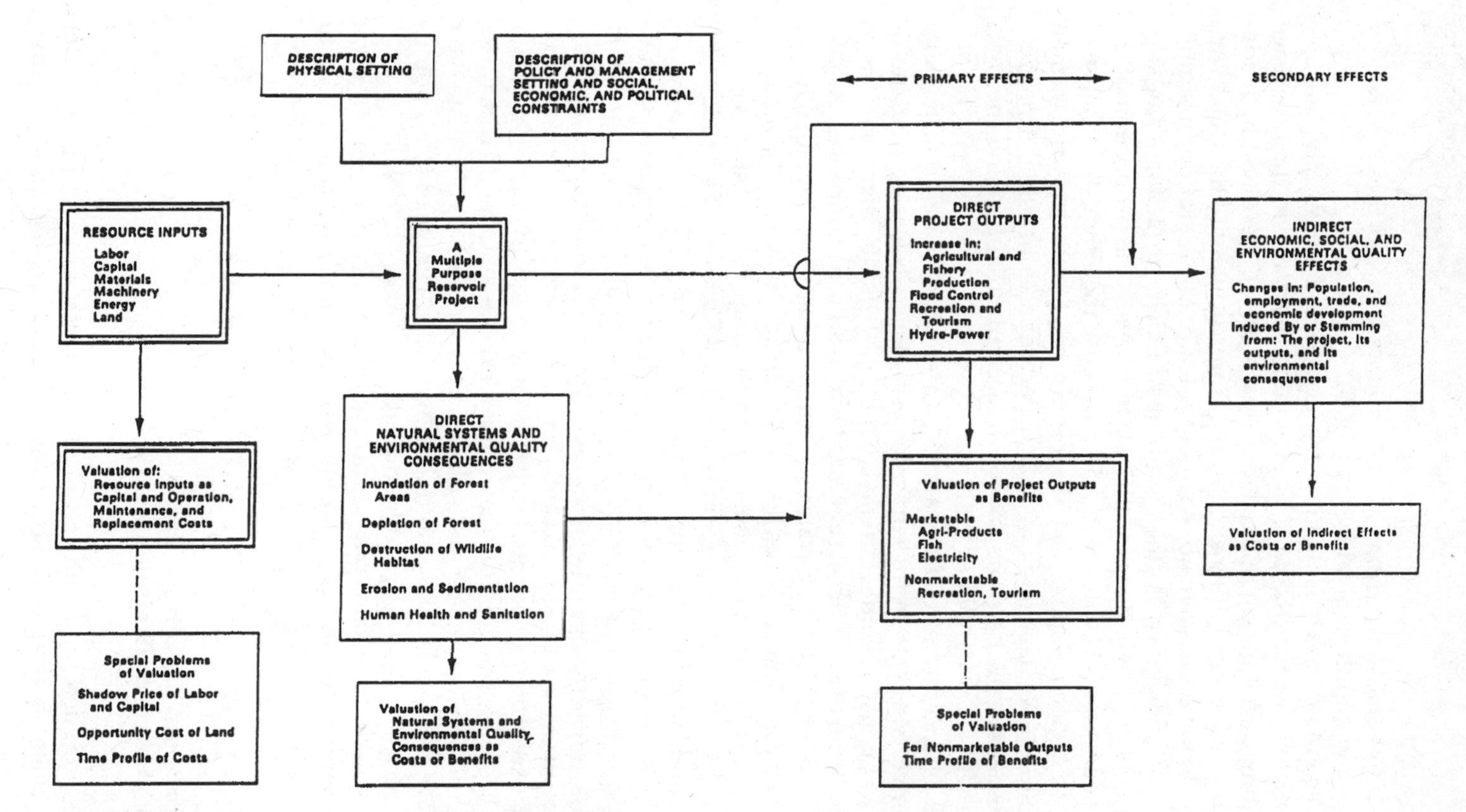

Figure 14.2 A comprehensive examination of benefits and costs of a typical development project. (*Hufschmidt et al., 1993, p. 51.*)

water purchased from urban water vendors than do those served by urban water supplies (Sivalingnam, 1995).

The economic reality is that if revenues received for supplying water to users do not cover the costs of providing the water in the first place (including development of needed infrastructure), there will be inadequate funds for the maintenance and/or improvement of the water supply. Inadequate maintenance ensures a steady deterioration in supply quality and reliability. Further, without sufficient funds for water supply expansion, it will be very difficult to increase the proportion of the population supplied with clean water. Because the single major cause of death on a global scale is waterborne disease, particularly for women and children, the lack of clean, safe water imposes health and production costs that must also be taken into account in appraising the costs and benefits of water policy options.

Lack of incentives to use water in an environmentally sustainable manner clearly abets inefficient water use. The costs arising from inefficient water use include forfeited benefits—benefits that could have resulted from using otherwise wasted water for other purposes. As previously noted, if irrigation water is used in an inefficient manner, it may mean that an urban area does not receive an adequate supply of needed water resources. Another related cost resulting from wasted water is that a higher level of funding for water infrastructure may be required. These funds could be used for other beneficial purposes.

At the same time, providing incentives for environmentally sustainable use of water means more than simply applying water prices to reflect the costs of supplying the water. Rather, the wider sectoral policy context also must provide incentives for environmentally sustainable use of freshwater resources. As previously noted, subsidies paid for agricultural production often encourage excessive or inefficient water use. Similarly, subsidizing pesticide or fertilizer use can encourage their excessive use, potentially resulting in increased runoff of these polluting materials to receiving water bodies (Aylward et al., 1995).

Taxes on pollution also provide an option for conserving and protecting freshwater resources. Nevertheless, the first priority must be to remove disincentives for the *efficient* use of these resources (Meister, 1995).

Institutional structures for management on drainage basin scale

The use of such tools and techniques requires an appropriate water management institutional structure—a structure requiring the integrated use of such tools and the capacity to implement them and their findings.

This institutional framework (Howe, 1995) should ensure

- Coordinated management of surface water and groundwater
- Coordinated management of water quantity and quality
- Provision of incentives for greater economic and physical efficiencies in water use
- Protection of in-stream flow values and other public values related to water systems

This framework requires water management institutions (Howe, 1995) with the following characteristics:

- Capability of coordinating water plans and management procedures with other functional agencies (e.g., agriculture, environment, economic planning, industry)
- Capability of considering a wide range of alternative solutions to water problems, including nonstructural measures and the use of economic instruments (e.g., pricing, taxes, tradable permits, subsidies)
- Separate planning and evaluation from construction and management functions (e.g., agencies responsible for building dams do not also have responsibility for watershed management)
- The multidisciplinary expertise to carry out multiple-objective planning and evaluation (i.e., the integrated approach discussed above)
- Observation of the "subsidiarity principle" in assigning responsibilities to agencies at national, provincial, and local levels (i.e., assign responsibilities to the lowest level consistent with the scale of the water issue)
- Expertise to involve all stakeholders in the planning operation from the beginning
- Incorporation of a reward structure that will stimulate creativity and innovation as well as learning through postconstruction analyses

Developing legislation to address environmentally sustainable management and use of water resources

The use of integrated tools and techniques, including establishment of an appropriate institutional structure for water management, ultimately requires enforcement of relevant laws. To this end, laws to address water and environmental issues ideally should incorporate the following elements:

- Management of water resources on the level of the watershed or drainage basin (or groundwater aquifer)
- Principles of sustainable water management and use
- Integrated management of water and environmental issues together
- Prevention of fragmentation of water allocation and use decision among multiple government departments
- Integrated economic and environmental policy and project appraisals
- Establishment of water management institutions as outlined above
- Establishment of enforceable incentives for environmentally sustainable management and use of water resources

Furuseth and Cocklin (1995) provide an example of legislation that incorporates most of the above-noted elements.

The passage and enforcement of laws incorporating the above-noted elements are necessary expressions of a government's will to achieve the environmentally sustainable management of river basins. In the absence of such legislation, piecemeal application of tools and techniques will continue, thereby worsening water problems over the long term on a global, regional, and subregional basis.

Population Growth and Water Resources

The consequence of uncontrolled population growth is a key element to consider in addressing environmentally sustainable water management, particularly for developing countries. As previously noted, the earth contains a fixed quantity of readily usable water; with increasing population, the per capita availability of this water is decreasing, to critical levels in some regions. Unfortunately, therefore, no matter how comprehensive the integrated economic and environmental management efforts may be, it is unlikely to compensate for population increases in some regions. In these regions, the issue of control of excessive population growth is critical to addressing the environmentally sustainable use of available water resources.

Water Resources and International Concerns

Some of the most difficult issues regarding management and use of water resources focus on international watersheds, which involve water resources shared by two or more countries. In the absence of watershed level of management of water resources, environmentally sustainable use is unlikely. Moreover, it also is unlikely that maximum human benefits can be gained from the available water resource. While integrated use of the tools and techniques discussed above also is applicable on the level of international drainage basins, another critical requirement is the development of binding agreements, outlining the commitments of all riparian governments to integrated watershed management.

Tools such as cost-benefit analyses and environmental impact assessments can assist governments in deciding how to share the environmental and economic benefits of watersheds in an equitable manner. Ultimately, however, management of international drainage basins is a diplomatic issue. A survey of different examples of international watershed management efforts confirms that a major factor in such situations is the commitment of all involved governments. Cooperation between the United States and Canada for the Great Lakes in North America is a relatively successful example of environmentally sustainable management and use of an international watershed. However, this example is an exception, not the rule.

Conclusions

The environmentally sustainable management and use of water resources can be largely based on integrated use of presently existing tools and techniques, rather than require development of new approaches. This, in turn, demands appropriate water management institutions and, ultimately, the legislation necessary to establish such institutions and to ensure that available tools and techniques are used in an appropriate manner. Such laws are an expression of the government's will to achieve environmentally sustainable management and use of water resources. In the absence of such will, little can be readily achieved.

If governments do not have the will to introduce appropriate institutional and legal frameworks, these tools and techniques can be used to highlight the costs of environmentally unsustainable water use to governments as well as to assist in creating the necessary public and government will. In some regions, however, uncontrolled population growth will work to ensure economic, environmental, and social costs of water resources that will be extremely difficult to address in an adequate manner.

It is obvious that there are real difficulties in implementation of an integrated approach of the type outlined above, and this chapter does not purport to diminish their importance. Nevertheless, without the use of such an approach, the world's water problems will bring increasing misery to billions of people.

(*Note:* The discussion presented above represents the opinions of the authors, and does not constitute a recommendation or policy statement of the United Nations Environment Programme (UNEP) or the United Nations system.)

References

Aylward, B., Echeverria, J., and Barbier, E. B. 1995. Economic Incentives for Watershed Protection: A Report on an Ongoing Study of Arenal, Costa Rica. CREED Working Paper No. 3, International Institute for Environment & Development, London, United Kingdom.

Barbier, E. B., Adams, W. M., and Kimmage, K. 1991. Economic Valuation of Wetland Benefits: The Hadejia-Jama'are Floodplain, Nigeria. Discussion Paper DP 91-02, Environmental Economics Centre, International Centre for Environment and Development, London, United Kingdom.

Barbier, E. B., Adams, W. M., and Kimmage, K. 1993. An Economic Valuation of Wetland Benefits. In G. E. Hollis, W. M. Adams, and M. Aminu-Kano (eds.), *The Hadejia-Jama'are Wetlands: Environment, Economy and Sustainable Development of a Shelian Floodplain Wetlands.* International Union for the Conservation of Nature, Goland, Switzerland. pp. 191–209.

Furuseth, O., and Cocklin, C. 1995. An Institutional Framework for Sustainable Resource Management: The New Zealand Model, *Natural Resources Journal,* vol. 35, spring.

Harrison, P. 1987. *The Greening of Africa.* Paladin Press, London, United Kingdom.

Howe, C. 1995. Guidelines for the Design of Effective Water Management Institutions Utilizing Economic Instruments. Report presented at Workshop on the Use of Economic Principles for the Integrated Management of Freshwater Resources, United Nations Environment Programme (UNEP), Nairobi, Kenya, June.

Hufschmidt, M. M., James, D. E., Meister, A. D., Bower, B. T., and Diaxon, J. A. 1993. *Environment, Natural Systems and Development—An Economic Valuation Guide.* Johns Hopkins University Press, Baltimore, MD, p. 50.

Meister, A. 1995. Additional Economic Instruments to Achieve Freshwater Management Objectives. Report presented at Workshop on the Use of Economic Principles for the Integrated Management of Freshwater Resources, United Nations Environment Programme (UNEP), Nairobi, Kenya, June.

Serageldin, I. 1995. Towards Sustainable Management of Water Resources. Technical Report, World Bank, Washington, DC, p. 10.

Sivalingnam, G. 1995. Pricing Water for Sustainable Development. Report presented at Workshop on the Use of Economic Principles for the Integrated Management of Freshwater Resources, United Nations Environment Programme (UNEP), Nairobi, Kenya, June.

United Nations Development Programme. 1995. The Aral in Crisis. Report presented at Tashkent, Uzbekistan.

United Nations Environment Programme. 1993. Diagnostic Study for the Development of an Action Plan for the Conservation of the Aral Sea. Technical Report, Water Branch, UNEP, Nairobi, Kenya.

World Bank. 1993. Water Resources Management. Technical Report, World Bank, Washington, DC.

World Bank. 1996. African Water Resources: Challenges and Opportunities for Sustainable Development. Technical Report, World Bank, Washington, DC, pp. 19–27.

15
Coastal Watershed Management

Jennifer A. Doyle-Breen
Metcalf & Eddy, Inc.
Wakefield, Massachusetts

Introduction

Coastal watershed planning and management have become an increasingly prominent feature in today's discussion of identifying and solving environmental problems. Traditionally, environmental management has focused on managing smaller-scale phenomena, such as wetland fill, point source discharges, and hazardous waste contamination. While each of these areas is an important source of harm to the environment, none of these exists in isolation from the rest of the environment. A coordinated approach which evaluates the effects and interactions of many elements in the landscape is needed to comprehensively and effectively manage natural resources and human activities, and ultimately to ensure that the two can coexist harmoniously.

Evaluating the landscape on the basis of watershed divisions is a natural and logical approach to environmental planning. Watersheds are natural hydrologic boundaries demarcating the area which contributes water to a particular receiving body. Thus, all events within a watershed may potentially affect either the quantity or the quality of water in the receiving water body. Clearly, it is necessary to evaluate activities within these boundaries in order to plan for and manage the quality of the receiving water body. This approach is much more desirable than arbitrarily dividing management activities along political boundaries, such as town lines, city boundaries, state limits, or even national borders. Furthermore, by considering the watershed as a whole, the interactions of all events potentially affecting the water body may be considered rather than individual activities being judged as isolated events which are not affected by, and do not impact, other activities and locations within the watershed.

In the coastal zone, the ultimate receiving water body is the ocean. The coastal watershed includes many fragile resources which exist in a delicate balance at the juncture of the inland and oceanic environments. Many social and economic activities are dependent upon the health of the ocean, including commercial fishing; recreational activities, such as swimming and boating; tourism; and municipal activities, such as management of wastewater. Similarly, drinking water management may be dependent on water bodies and groundwater which are elements of the coastal watershed. All these activities concern not only the health of the environment, but also the economic health of the coastal zone community. The natural resources of the coastal watershed are often threatened by the very uses which depend upon them. Maintaining a balance between the desired uses of the coastal watershed and the delicate coastal environment is a difficult, but necessary task.

Prudent management of coastal watersheds must incorporate many different elements and must consider the vast landscape that contributes to its health. This chapter discusses one possible approach to developing an effective coastal watershed management plan and the benefits which can be realized by considering the entire coastal watershed when management decisions are made. To illustrate the implementation of many of the concepts described, this chapter also discusses the development and implementation of effective, comprehensive coastal watershed management along the southeast coast of the United States: the Georgia coastal zone management program.

Components of Successful Coastal Watershed Management

Developing and implementing a coastal watershed management plan involve a number of components. First, some background work is necessary to establish baseline conditions within the watershed, including habitat mapping and pollution surveys. Next, goals must be established for the watershed. Once a determination has been made regarding the current and desired future characteristics of the watershed, then an overall program must be designed which outlines the specific strategies to be implemented to achieve those goals.

Establishing baseline conditions

The first step in developing a coastal watershed management plan is to assess the existing resources within the watershed. This involves a comprehensive effort aimed at identifying vegetation cover types, including both upland and wetland habitats; wildlife species present in the watershed and their habitat requirements; important fishery areas in the coastal zone; sensitive habitat areas along the coastal zone, including coral reefs, estuaries, or other restricted habitat types; critical habitat areas for threatened or endangered plants or animals; soil and geological features; groundwater locations and quality; and current land uses throughout the watershed. It is also useful to demarcate subwatersheds within the watershed for which planning efforts are being implemented. This baseline data collection lays the groundwork for the planning effort by clearly identifying the important resources within the watershed which should be protected and the current land uses throughout the watershed.

The second step of the early planning phase of coastal zone management planning is to identify pollutant sources throughout the coastal watershed. Sources of contamination in a watershed are numerous and varied, and they depend upon the particular communities and land uses within the watershed. Identifying pollutant sources requires evaluating land uses with the watershed, surveying businesses and communities, and correlating areas of poor water quality with potential upstream contamination sources.

One prominent source of coastal contamination is wastewater discharge. Many coastal communities discharge treated wastewater from a water pollution control plant into the ocean or rivers or estuaries within the coastal watershed. Although these facilities are regulated under the U.S. Environmental Protection Agency's National Pollutant Discharge Elimination System permit program, facilities in some coastal communities may periodically exceed their discharge limits. In some coastal communities, untreated wastewater is discharged to coastal receiving waters during storm events, when flow in combined stormwater and sanitary sewers exceeds their carrying capacity and overflows into rivers, estuaries, or bays. In addition to point sources of wastewater discharge, failing residential septic systems present a nonpoint source of contamination for coastal waters in many communities.

Nonpoint source stormwater runoff within the coastal watershed may introduce contamination to water bodies and waterways.

Management practices utilized at landfills, agricultural lands, dairies, and industrial sites all involve chemicals which may be introduced to receiving waters during storms. Even road maintenance practices, particularly application of sand and salt, can introduce contaminants into the watershed.

Establishing baseline conditions within the coastal watershed should also involve assessing the existing water quality of the ocean, bays, estuaries, rivers, and other water bodies within the coastal zone. Some portions of the coastal zone may be experiencing poorer water quality than others. Identifying these areas will aid in focusing coastal watershed improvement efforts on those portions of the coastal zone which are most in need of help. Conversely, the assessment of baseline water quality conditions may demonstrate that existing water quality is relatively good. In either case, establishing the existing quality of receiving waters within the coastal watershed will serve an important role in developing goals for the coastal zone management plan.

The collection of baseline data throughout the coastal watershed will require the coordination of many different individuals with varying areas of expertise, such as botanists, zoologists, hydrogeologists, petroleum geologists, engineers, fishers, mariculturists, environmental pollution experts, cartographers, geographical information system (GIS) specialists, and water resource management experts. In addition, coordination with many different institutions, organizations, and individuals will be required in order to locate existing sources of information. Universities, government agencies, conservation groups, local planning groups, and private landowners are all potential sources of baseline information.

The information collected during the baseline investigation phase of the coastal watershed planning endeavor can be summarized in a number of ways. Baseline information describing physical features of the watershed landscape can be depicted on maps. For example, a GIS can be used to generate comprehensive maps summarizing the information collected. In addition, this information could be summarized in a database, which allows easy access to particular pieces of information as needed. A comprehensive and efficient approach to summarizing existing potential contaminants in the coastal watershed and existing water quality conditions is to spell out these factors in a checklist (see Table 15.1). Use of a checklist facilitates the identification and development of management practices for all known significant sources of contamination that may be responsible for deteriorated water quality in some or all parts of the coastal watershed. This approach also results in a savings of time and money by focusing future planning efforts on problematic issues and locations, and avoiding further consideration of contaminant sources of

Table 15.1 Watershed Survey Checklist

Information	Significant	Not significant	Unknown
I. General conditions			
A. Changes in available water quantity?			
B. Construction of water diversion or reservoir projects?			
C. Relocation of intakes?			
II. Contaminant sources			
A. Wastewater treatment			
1. Treatment plant effluent discharges			
2. Storage, transport, treatment, disposal to land			
3. Residential septic systems			
4. Commercial/industrial septic systems			
B. Reclaimed water			
C. Urban areas			
D. Agricultural crop land use			
E. Pesticide/herbicide use			
F. Grazing animals			
G. Concentrated animal facilities (feedlots, zoos, etc.)			
H. Wild animal populations			
I. Mines			
1. Active			
2. Inactive			
J. Disposal facilities			
1. Solid waste			
2. Hazardous waste			
K. Logging			
L. Recreation			
1. Reservoir body contact			
2. Reservoir nonbody contact			
3. Watershed activities			

Table 15.1 (*Continued*) Watershed Survey Checklist

Information	Significant	Not significant	Unknown
M. Unauthorized activity			
1. Illegal dumping			
2. Underground storage tank leaks			
3. Other			
N. Traffic accidents/spills			
1. Transportation corridors			
2. History of accidents/spills			
O. Groundwater discharges			
1. Natural discharges			
2. Gas, oil, geothermal wells			
P. Seawater intrusion			
Q. Geologic hazards			
1. Landslides			
2. Earthquakes			
3. Floods			
4. Other			
R. Fires			
III. Growth			
A. Population/general urban area increase			
B. Land-use changes			
C. Industrial-use increase			
IV. Water quality			
A. Changes in raw water quality			
B. Difficulty meeting drinking water standards			

land uses demonstrated to be insignificant. This approach also provides a clear identification of future studies that may be needed to understand the potential significance (if any) for all items checked as unknown.

Watershed goals

An extremely crucial element of any successful coastal watershed management plan is the adequate identification of goals. Goals for the coastal watershed will depend on the needs of the particular community or communities within the area. However, many goals are likely universal. Recreational boating, swimming, fin-fishing, and shell-fishing are common desired uses of many coastal and estuarine waters. Some of these uses may be currently provided in the coastal zone, while some desired uses may be limited due to contamination. Identifying which uses are desired in which portions of the coastal zone is extremely important, since some of these uses may not be compatible.

In addition to recreational uses, communities should identify the specific commercial and industrial uses of the coastal watershed that are desirable and should be encouraged. These uses may bring economic benefits to the communities within the coastal watershed, but can also result in adverse environmental impacts if not properly managed. To identify whether industrial and/or commercial uses of the coastal zone are appropriate goals for the coastal watershed zone, and the desirable locations for these endeavors, is an important element of coastal watershed planning. For example, it may be beneficial to locate industrial activities, which have the potential to introduce contamination to the environment, within subwatersheds that are not planned to be used for swimming. Similarly, agricultural activities and dairy farms should preferably be located in areas where runoff containing pesticides, herbicides, or fecal coliform does not reach recreational or commercial shellfish beds.

Habitat protection is another goal to consider for the coastal watershed. Critical habitat for endangered or threatened species may be present in the coastal watershed; protection of this habitat may be a goal of coastal watershed management. Historically, one goal of environmental management has been wetland protection. Wetland soils and vegetation provide wildlife food and shelter; retain toxicants and heavy metals; retain floodwaters; retain sediments and prevent erosion; provide aesthetic viewing opportunities; and provide recreational opportunities. However, wetland protection by itself may not guarantee protection of all wetland functions if all areas around wetlands are allowed to be developed for residential, commercial, or industrial uses. Upland habitat provides a buffer zone around wetlands, which permits them to function in a healthy manner. In addition, upland habitat corridors between wetlands allow wildlife to travel between wetland habitats. Furthermore, many wildlife species require upland habitat rather than, or in addition to, wetland habitat. Upland habitats also provide many valuable functions similar to those provided by wetlands, such as aesthetic and recreational opportunities. Thus, protection of upland habitats is equal in importance to protection of wetland habitats, and should be considered when goals for coastal watershed management are developed.

Developing strategies for achieving goals

Once the baseline conditions and goals for the coastal watershed have been established, it is necessary to develop strategies for achieving those goals. The overall coastal watershed management program may involve a number of strategies for achieving each of the goals outlined during the planning process. For example, if one goal is to reopen shellfishing beds in an area where the water quality is currently poor due to herbicide application on a nearby golf course, then one strategy may be to require the golf course to investigate alternative methods of controlling weed growth on its property. Similarly, a strategy to protect commercial fishery resources in a particular coastal area would involve placing limits on the number of fish which each fisher can catch in a certain period. Protecting a particular large tract of undisturbed habitat may require establishing a wildlife sanctuary in this area or prohibiting development within and immediately adjacent to this area.

Clearly, the strategies needed to attain the goals for a particular coastal watershed depend on the goals themselves, and will therefore vary from community to community and watershed to watershed. However, all the above examples of strategies which may be implemented to achieve coastal watershed goals have a common element: the need for adequate legislative mechanisms to control activities and protect resources. Although specific goals and strategies may vary among communities and among coastal watershed areas, all will require an assessment of the existing regulatory institutions. The assessment of existing regulations should evaluate whether there is sufficient regulatory power to implement strategies which will aid in achieving the watershed goals and which will prevent actions that thwart attainment of these goals. Examples of legislative actions which may be necessary include establishing wildlife sanctuary areas, developing new zoning ordinances, and developing adequate stormwater management programs. In addition, it may be necessary to introduce new regulations, such as those providing the following functions:

- Limitation of alteration to wetlands and buffer zones
- Protection of endangered-species habitat
- Limitation of fill and construction in flood zones
- Management of hazardous materials in the watershed
- Management of pesticide and herbicide application

Other strategies for addressing problems within the watershed may not involve creating new regulations, but may instead require municipalities to implement new management techniques for public works projects. For example, it may be desirable to cease the application of road salts adjacent to particularly critical habitat during winter months, and instead to post signs warning of potentially slippery road conditions. Another strategy might entail implementing a recycling program to cut down on the amount of waste being deposited at the local landfill. The potential strategies for achieving coastal watershed goals are as varied as the goals themselves.

It is important to consider financial realities when strategies are formulated for obtaining watershed goals. The ideal solution to a problem within the watershed will achieve nothing if it is so expensive to implement that budgetary constraints prevent its realization. Often simple, inexpensive measures result in the same water quality benefits as more costly measures, if implemented throughout the watershed. For example, installing grates over drains to prevent debris from entering streams and planting grass strips around impervious surface to collect sediments are simplistic actions which can result in substantial improvements in downstream water quality.

Another consideration regarding the development of strategies is phasing or prioritization of the various components of the overall coastal watershed management plan. There may be some components of the plan which can be implemented immediately, without waiting for legislative approval of funds or regulations. The implementation of these strategies can be initiated first, and this may achieve benefits before later strategies are finalized and approved. It may be prudent to pursue smaller goals first, so that some benefits are achieved while more difficult issues are still being addressed. For example, installation of grates over drains would require little prior planning or funding allocation, whereas identifying the source responsible for contamination of swimming waters may require years of study and planning before a solution is identified and then implemented. Similarly, goals for the coastal watershed should be prioritized. If funding is limited, it is worthwhile to concentrate large sums of money in areas where they will achieve the greatest benefit.

Public outreach

Public outreach is a key component of any successful coastal watershed management program. First and foremost, this component is important because the watershed and its health are the responsibility of all residents. Soliciting input from the public is important for identifying goals and developing realistic strategies for problem solution. Often, local communities and residents are more knowledgeable about issues in the land surrounding them than regulators or public officials are. Once the public understands the importance of coastal watershed management and the economic and environmental benefits that can be achieved by addressing watershed problems, a sense of environmental stewardship will be fostered which will aid in achieving the goals established for the watershed. Furthermore, educating the public about watershed goals and problems will help to build consensus regarding the strategies that should be implemented to achieve goals and resolve problems. Achieving this consensus early in the planning process may avoid delays in implementing solutions later.

Monitoring

The final component of successful and effective coastal watershed management is monitoring. Implementing coastal watershed protection measures is worthless if there is no assessment of the success or failure of these measures. The goals established at the commencement of the coastal watershed management process should be evaluated periodically, and an assessment should be made of the efficacy of the implemented strategies in achieving these goals. Strategies which do not appear to be solving problems or achieving the set goals should be reevaluated. This process may require modifying strategies or developing new strategies altogether. Similarly, the goals for the coastal watershed may themselves change over time. The coastal zone management (CZM) process is a dynamic one that requires the flexibility to modify plans as time goes on. For each strategy that is established, an appropriate time frame and mechanism should be established that will permit a timely and comprehensive review of the achievements to date.

Case Study: Georgia CZM Plan

To illustrate the complex and intricate nature of watershed management, it is instructive to evaluate the efficacy of the coastal zone management program developed by the state of Georgia. The watershed-based Georgia CZM program included

- Marine fisheries management
- Petroleum exploration and production coordination
- Development of regulations to protect Georgia's valuable marine and estuarine renewable biological resources
- Promotion of mariculture
- Implementation of public education activities to facilitate ecotourism success

As the CZM program developed (in the early 1980s), Georgia faced increased pressure to develop and produce new offshore petroleum reserves to provide revenues for the state. The near-shore waters of Georgia are among the most biologically productive and biologically diverse in the United States. These fragile, highly productive areas had been threatened by coastal tourism development, coastal energy production, and excessive fishing from both marine and sport fishing interests.

An assembly and integration of baseline data that had been developed over many years allowed Georgia to recognize the fragile nature of renewable resources. The initial evaluation of the data resulted in the formulation of a monitoring and evaluation effort. During the development of the supplemental field data, Georgia initiated development of a draft CZM plan. Final adoption and implementation of the written plan included a wide diversity of components.

Key components of the CZM plan included proposed regulations for all the sand-sharing system (the dynamic sand dunes and shoreland) along Georgia's coastline. The state also adopted goals, policies, and regulations aimed at establishing fishing seasons and catch limits. Furthermore, protection zones were delineated around the estuarine bays, which serve as a nursery for many species of marine life. Georgia's CZM plan also involved effective public education and included measures to promote ecotourism and coastal preserves. The Georgia CZM effort resulted in the creation of numerous wildlife sanctuaries and the development of new state and federal regulations specifically aimed at protection of the coastal zone (see Table 15.2).

To accomplish this collection of watershed-based management strategies, a diverse team of planners, scientists, engineers, developers, petroleum geologists, marine and estuarine parks development staff, environmental economists, mariculturists, and environmental pollution experts worked closely with the public, as well as with many government agencies, to develop and implement an acceptable plan. The diverse nature of the program required expertise in conversing with many people with many different levels of education in order to develop a consensus on the need for environmentally effective CZM.

Table 15.2 Accomplishments of the Georgia Coastal Zone Management Program

Sanctuaries and Preserves Created
■ Gray's Reef National Marine Sanctuary
■ Duplin Estuary National Estuarine Sanctuary
■ Richard J. Reynolds Wildlife Preserve on Sapelo Island
■ Ossabaw Island Project Foundation Reserve on Ossabaw Island, which is a private preserve for both upland and estuarine plants and animals and their aesthetic appreciation
■ Larkin Foundation Preserve on St. Catherines Island, which is a private preserve to promote the continued breeding success of warm-blooded, terrestrial animals threatened for extinction
■ Creation of the Cumberland Island National Seashore (a part of the U.S. National Parks System)
Legislation Developed
■ Legislation limiting the amount of shrimp that could be caught, the locations where finishing was prohibited, and the seasons acceptable for fishing elsewhere
■ Legislation requiring restrictive permitting for any planned development activities in the sand-sharing system
■ Legislation requiring restrictive permitting for any activity in any coastal, estuarine, or marine wetland
Other Accomplishments
■ Development of a state policy to manage potential energy (oil and gas) development so as to prevent accidental pollution of coastal areas within Georgia's coastal zone
■ Development of federal restrictions related to commercial fishing gear to prevent the accidental catch of endangered sea turtles by shrimp fishers
■ Establishment of policies to promote ecotourism development
■ Creation of two coastal museums to interpret early coastal history and human settlement
■ More strictly regulated tourism and industrial development
■ Development of policies promoting mariculture
■ Publicly accepted expansion and development of coastal ports

Benefits of Successful Management

The benefits of prudent coastal watershed management are numerous. First, many of or all the goals for the watershed will be realized. People may enjoy safe swimming areas which were previously absent. Industrial or commercial endeavors may be encouraged in certain portions of the watershed, resulting in economic benefits for the community, while important resources are simultaneously protected. Overall, natural resources will be safeguarded for future generations, and a healthy environment will be encouraged and maintained.

Financial savings will almost certainly result from carefully planned management of the coastal watershed. Identification and resolution of watershed problems at their early stages are much less costly than the development and implementation of programs to fix full-fledged environmental catastrophes. Often, simple nontechnological approaches will serve to reverse negative trends in the watershed at the early stages. In addition, thoughtful planning will likely eliminate the occurrence of some problems altogether. Prevention is the least costly way to avoid many problems and maintain desired uses within the watershed.

Other important outcomes of successful coastal watershed management include the creation of a sense of environmental stewardship and bolstering of stakeholder confidence in the political infrastructure responsible for protecting natural resources. Educating and involving the public will foster public recognition that everyone is an important component of the process that protects the environment, and that everyone shares the responsibility to protect the environment. As the public recognition of these facts increases, the likelihood of attaining watershed goals increases as well.

Conclusions

The coastal environment includes many fragile natural resources and is also the site of many industrial, commercial, and tourism endeavors which provide economic benefits for coastal communities. To balance these uses of the coastal watershed with the very resources upon which these uses depend is a difficult task.

It is essential that management of coastal natural resources integrate the many different components in the coastal watershed rather than focus on individual, isolated elements in the landscape. Focusing on individual activities within the watershed rather than viewing them as a whole will result in a patchwork approach to environmental protection, which achieves few positive results and does not lend itself to achieving watershed-wide goals.

Effective, comprehensive coastal watershed planning can be accomplished in an organized manner by gathering baseline information, establishing goals, and devising strategies for achieving those goals. Identifying goals and strategies requires the input of many different partners within the watershed, including scientists, residents, fishers, businesses, municipalities, legislators, developers, and government officials. Periodic assessment of the coastal watershed program allows adjustments to be made when and where necessary, so that ineffective strategies are replaced by new, innovative approaches in order to attain the goals for the watershed.

Proceeding through the coastal watershed planning process in a logical manner, such as that outlined here, will ensure that activities within the coastal watershed are coordinated so that their synergism results in an environment in which natural resources and human activities can continue to coexist for many years to come. The effort required to plan for management of coastal watersheds is well worth the result: a healthy environment that can provide enjoyment while also supporting human uses.

16

Anatomy of a Drought

Gerald J. Kauffman
Water Resources Engineer
Water Resources Agency
for New Castle County
Newark, Delaware

Introduction

The drought of 1995 was a significant event which severely reduced the availability of public water supplies in northern New Castle County (Delaware). The year-long drought was characterized by an extended period of deficit precipitation which resulted in near-record low stream flows in the Christiana watershed, declining groundwater levels, and a depleted Hoopes Reservoir (see Figs. 16.1 and 16.2). The dry conditions reduced the availability of stream flows for water supply purposes, nearly resulting in water shortages just after Labor Day in 1995. Fortunately, record rainfall during October averted the water emergency, and water conditions returned to normal.

The drought created hardships for the citizens and businesses of New Castle County, yet it highlighted opportunities for improvement of the regional water supply system. The following summary discusses the chronology of the drought and various drought management activities by state, county, local, and private water agencies. More importantly, this chapter discusses the lessons and actions which are needed to improve watershed management activities for future droughts.

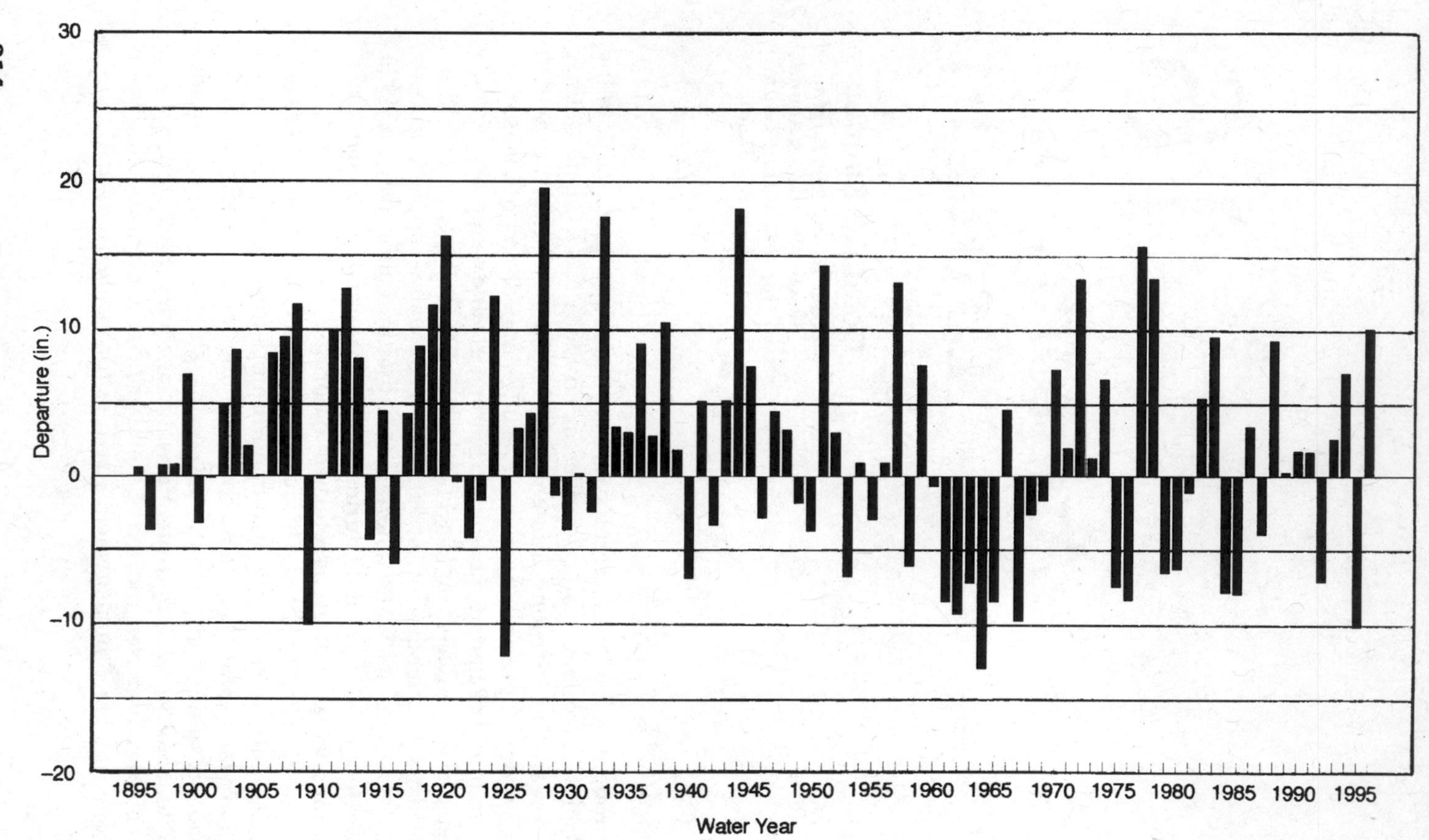

Figure 16.1 Average annual precipitation in northern New Castle County, 1895 to 1995.

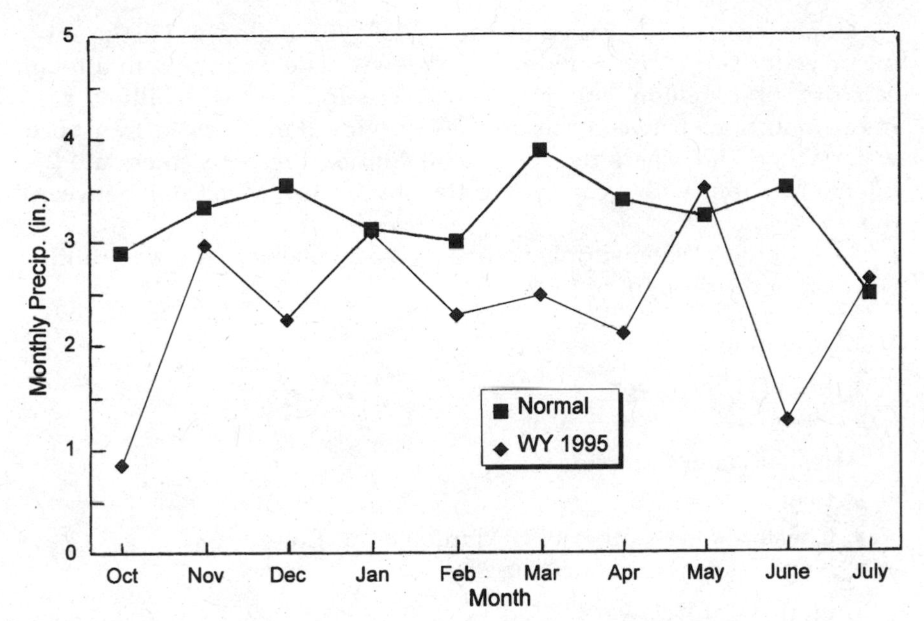

Figure 16.2 Precipitation in northern New Castle County, water year 1995. (*U.S. National Weather Service.*)

Public water supply system

Public water supplies in northern New Castle County are provided by five major public and private utilities. The city of Newark, city of Wilmington, and city of New Castle operate public water utilities which supply water in and around their respective municipalities. The Artesian Water Company and United Water Delaware are private investor-owned utilities which supply water to the suburbanizing areas of unincorporated New Castle County. Average water production for the five water suppliers ranges from 61 Mgal/day in March to 78 Mgal/day in June.

The five utilities obtain water from surface water in the Christiana drainage basin, groundwater, and interconnected supplies. Surface water supplies provide over 70 percent of the public water to New Castle County from Brandywine Creek, Red Clay Creek, White Clay Creek, and Christina River. Minimum 7Q10 in-stream flow requirements are in place along White Clay Creek at the Newark water treatment plant (WTP) and the United Water Delaware Stanton Filter Plant. Groundwater provides 30 percent of the public water supplies from the Cockeysville, Columbia, and Potomac aquifers. Supplemental supplies are provided to Artesian and United Water via interconnections.

Hoopes Reservoir is owned by the city of Wilmington and is the only major water supply impoundment in New Castle County with a total capacity of 2 billion gal and a usable supply of 1.5 billion gal. Interconnections between the utilities provide the ability to transport water when and where needed in the county. Presently, there are 23 interconnections with a capacity to transfer up to 8 Mgal/day between the utilities.

The following water supply sources were available to the water utilities during the drought of 1995:

1. Wilmington
   - Brandywine Creek
   - Hoopes Reservoir
2. Artesian Water Company
   - Wells
   - Chester Water Authority (PA) interconnection
   - New Castle interconnection
3. United Water Delaware
   - White Clay Creek and Stanton WTP—Hoopes Reservoir
   - Christina River and Smalley's Pond
   - CWA (PA) interconnection
   - Artesian Water Company connection
   - Wilmington interconnection
4. City of Newark
   - White Clay Creek and surface WTP
   - Wells
   - United Water interconnection
   - Artesian Water interconnection
5. New Castle Board of Water & Light
   - Wells

Chronology of the Drought

October 1994 through May 1995—Prologue

The drought of 1995 actually started in the fall of 1994 when, after a wet summer, for several consecutive months there was less than normal precipitation. The dry conditions continued through the winter of 1995, and only one snowstorm occurred during January. The lack of snow

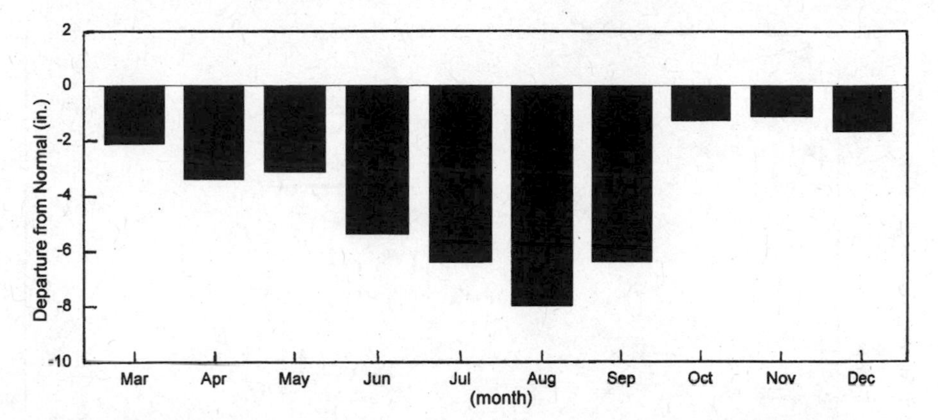

Figure 16.3 Precipitation at New Castle County Airport, 1995. (*U.S. National Weather Service.*)

resulted in decreased groundwater recharge and depleted stream baseflow levels during the spring snowmelt period. The lack of rain and snow resulted in April mean flows for the Red Clay and White Clay Creeks which were the lowest and third-lowest on record, respectively. By Memorial Day, the monthly precipitation at New Castle County Airport measured below normal for seven of the eight previous months of water year 1995, which began in October 1994 (see Fig. 16.3).

June 1995—decline

During June 1995, the low-water conditions continued. The cumulative precipitation deficit for water year 1995 approached 7 in, which was comparable to the drought year of 1966. The city of Newark considered the possibility of voluntary water restrictions due to declining stream flows at the White Clay Creek surface water treatment plant. The Department of Natural Resources and Environmental Control (DNREC) and the Delaware Geological Survey (DGS) were requested to prepare a press release notifying the public about the continuing dry conditions.

July 1995—advisory

During July 1995, drought management activities accelerated due to the continuing low-water conditions. The hottest day ever in Wilmington occurred on July 15, when the public water suppliers recorded a peak water demand of 87 Mgal/day. The governor's drought advisory committee convened and recommended on July 19 that the governor issue a

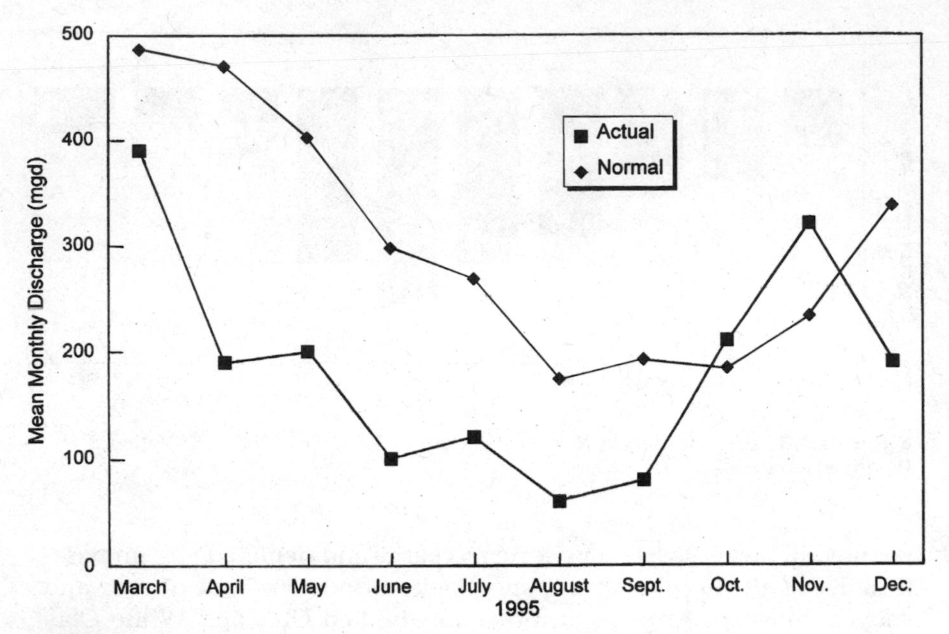

Figure 16.4 Mean monthly discharge (Mgal/day) at Brandywine Creek at Wilmington, Delaware. (*U.S. Geological Survey.*)

drought advisory with voluntary water restrictions for northern New Castle County. Water conditions continued to decline at the end of July when the cumulative precipitation deficit for water year 1995 exceeded 10 in and the stream flows reached 7Q5 (see Fig. 16.4). According to DNREC indicators, northern New Castle County was officially in "drought" status.

August 1995—warning

During August 1995, water conditions continued to deteriorate. On August 18, the city of Newark ordered voluntary water restrictions in its service area. The governor's drought advisory committee, on August 22, recommended that the governor declare a drought warning with possible mandatory water restrictions in northern New Castle County after a 20-day public notice period. Public water suppliers were requested by the governor's drought advisory committee to review their emergency contingency plans. On August 28, the city of Newark ordered mandatory water restrictions due to declining stream flows and the prospect of increased water demand during the new semester at the University of Delaware.

August concluded with worsening water conditions. Ten of the preceding 11 months recorded deficit precipitation. White Clay Creek and Brandywine Creek reached 7Q10 and recorded the third-lowest flows in 40 and 50 years, respectively (see Fig. 16.4). Hoopes Reservoir was observed at 65 percent of capacity. The city of Wilmington continued to release water from the reservoir into Red Clay Creek to supplement flows at United Water Delaware's Stanton Filter Plant. Groundwater levels in shallow monitoring wells continued to decline due to lack of recharge.

September 1995—emergency

September 1995 represented the trough or most critical period of the drought (see Fig. 16.5). On September 1, United Water Delaware ordered mandatory water restrictions due to declining White Clay Creek flows. With the drought emergency declaration, the minimum flow standard at the United Water Delaware Stanton Filter Plant was waived by the Delaware River Basin Commission (DRBC). On Labor Day, September 4, the city of Wilmington ordered mandatory water restrictions. On the same day, the governor of Delaware declared a drought emergency with the intent to declare mandatory water restrictions in northern New Castle County after a 7-day public notice period. Artesian Water Company followed with mandatory water restrictions

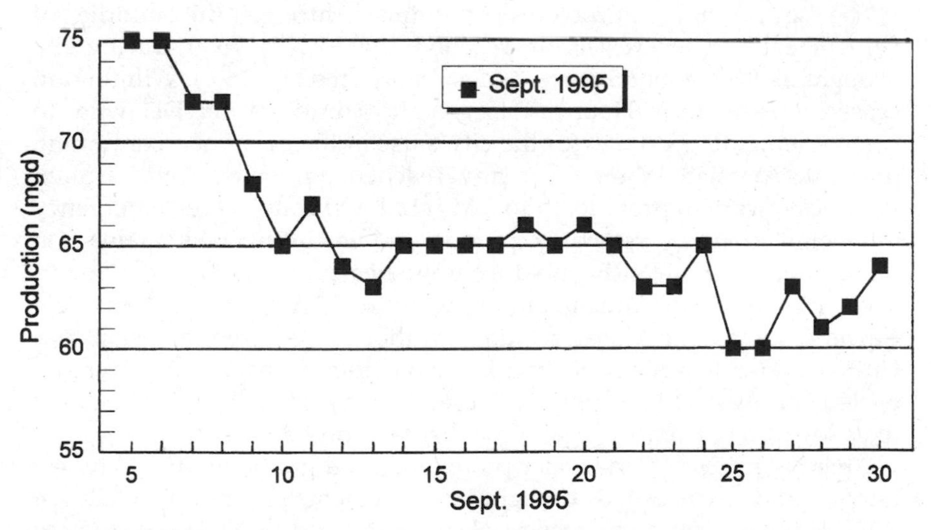

Figure 16.5 Public water supply production, northern New Castle County, Delaware. (*Water Resources Agency for New Castle County.*)

on September 5. On September 7, during a governor's drought advisory committee meeting, the Delaware Geological Survey reported that the water conditions index was in the "water shortage" range. DNREC requested industries to submit conservation plans to reduce water use by 25 percent. The Water Resources Agency for New Castle, Delaware (WRANCC) and United Water Delaware (UWD) petitioned Wilmington to continue releases from Hoopes Reservoir for use by UWD. The WRANCC was requested to track daily supply and demand of the five water suppliers. United Water reported that unless rainfall or supplemental flows from Hoopes Reservoir were received, a water shortage in its service area was imminent.

The week of September 8 through 15, 1995, was the nadir of the drought. Some areas had not received rain for 3 to 4 weeks. Brandywine Creek approached the 7Q10. White Clay Creek approached low levels not seen since the drought years of the early 1960s. Wilmington experienced difficulty in capturing the historically low stream flows along Brandywine Creek to meet water demand in the city. United Water Delaware curtailed water withdrawals to 30 percent of allocated capacity and constructed a low dam along White Clay Creek to capture tidal flow at the Stanton Filter Plant. Due to lack of stream flows, sodium and chloride levels in White Clay Creek continued to rise above the allowable limits set by the Delaware drinking water standards. The city of Newark ceased withdrawals at its White Clay Creek water treatment plant. The cumulative precipitation deficit for the water year reached −12 in, which had been exceeded only three times in 100 years.

The severe water problems continued through the middle of September. Hoopes Reservoir reached the lowest level during the drought at −7.5 ft, or 55 percent of capacity (see Fig. 16.6). Wilmington ceased releases from Hoopes Reservoir to United Water Delaware to reserve remaining storage for the city's use in the event of a continuing drought. Artesian Water Company reached agreement with United Water Delaware to provide up to 2 Mgal/day through three emergency interconnections by reversing pumps and installing new mains. On September 11, the DNREC held a public hearing regarding the governor's intention to declare mandatory water restrictions in northern New Castle County. The Chester Water Authority required Artesian and United Water to reduce their takes from the Pennsylvania interconnected supply by 10 percent. By September 15, water demand declined to 65 Mgal/day from 75 Mgal/day observed on Labor Day.

After September 18, the water problems eased gradually due to timely rainfall and coordinated drought management actions. The city of Wilmington approved emergency releases to United Water Delaware from a refilling Hoopes Reservoir. Industries utilized emergency surface water

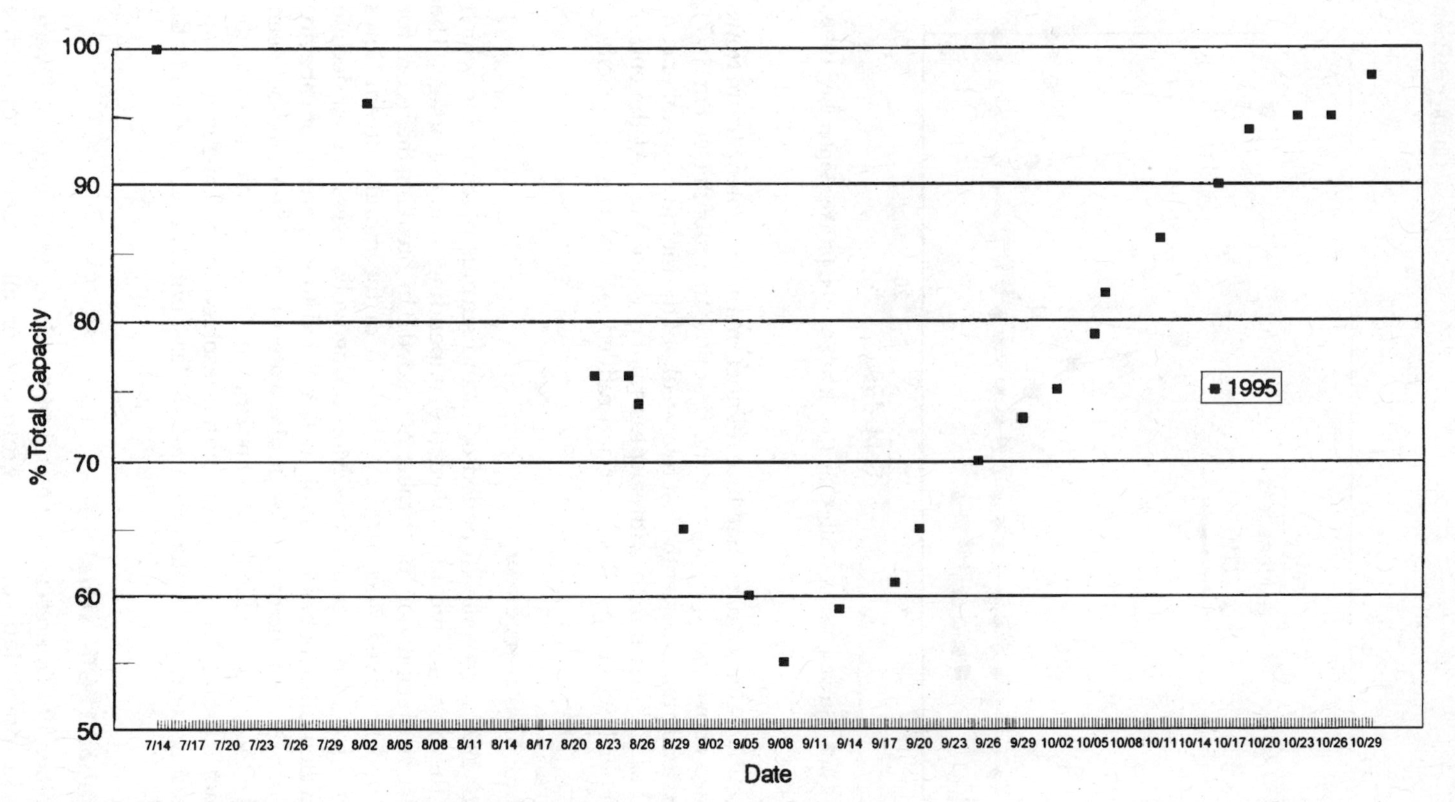

Figure 16.6 Capacity of Hoopes Reservoir for July through October, 1995. (*City of Wilmington and Water Resources Agency for New Castle County.*)

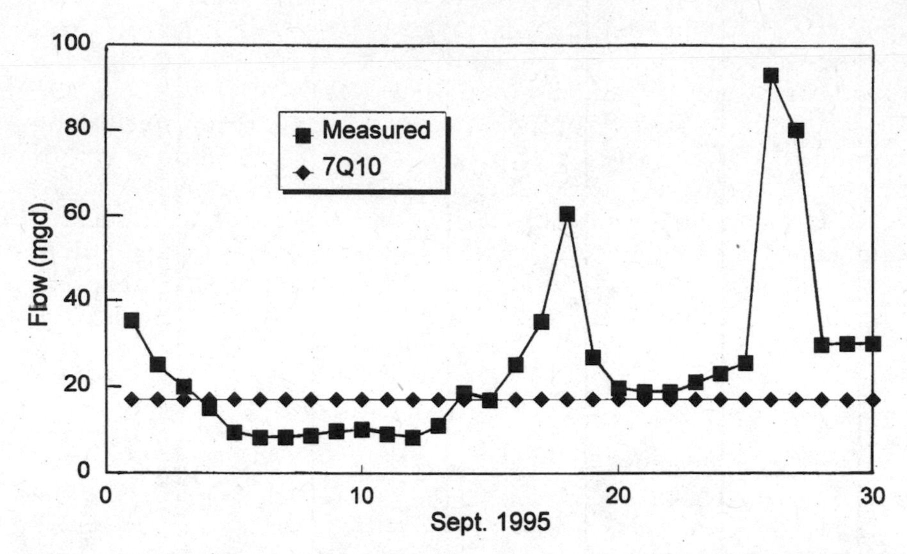

Figure 16.7 Stream flow at White Clay Creek at Stanton intake, September 1995.

and groundwater supplies to reduce reliance on public water. Stream flows increased above the 7Q10 flow for the first time in a month (see Fig. 16.7). Due to steady rainfall over the last two weeks of the month, September concluded as the first month with surplus rainfall since May. At the end of September 1995, Hoopes Reservoir had refilled to 73 percent of capacity.

October 1995—recovery

October 1995 represented a period of record surplus precipitation which eased the drought and forestalled the impending water shortage. The cumulative precipitation for October exceeded 8 in, breaking the record for the period observed since 1895. At the end of the month, stream flows increased to nearly normal levels. Monitoring wells reversed their decline and leveled off. Hoopes Reservoir was refilled to 95 percent of capacity. Water demand for the five major public water suppliers hovered between 60 and 62 Mgal/day, which is about normal for October. On October 26, the governor's drought advisory committee resolved to continue monitoring the steadily improving water conditions at least until the end of the month.

November 1995—relief

November 1995 represented the end of the drought emergency. Due to improved precipitation, stream, groundwater, and reservoir levels, the governor's drought advisory committee on November 3 agreed to recommend suspension of the drought emergency and relaxation of the mandatory water restrictions in northern New Castle County. On November 6, 1995, Governor Carper convened a press conference along the banks of Brandywine Creek. At the press conference, the governor signed an executive order suspending the drought emergency and relaxing mandatory restrictions. November concluded with surplus rainfall with a drought warning and voluntary water restrictions still in effect for northern New Castle County. By Thanksgiving, the public and private water suppliers had resumed normal operations.

January 1996—epilogue

An appropriate epilogue to the drought of 1995 occurred with the blizzard and floods of January 1996. On January 8, 1996, the "blizzard of the century" dropped a record-tying 22 in of snow at Wilmington, representing millions of gallons of equivalent moisture. Ten days later, a combination of balmy weather and rainstorms melted the snowpack, resulting in flooding along the streams in Christina Basin. The above-normal precipitation from these events reversed the decline in water conditions, resulting in recharge and recovery of groundwater levels (see Fig. 16.8).

On February 9, 1996, the governor's drought advisory committee convened and recommended terminating the drought warning, which had

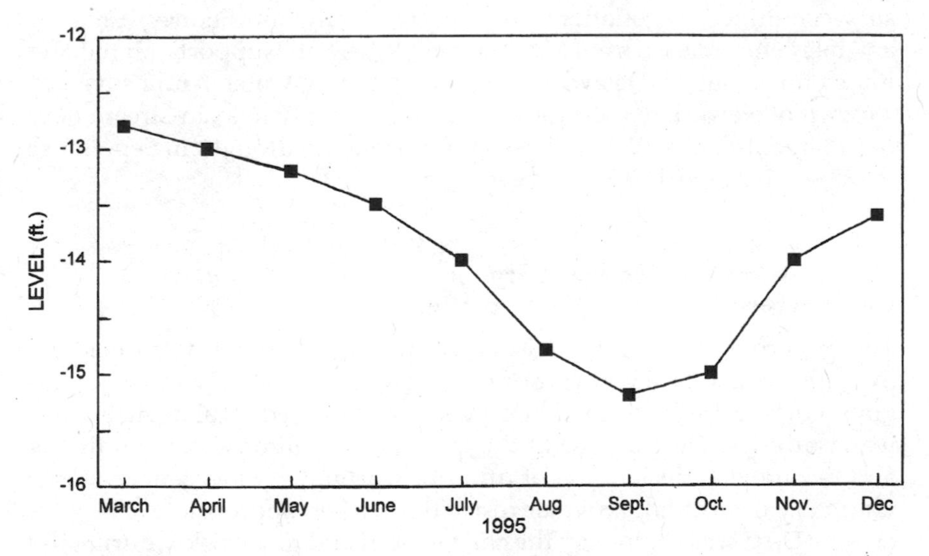

Figure 16.8 Level of water table well in New Castle County, Delaware, for March to December 1995. (*Delaware Geological Survey*.)

been in effect for the last 7 months. At the end of February, hydrologic indicators registered above normal, indicating the end of the drought.

Drought Coordination

Drought coordination activities were administered by three committees which worked to avert a water shortage in New Castle County. The responsibilities of these committees included monitoring of water conditions, coordination of water supplies between the utilities, and declaration of drought warnings and water restrictions. The activities of the drought coordination committees are described below.

Christina Basin Drought Management Committee (CBDMC)

This committee consists of state and county agencies and public and private water utilities in the Pennsylvania and Delaware portions of Christina Basin. The purpose of the CBDMC is to coordinate interstate drought management activities within Christina Basin in Pennsylvania, which provides 40 percent of the water for Chester County, and Delaware, which provides 70 percent of the water for New Castle County. Significant actions by the CBDMC during the drought of 1995 included recommendations to convene the governor's drought advisory committee, negotiations between Delaware and Pennsylvania for an emergency release from Marsh Creek Reservoir, support of interstate interconnections to Delaware from the Chester Water Authority, and support of Newark request to DRBC to temporarily relax in-stream flow requirements along White Clay Creek during the drought in September (see Figs. 16.9 and 16.10).

Governor's Drought Advisory Committee

The governor's drought advisory committee (DAC) was formed by executive order by the governor of Delaware. The DAC consists of the governor's office, state cabinet officials, and invited local agencies and water utilities. The purpose of the DAC is to monitor water conditions and recommend declaration of drought warning or emergency and/or voluntary or mandatory water restrictions when appropriate. The governor's DAC was chaired by the chief of staff and met weekly during the drought of 1995.

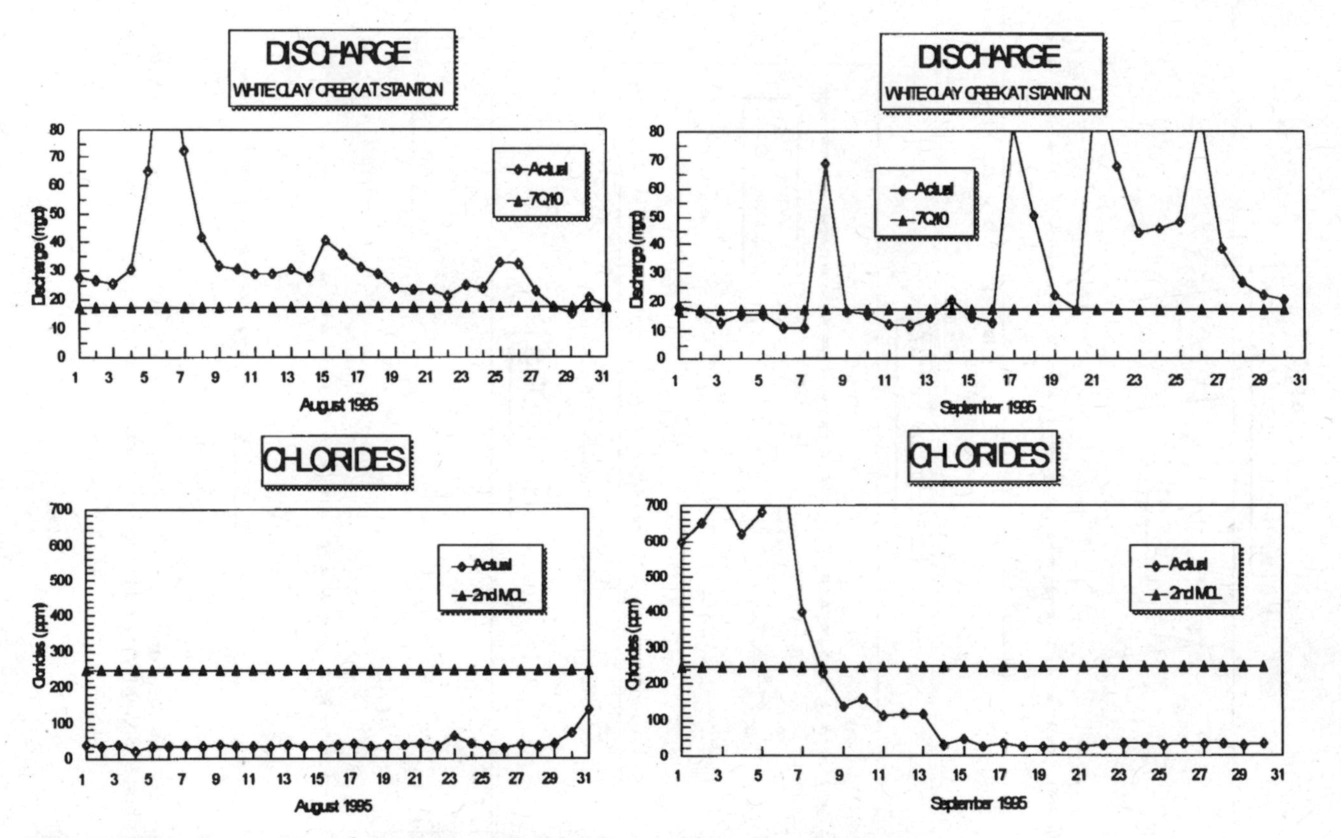

Figure 16.9 Comparison of discharge and chlorides at White Clay Creek at Stanton, August and September 1995.

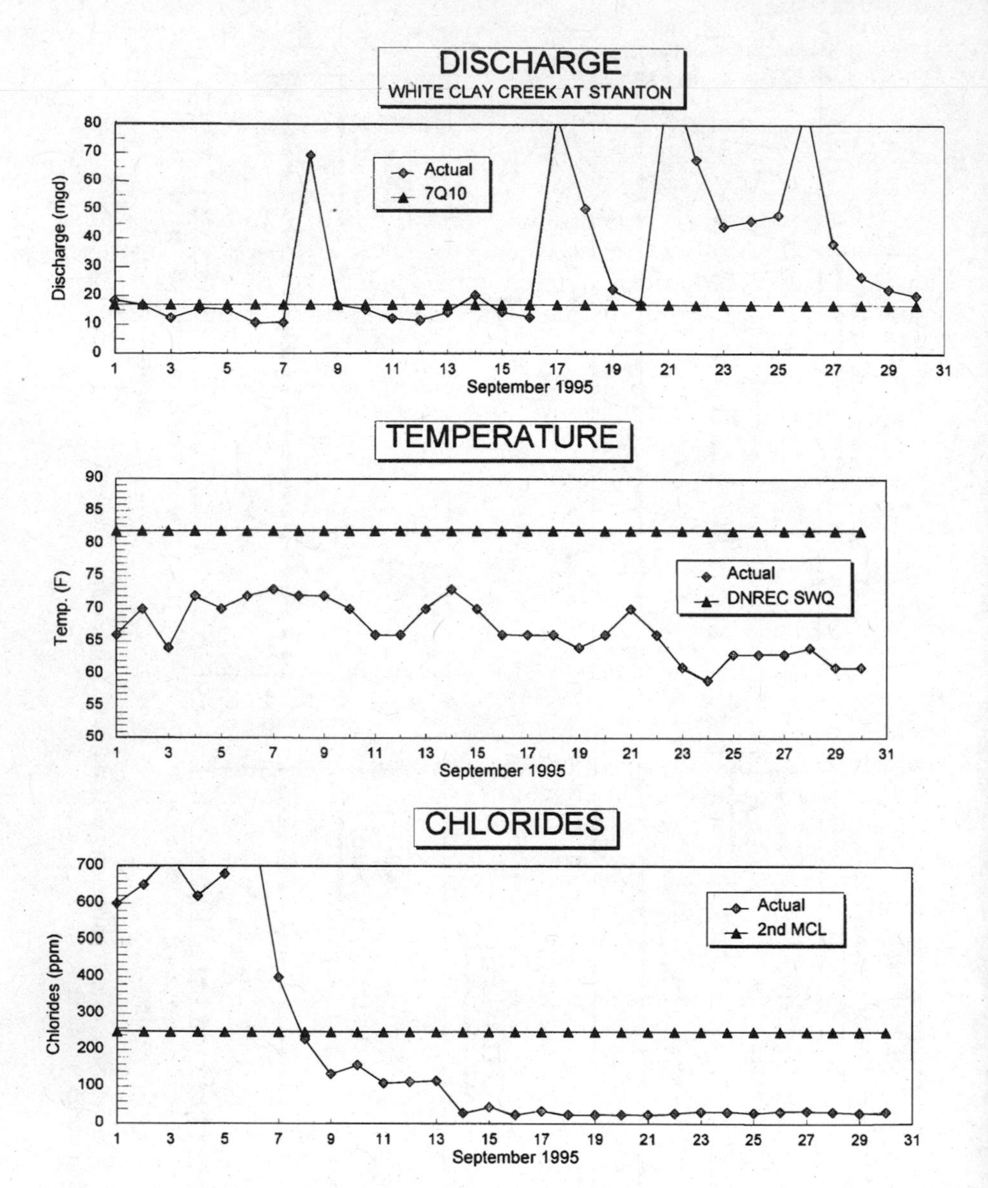

Figure 16.10 Measurements of pollutants at White Clay Creek at Stanton, September 1995, showing relaxation of in-stream flow requirements.

Water Resources Technical Coordinating Committee

The Water Resources Technical Coordinating Committee (WRTCC) met almost weekly to coordinate the technical aspects of the drought. The WRTCC was convened at the offices of the Water Resources Agency for New Castle County and consisted of the state, county, and public and private water utilities. At the weekly technical meetings, the agencies and utilities shared information on water conditions involving precipitation, stream flow, groundwater levels, and reservoir levels. Water supply production and demand summaries were prepared daily and distributed for drought coordination. In addition, the WRTCC conducted activities including recommendations regarding water restrictions by individual utilities, evaluations of Hoopes Reservoir, development of emergency water supply contingencies, and evaluation of worst-case drought scenarios.

Lessons and Actions

The drought of 1995, while it caused many water supply hardships, provided a learning experience which can enhance our abilities to cope with future droughts. The drought provided certain lessons and actions which can be logically grouped into watershed-based physical, institutional, policy, and public education categories. The water resources community in Delaware was urged to consider the following recommendations to improve drought management activities in the future.

Physical

Need for additional water supply. The drought emphasized the need for an additional, permanent, reliable, in-state source of sustainable water supply to meet the recognized deficit of 20 Mgal/day in northern New Castle County. This area lacks adequate storage to replenish surface water supplies during periods of low stream flow. The Churchmans Environmental Impact Statement (EIS) should proceed with all due speed to permit, design, and construct a reservoir by the end of this century.

Brandywine–Hoopes–Red Clay system. There were concerns during the drought regarding the adequacy of the Brandywine–Hoopes Reservoir–Red Clay system. The city of Wilmington experienced difficulty withdrawing water from Brandywine Creek during record low stream flow levels. There were questions regarding the actual capacity and conditions of pumps which withdraw water from Brandywine Creek and fill and drain Hoopes Reservoir. An infrastructure assessment should be conducted to evaluate the capacity and condition of gates, canals, tailraces, pump stations, and pipelines in this vital system.

Hoopes Reservoir operating plan. During the drought, there were questions about the actual capacity of Hoopes Reservoir and whether sufficient storage was available for Wilmington and United Water Delaware through the end of the drought. There were questions regarding the capacity of the reservoir depending on the condition of the outlet gates, accumulated sediment in the impoundment, and water quality based upon predicted high levels of manganese at lower depths. When Hoopes reached 55 percent of capacity, the city ceased releases to United Water Delaware to reserve the balance of storage for the city during continuing drought. A Hoopes Reservoir operating plan should be developed which addresses the questions about actual volume, outlet structure condition, and water quality. The plan should specify written operating guidelines, similar to those for Marsh Creek Reservoir, which dictate storage commitments and required releases for the city of Wilmington and United Water Delaware during normal and drought periods.

Interconnected network. The interconnected network was one reason that a water shortage was averted in northern New Castle County. During the critical dry period in early September 1995, up to 5 Mgal/day was transferred between the utilities to meet demand in service areas of short supply. Artesian Water Company installed three emergency interconnections which conveyed 2 Mgal/day to United Water by reversing pumps and laying new mains. A hydraulic evaluation should be conducted to identify opportunities for increased interconnections between the utilities during normal and drought emergency periods. Also, formal interconnection agreements should be prepared by the utilities which specify capacity, duration, and financial arrangements during normal and drought periods.

Interim water supply alternatives. The drought highlighted the need to develop all possible water supplies to meet deficits over the next few years. The following projects should be developed to maximize water supplies over the next several years until a regional water supply alternative is implemented by the Churchmans EIS:

- Continued CWA interconnections from Pennsylvania
- UWD tidal enhancement structure along White Clay Creek
- Water treatment plant at Newark south well field

Stream gauging. During the drought, the media incorrectly reported problems at the gauge along Brandywine Creek at Wilmington during record low stream flows. The Delaware Geological Survey performed outstanding work by continuing difficult stream gauging measurements during record low flows when levels dropped "off the chart."

Institutional

Drought emergency preparedness plan. The critically low water supply conditions emphasized the need to prepare a statewide drought emergency preparedness plan. Such a plan should be prepared by the Delaware Emergency Management Agency (DEMA) and should describe emergency water supplies (tankers, desalting plants, etc.), priority service areas (nursing homes, hospitals, etc.), and a chain of command for drought coordination during an emergency. The drought emergency plan should be similar in scope and scale to the plans prepared for floods, fires, snowstorms, and other natural disasters.

Marsh Creek Reservoir accord. During the critical drought period in early September, Delaware requested Pennsylvania to provide an emergency release from Marsh Creek into Brandywine Creek for capture by the city of Wilmington. While the terms of the release were being negotiated, the drought eased, stream flows increased and the emergency release from Marsh Creek was deemed no longer necessary. While in a nondrought period, the two states should finalize the paperwork regarding an emergency release from Marsh Creek, should it be needed in the future. The agreement should specify that a release would be requested only as a contingency, in an emergency, and only as a last resort to supplement low flows in Brandywine Creek at Wilmington.

Water master. During the drought emergency, there was debate and difficulty allocating stream flows, withdrawals, and reservoir storage for use by water suppliers with the greatest need. In the Delaware River basin, a "river master" is appointed as an impartial party to equitably apportion stream flow, withdrawals, and reservoir storage in accordance with negotiated agreements. A similar impartial position, perhaps a "water master," should be considered to allocate water supplies during a drought emergency in Christina basin. The water master would be appointed as a position within DEMA with duties specified through negotiated agreements among public agencies and the water utilities.

Relaxed water standards. A water shortage in northern New Castle County was averted partially through the relaxation of water standards during the drought emergency. Water utilities were permitted to temporarily exceed well allocation standards. Drinking water standards for sodium and chlorides were exceeded for two weeks in White Clay Creek due to lack of freshwater flow. Minimum in-stream flow standards were waived during the duration of the drought emergency along White Clay Creek at the Newark WTP and the UWD Stanton Filter Plant. Additional storage is needed to supplement existing water supplies so water quantity and quality standards are not compromised during future droughts.

Policy

Three-phase drought declaration. The drought outlined the need to update the DNREC drought management plan. The current plan specifies a two-phase drought declaration (warning and emergency) which provides little opportunity for an early public "advisory" and is inconsistent with the existing three-phase drought declaration policy in use by neighboring Chester County, Pennsylvania. The following three-phase drought declaration policy should be developed in Delaware depending on the severity of water conditions:

- Advisory (voluntary restrictions)
- Warning (voluntary restrictions)
- Emergency (mandatory restrictions)

Public notice for drought emergency and mandatory restrictions. The current Delaware drought management plan specifies a minimum 20-day public notice period before a drought emergency and mandatory restrictions can be declared by the governor. The recent drought indicated that water conditions can decline from warning to emergency status more rapidly than 20 days. A shorter public notice period, say 7 days, should be instituted to permit a more expedient declaration of drought emergency and mandatory restrictions in the event of rapidly declining water conditions.

Drought forecast index. The current DNREC drought forecast index (bowling chart) measures only Brandywine Creek and precipitation levels as indicators of water conditions in northern New Castle County. The recent drought exemplified the need to monitor other water conditions such as groundwater and reservoir levels as indicators of impending drought. The DNREC drought forecast index should be revised to incorporate factors based on

- Precipitation
- Soil moisture
- Stream flow
- Groundwater level
- Reservoir level

Drought emergency expiration. While the policies for entering a drought are well defined, there is no clear policy for expiration of a drought emergency or warning. An automatic expiration date should be included in a governor's drought emergency executive order to avoid the embarrassment of calling off a drought during a wet period.

Public education

Drought education. The recent drought illustrated public misconceptions about the need to order mandatory restrictions and declare drought emergency. A public education program should be enacted to inform the public about watershed-based drought management and water conservation activities.

Media visibility and relations. Considerable television and newspaper coverage was provided during this drought (see Fig. 16.11). Most of the media coverage was desirable, to communicate the need to conserve water. However, certain articles were not factual and created a false public perception regarding the competence of drought management activities by public agencies and the utilities.

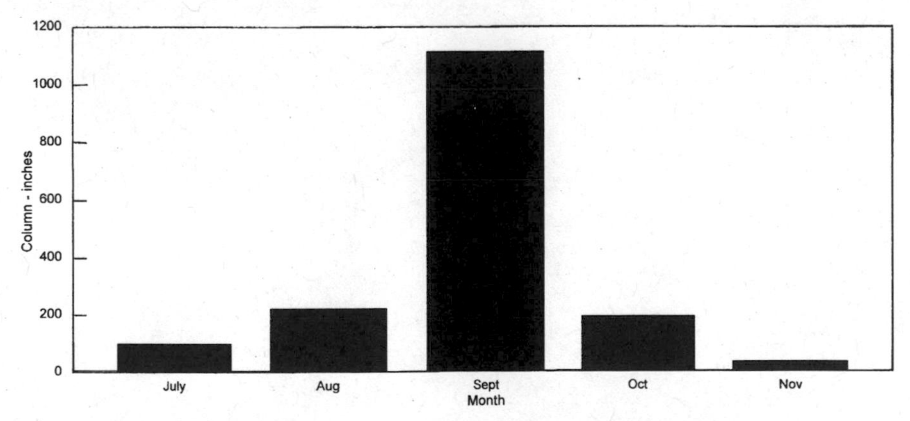

Figure 16.11 Extent of media coverage during the drought of 1995. (*The News Journal, Wilmington, Delaware.*)

17

Integrated Management and Water Resources Protection

The Case of the Paris (France) Area

T. Vandevelde and N. Fauchon
Compagnie Générale des Eaux
Paris, France

Abstract

Following an overview of the problems of the drinking water supply in France and then a quick description of the various players and administrative and financial structures, the case of the water supply of the Paris urban area is taken to describe an example of an integrated watershed approach applied to the protection of the water resources of a large conurbation. The system for combating accidental pollution that has been set up is based on studying the potential risks, monitoring raw water, and simulating the impact of a pollution event. In the Paris area, since 1990, the Compagnie Générale des Eaux and the Syndicat des Eaux d'Île-de-France have implemented diffuse pollution control programs for the protection of water resources in the Paris area, according to a dual approach of treatment and prevention.

Introduction

In France, surface water is a resource vital to the supply of drinking water. As a matter of fact, it provides more than 40 percent of needs. The largest French conurbations, e.g., Paris, Lyons, and Toulouse, basically depend on surface water for their drinking water supply. Most of the time, this surface resource is much more vulnerable than the underground resource and finds itself continually threatened by numerous instances of chronic or accidental, localized, dispersed, or diffuse pollution.

Industrial sites situated far upstream, several hundred kilometers from a water intake, may thus potentially be a factor in significant deterioration and lead to an interruption in the drinking water supply. Furthermore, multiple use of the water course may be a source of conflict and entail changes in its quantity or its quality capable of seriously affecting water production.

Consequently, surface water is a fragile and coveted asset which is best managed by taking various interests into account. For a long time water management has been carried out on a use-by-use basis. This sector-oriented approach, however, quickly shows its weaknesses as soon as the resource becomes more limited. It has therefore been necessary to conceive a more global approach in order to go beyond the stage of sector-oriented use-by-use management (which is only possible in the very rare instances of an inexhaustible, pollution-proof resource) and thus strive for integrated management, i.e., management that seeks to reconcile the satisfaction of the various uses with the preservation of resources and the natural heritage.

In fact, the term *integrated management* has several meanings which may be different. Some authors regard it as necessary that it take into account the spatial dimensions of a problem; others, the administrative consistency; yet others, a functional integration. With regard to the protection of water resources, each of these dimensions is obviously important, and a multiuse approach, on the watershed or catchment basin scale, which takes into account the various players is the only really effective one. See Fig. 17.1.

General Data on Water Supply in France

Water needs

The average per capita consumption in France, for domestic uses, is estimated at approximately 140 L/day of drinking water. This consumption

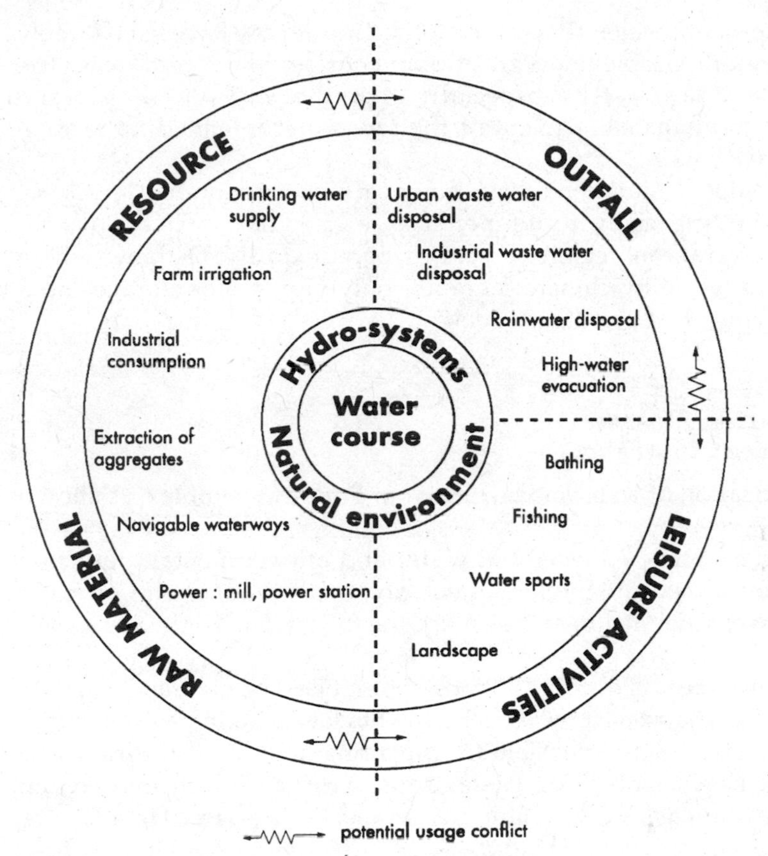

Figure 17.1 Uses and functions of surface water.

places France in the middle of the group in relation to the other European countries—Great Britain, 135 L/day; Italy, 220 L/day; Switzerland, 264 L/day, according to the Water Authorities Association (1988). Tapping water for the human drinking water supply, of course, is only a small percentage of the raw water drawn off. Thus the amounts of water drawn off (net consumption) for agriculture or cooling thermonuclear power stations are greater by far, as shown in the table of net consumption below.

Agriculture	2.9 km^3/yr in 1992
Electricity	1.5 km^3/yr in 1992
Drinking water	0.7 km^3/yr in 1992
Industry	0.25 km^3/yr in 1992
Others (ship canals, etc.)	2.6 km^3/yr in 1992

All these water needs, all uses combined, until now have been largely met by France's sizable underground and surface water resources. The position of surface water is especially important in France: It actually represents more than 40 percent of the raw water intended for human consumption.

Critical supply situations may, however, be encountered locally in some regions, especially in summer. This, e.g., is the case in the dry climate Mediterranean regions or more paradoxically Brittany, where despite a rather rainy climate there are only very minor storage capabilities as a result of granite bedrock.

Organization of water management in France

The organization of water management in France is complex, as shown in Fig. 17.2.

It is at the watershed level that water and environmental protection actions are planned and implemented. Among the bodies whose area of activity covers the catchment basin are the following:

Basin committee. For over 20 years, committees called *basin committees*, one for each of the six major watershed catchment basins into which France is divided, have been responsible for preparing and implementing water resource management policies. These committees are made up of representatives from the state, local authorities, and various users (see Fig. 17.2).

Water agency. One financial establishment per basin, called a *water agency*, under the supervision of the basin committee and the dual supervision of the Ministries of the Environment and Finance, is responsible for implementing this policy. The water agencies are in charge of collecting dues which are then redistributed in the form of loans for assisting in the financing of pollution control works and resource protection programs. Two broad principles are applied: The polluter pays, and whoever cleans up receives assistance.

Regulatory aspects. The new water law (January 1992) has emphasized the global nature of water management. It replaces the former sector-oriented logic in order to achieve a more integrated approach to the water environment as well as satisfy legitimate economic water use. The first two articles indeed declare water to be a common national heritage and therefore define the objectives of balanced water resource management, i.e., ecosystem conservation and maintenance, use reconciliation, water resource protection and development, and so on.

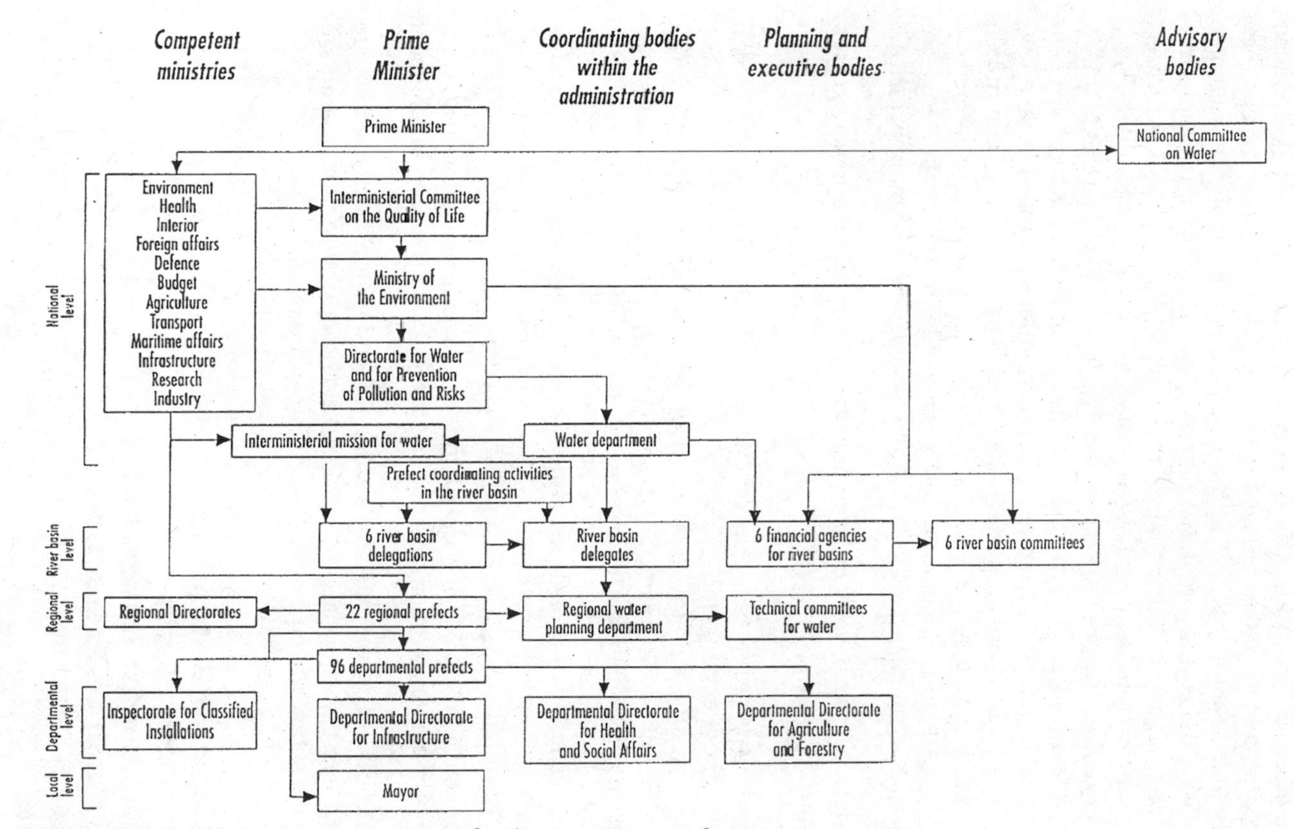

Figure 17.2 Administrative structures for the management of water resources in continental France. (*Sadave, October 1991.*)

The essential innovation is the creation of new management tools, extending coordination at the local level: the master plan for water development and management (SDAGE) and the water development and management plan (SAGE). Recalling the town planning system, SDAGE is supposed to provide a reference and organizational framework for SAGE, which is a local plan; they ensure the consistency of water policy throughout the basin. Water agencies are currently giving priority to master plan completion.

The master plans for water development and management are compulsory and must be implemented and completed before January 1997. Initiated by a coordinator prefect, they are prepared by basin committees in association with state representatives, general and county councils, as well as water users. They set the basic guidelines for balanced water resource management in the six French basins, with two main principles:

- Progression from water management to water environment management
- Priority given to collective interest

Any administrative program or decision should be compatible with these in the area of water.

They usually contain reports and diagnoses, objectives, and definition of the means to meet them over a 15-year period. A dual approach has been implemented for carrying out this work

- A geographic approach identifies the main issues within each subbasin and forecasting changes.
- A thematic approach enables us to deepen our knowledge and approaches to qualitative and quantitative resource management, water environment protection, and risk and large-scale planning management.

Currently, all basin master plans for water development and management are in the last phase—the approval and adoption phase—and will meet the schedule requirements.

Example of Watershed Catchment Basin Management: Integrated Approach to Protection of Water Resources in Paris Area

Surface water provides more than one-half of the water requirements of the urban area of Paris (with more than 10 million inhabitants). For a long time, underground water has not been sufficient to meet the needs of Paris and its suburbs, and the spread of the urban area is dependent on surface water courses. In the fast-growing suburban areas, surface water has come to represent almost all the available resources; thus the Syndicat des Eaux d'Ile-de-France (Ile-de-France Water Syndicate, SEDIF), a large structure grouping together 144 suburban communes and supplying over 4 million people, is more than 95 percent dependent on surface water. The main water intakes are situated in the lowest parts of the catchment basin at the confluence of the rivers Seine and Marne and also on the Oise River. Human activities abound upstream from the water intakes and can be sources of significant deterioration in water quality. There is a great deal of chronic industrial and domestic waste disposal. Accidental pollution is relatively frequent, and pollution of agricultural origin also presents problems.

A dual approach has been developed to face up to these fragile and deteriorated resources. On one hand, plant treatment processes have been continuously improved. Processes for the elimination of pesticides by coupled ozonization, e.g., have recently been installed in plants; nanofiltration technologies are in the process of being implemented. In parallel with this, significant work has been done on risk prevention.

It is within the framework of this preventive approach that a policy of protection areas has been developed. French regulations allow the definition of so-called protection areas for all drinking water catchment zones, in which certain pollutant activities are restricted or even completely prohibited. These regulations are relatively well suited to the case of underground water where, by acting over relatively small surface areas, we may hope to control most of the pollution risks. This is not completely so with regard to surface water, where activities very remote from a water intake on a river may still have a significant impact.

The approach via protection areas has therefore been adapted with respect to surface water. The fact must then be accepted that "zero risk" does not exist, and to better combat them, we must seek to identify all the potential risks and set up a monitoring and antipollution system. Several stages can be identified in such an approach: monitoring of resources, studies of potential contamination risks, diagnosis of resource vulnerability, use of crisis management tools after a pollution event, and identification of risk prevention measures.

Concrete examples of this approach are given later, based on what has been done in the Paris area with regard to the fight against accidental pollution, wastewater treatment, and the control of pollution of agricultural origin.

Combating Accidental Pollution

The system for combating accidental pollution that has been set up is based on (1) studying the potential risks, (2) monitoring raw water, and (3) simulating the impact of a pollution event.

Studying the risks

Risk studies are regularly carried out for all the basins upstream from the water intakes of the Paris area, and a central database contains information on more than 600 potentially dangerous industrial sites. This information is periodically validated by discussions with industrialists. Government departments participate in this work in cooperation with water suppliers.

Monitoring raw water quality

A network of automatic stations continually monitors the quality of the raw water upstream from water intakes. Analyzers measure the standard pollution indices (TOC, nitrates, nitrites, etc.) but also more complex pollutants such as hydrocarbons or heavy metals. They have the sensitivity threshold required for complying with the most stringent European directives. Each device performs an average of 50 operations per 24 h.

Controlling the impact of pollution

Various mathematical models have been developed to simulate the variations in water quality parameters in the event of accidental pollution. The "Disperso" model, e.g., determines the characteristics of an accidental discharge upstream from a water intake and forecasts the arrival time of polluted water at the water intake, as well as pollution duration and concentration data. These models form valuable decision support tools, enabling the best use to be made of the means of combat.

Wastewater treatment

The principle set out in the new water treatment master plan for the Paris area consists of favoring the treatment of wastewater as close as possible to its collection area. The act of treating at the source avoids the transfer of water from upstream of towns in a downstream direction, which in addition to a high transport cost has a significant ecological impact by depriving the whole section of river of water.

Thus, several new wastewater treatment stations are being built, or are about to be built, in the urban area in order to seek short cycles between collection and treatment and relieve the huge purification plant situated at Achères downstream from the Paris conurbation. This station (2,100,000 m^3/day), dating from the middle of the century, has become unmanageable owing to its size, and its discharges have had a major impact on water quality downstream from Paris.

Recent technological progress, which is expressed in an improvement in the efficiency of purification, a reduction in pollution (noise, smell), and a reduction in the size of structures, is opening up new possibilities. Recent constructions in Monaco, Antibes, and Marseilles demonstrate that it is now possible to build treatment plants right in the middle of town, sometimes completely underground, while respecting the environment and meeting the most demanding treatment standards. In Paris, these new technologies have enabled the construction of several purification plants right in the urban area. The cost of these new plants can be partly offset by the savings associated with the reduction in the length of the collecting systems.

Controlling Diffuse Pollution of Agricultural Origin

European regulatory provisions on drinking water quality over the last 10 years have led to profound changes, not only with regard to drinking water plant treatment and production conditions, but also with regard to resource management and protection. These provisions, applied in France from 1989, and in particular the maximum acceptable concentration for pesticides in drinking water set at 0.1 μg/L, have necessitated the setting up of major research and development programs on the subject of agricultural micropollution.

Need for a policy of prevention

For a number of years, the producers of drinking water have been confronted by the pollution of resources by nitrates, but the questions posed by pesticides have sharply revived the problem of diffuse pollution of agricultural origin.

The control of diffuse pollution, as it manifests itself in surface water in the Paris area (see Figs. 17.3 and 17.4), necessarily ranges from the definition and placing in service of additional treatment devices in drinking water plants, to preventive actions aimed at limiting the contamination of the natural environment by fertilizers and pesticides.

Curative solutions to the problem of pesticides are, in fact, limited from both a technical and a financial point of view. On the technical level, considerable development efforts have been carried out for eliminating pesticides in the plant production lines. With regard to the treatment of atrazine, which in France still remains the most frequently encountered agricultural micropollutant in underground water and surface water, the processes available are generally effective and economically viable only for pollutant concentrations not exceeding 1 to 1.5 μg/L; such concentrations are frequently observed on many sites and are even exceeded in surface water during certain times of the year (see Fig. 17.5). And on the financial level, a large number of local authorities, especially in rural areas, have limited means and hesitate to make investments liable to entail a significant increase in the price of drinking water.

The implementation of actions for preventing agricultural pollution therefore appears essential, less for saving on a specific treatment device

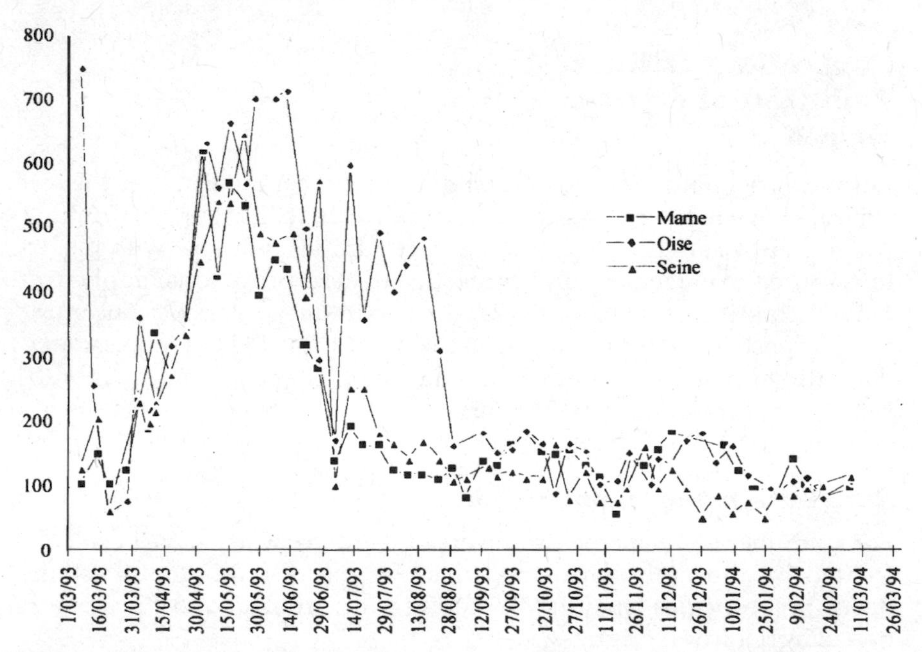

Figure 17.3 Changes in atrazine concentration in the rivers Seine, Marne, and Oise, March 1993 to March 1994.

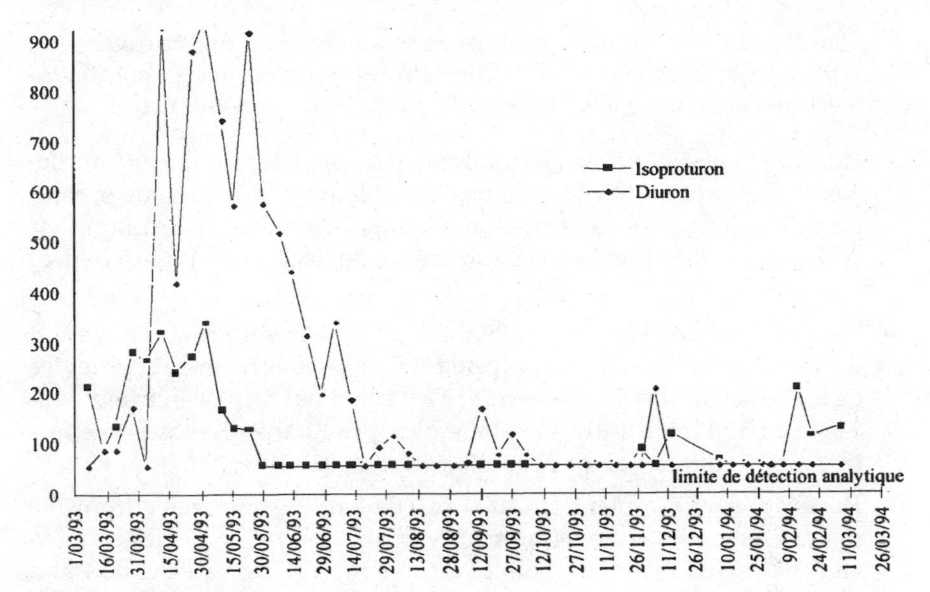

Figure 17.4 Changes in isoproturon and diuron concentrations in the river Marne, March 1993 to March 1994.

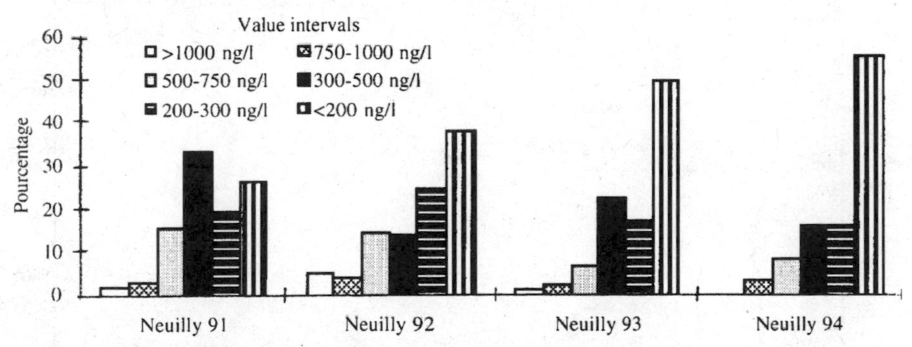

Figure 17.5 Distribution of atrazine concentration values noted in the Marne River in 1991, 1992, 1993, and 1994. Distribution by value intervals, corresponding to technical thresholds and specific treatment constraints in the Neuilly-sur-Marne treatment plant: reinforced ozonization, ozonization coupled with hydrogen peroxide, use of powdered activated carbon, and so on.

which may be necessary anyway on many sites in the short term, than for completing and protecting resources in the medium and long terms.

In the Paris area, since 1990, the Compagnie Générale des Eaux and the Syndicat des Eaux d'Île-de-France have implemented diffuse pollution control programs for the protection of water resources in the Paris area, according to this dual approach of treatment and prevention.

Considerable investments have been made in terms of improving the treatment lines of the Syndicat's three plants with regard to plant-protective products, and more especially triazines and phenyl ureas:

- The Choisy-le-Roi plant (800,000 m^3/day, using raw water from the Seine), equipped with two stages of biological filtration, over sand and granular active carbon, as well as powdered active carbon, has been fitted with a reinforced ozonization device, coupled with hydrogen peroxide.
- The Neuilly-sur-Marne plant (800,000 m^3/day, using raw water from the river Marne) has been equipped with a similar ozonization device coupled with hydrogen peroxide. This plant has also been fitted with a second biological filtration stage over granular active carbon effective in the elimination of pesticides such as triazines.
- The Méry-sur-Oise plant (180,000 m^3/day, using raw water from the river Oise) has also been equipped with an ozonization device coupled with hydrogen peroxide. This plant is currently being extended and modernized, which will bring its capacity up to 400,000 m^3/day; the treatment will be based on the nanofiltration process which is very effective in the elimination of organic molecules such as atrazine (see Fig. 17.6).

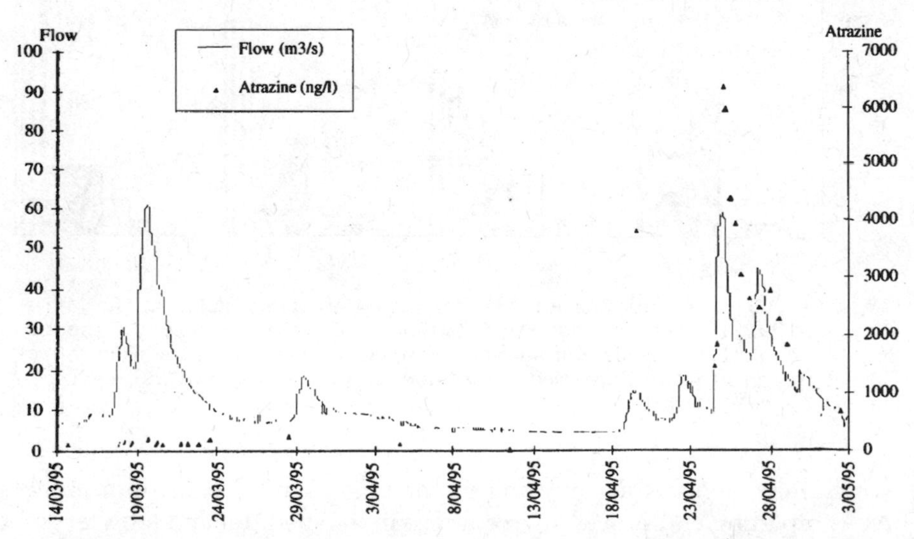

Figure 17.6 Analysis of the transfer mechanisms of pesticides requires detailed monitoring of water quality in conjunction with hydrology. In this log recorded on the Grand Morin River, at the time of the first flood episode, atrazine remained at the level of basic contamination (approximately 100 ng/L); at the time of the second flood episode, which occurred after the application of atrazine on the maize crops, we observe a sharp deterioration in the surface water associated with surface runoff.

In parallel with the creation of these installations, major study programs have been set up. On one hand, systematic sampling of watercourses has been carried out since 1990, with the aim of optimizing the treatment of water in the production lines and showing up any new contaminants as well as the principal mechanisms for the transfer of these contaminants in the natural environment. New multiresidue analytical techniques have been developed by the Compagnie Générale des Eaux's research center, in order to be able to proportion plant-protective products at concentrations compatible with the regulation values.

On the other hand, work has been started in cooperation with research establishments specializing in agronomy and hydraulic modeling, as well as with manufacturers of agropharmaceutical specialty products. The objective is to end up with a better knowledge of the transfer conditions of plant-protective products outside cultivated areas and in different sectors of the natural environment over the catchment basins.

In terms of prevention, the Compagnie Générale des Eaux's interest in the study of catchment basins and contaminant transfer mechanisms in the natural environment lies in being able to contribute to establishing a coherent diagnosis of the state of resources in connection with the additions of active matter over the catchment basins. This diagnosis should be capable of being expressed by the definition of an acceptable pollutant pressure according to the vulnerability of the environment.

Testing methods of diagnosing the impact of farming practices

The diagnostic study approach is based on the interpretation of surface water quality data. In the case of water intakes situated on the watercourses of the Paris area, i.e., downstream from urban or semiurban districts, the impact of urban runoff may be superimposed on that of farming activities:

- The data relating to atrazine are explained by contamination of underground water which is the result of heavy use of this herbicide in agriculture for long periods.
- The effects of seasonality, linked to surface runoff phenomena, are the direct consequence of applying active matter on cultivated areas in spring. Notably in the case of diuron or simazine, they may also have an urban origin, since road treatments are also carried out at the same period.
- The origin of brief episodes of strong concentrations may similarly be agricultural or urban in origin. A short peak of concentration will

originate from a source close to the point of measurement, i.e., from the urban area, while a longer peak corresponding to a distant source will be attributable to a badly monitored agricultural application or one carried out under unfavorable weather conditions.

These different elements have led the Compagnie Générale des Eaux to define a program of studies on a subcatchment basin of the river Marne, the catchment basin of the Grand Morin, situated 40 km east of Paris; this is a catchment basin of 1200 km^2 extending across Brie, which is an area of major cereal cultivation.

This program covers (1) the monitoring of water quality, (2) the acquisition of data relating to farming practices, and (3) the definition of vulnerable areas according to different approaches:

- Careful monitoring of water quality has revealed the main characteristics of the transfer of certain plant-protective products in the natural environment, and in particular the relationships existing between changes in concentrations in surface water and the hydrology of the basin (Fig. 17.6). Different multiresidue methods of determining the proportioning of micropollutants are applied or tested on the basin (triazines, phenyl ureas, carbamates, etc.).
- Surveys of farm practices have been carried out on representative samples of farmers in the basin, to discover the precise nature of the plant-protective products used and their conditions of application on crops (dates, doses, etc.). Remote detection has been used for updating the various cultivated areas.
- The methods of defining the basin's vulnerable areas are applied with regard to the risks of contamination of underground water and surface water by plant-protective products. Methods (DRASTIC, AF, etc.) based on mapping physical environment criteria involved in the transfer mechanisms (nature of the soils, size of the unsaturated zone, etc.) have been tested, but the approaches that have been favored depend on mechanistic simulations of the physical phenomena.

This work is based on simulating the water cycle on the scale of the whole basin, with the object, first, of modeling the mechanisms of surface runoff and groundwater table supply and, second, of simulating the impact on water quality in scenarios (see Fig. 17.7) relating to land use or certain cultivation practices (dates of applying herbicides, effects of artificial drainage, etc.).

The integrated control of the technical aspects of diffuse pollution, on the physical, chemical, and hydrological levels, forms a mediation tool with the agricultural professionals, as dialogue with the farmers is

Figure 17.7 Nanofiltration at Méry-sur-Oise.

based on advice on the use of plant-protective products suited to the constraints of the physical environment.

A concerted approach for sustainable change

Diffuse pollution of agricultural origin usually results from contamination by nitrates and pesticides of a major part of the different sectors of the natural environment involving water resources—soils, aquifers, and surface water. In large catchment basins, prevention may therefore concern extensive cultivated areas, covering hundreds, even thousands of square kilometers, and thus involve a large number of farmers.

For sustainable control of agricultural pollution, prevention programs must necessarily be based on the voluntary support of the farmers concerned, who must collectively participate in the program and actually change their individual farming practices. Such support is essential in an agricultural country like France, where the economic and political weight of the farmers is still very strong.

For this reason, and despite the existence of restrictive legislation, regulating or prohibiting the use of individual plant-protective substances, the principle of current prevention policies, rests on the idea that it is possible to introduce environmental practices into agriculture and to improve the use of fertilizers and pesticides, without, however, causing a fall in farmers' incomes.

This support from the farmers is all the more necessary since controlling contamination of the natural environment calls for profound changes in farming practices, affecting not only the use of fertilizers and pesticides according to specific vulnerability criteria of the natural environment, but also the year-to-year management of cultivated areas, rural development, and farm drainage. The prevention of pollution by nitrates in particular demands not only improvement of the conditions of fertilizing itself, but overall management of the whole nitrate cycle on the scale of the catchment basin, including the additions due to mineralization of the organic matter originating from crop residues. The prevention of pollution by pesticides calls for original methods taking into account the very small proportion of active matter contaminating the environment in relation to the mass of this active matter applied on cultivated areas. In the basin of the Seine in the Paris area, the annual flow of atrazine passing through the surface water has been estimated at between 1 and 2 percent of the total quantity of this herbicide applied to maize crops during the corresponding agricultural season. This flow causes concentrations on the order of 0.3 to 0.5 μg/L for 4 to 5 months in these watercourses, an effect to which must be added a permanent concentration of approximately 0.1 μg/L originating from the drainage of contaminated aquifer groundwater.

Scrutiny of the farming community's capacity for change led the Compagnie Générale des Eaux and the Syndicat des Eaux d'Île-de-France in the first instance to undertake nitrate pollution control operations, since the problem of pesticides is not yet considered to attract sufficient support from farmers in the Paris area.

In 1995, the drinking water suppliers therefore engaged in a program for reducing nitrate pollution in the basin of the river Marne; the main area of action is the catchment basin of the Grand Morin in which there are over 1000 farmers working. See Fig. 17.8.

The various groundwater aquifers of this basin display nitrate concentrations often higher than 50 mg/L, even reaching 100 mg/L on certain sites; the permanent nitrate concentrations in surface water are on the order of 30 mg/L, but in winter, during periods of excess water balance, they exceed 50 mg/L and may reach 100 mg/L during flood periods.

The program, which has been launched over an initial 4-year period, comprises the following stages:

- A diagnosis is made of each of the hydrographic units of the area studied, aimed at discovering the relative magnitude of the various

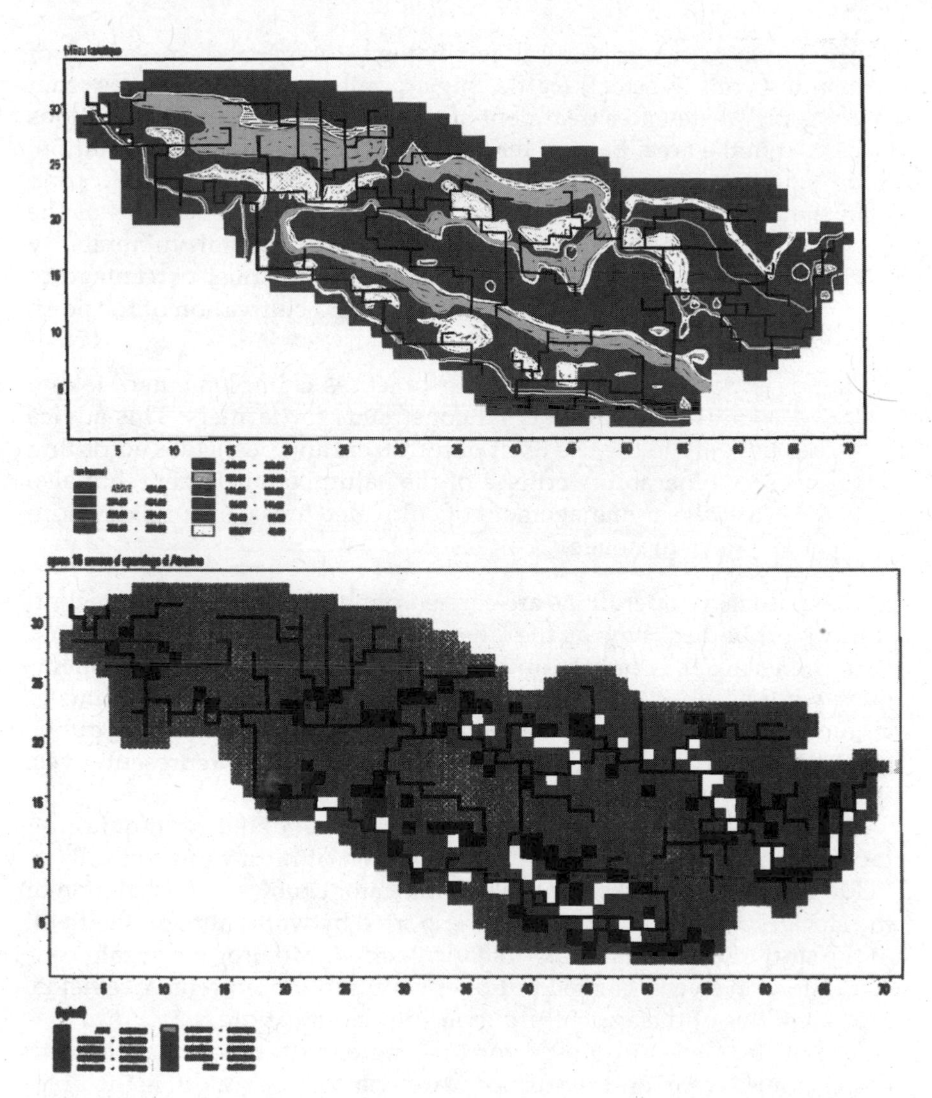

Figure 17.8 The basin of the Grand Morin River has been the subject of in-depth studies based on the simulation of pesticide transfers in underground wastewater and surface water. The water cycle in this catchment basin with an area of 1200 km^2 was modeled by the European hydrological system SHE, which is a physically based distributed modeling system. Pesticide transfer in soils and the unsaturated zone was modeled by the LEACH P model; the calculation for transport in the saturated zone was modeled by the MIKE SHE AD model. The first illustration shows the period of atrazine transfers in the basin's karstic aquifer; the second represents the level of contamination of the same aquifer after the application of atrazines onto maize fields, over a period of 15 successive years. These studies were carried out by the Laboratoire d'Hydraulique de France (LHF, Hydraulics Laboratory of France) on behalf of the Syndicat des Eaux d'Île-de-France.

nitrate sources: agricultural fertilizing, storage and spreading of manure from livestock (cattle, pigs, poultry), and discharges from domestic wastewater treatment stations of the various conurbations located in the area. Each of these sources requires a specific solution; the actual agricultural part of the diagnosis relates to a classification of the crop systems, taking into account the dominant crops of the various agricultural holdings and the environmental vulnerability criteria that have been adopted (nature of the soils; percentage of clay, sand, and loams; plateau cultivation and cultivation of hillsides, slopes, etc.).

- An advisory operation on farming practices is implemented, taking into account the environmental constraints for farmers. This advice relates not only to the use itself of fertilizers and pesticides according to specific vulnerability criteria of the natural environment, but also to the year-to-year management of cultivated areas, or rural development and farm drainage.

These advisory operations are carried out by farmers' normal technical support bodies, such as the Chamber of Agriculture, in association with the fertilizer suppliers and manufacturers and with the cooperatives whose task is to market farm produce. Advice is customized according to the previously defined classification of holdings. A public relations plan is drawn up, involving the farmers' usual representatives, the farming press, and the local authorities.

The operation is continually assessed and validated by monitoring indicators such as the residual concentration of nitrates in the soils of fields at the beginning of winter; this indicator enables the estimation of the stock of nitrates available to be exported by water outside the field, at the start of the excess water balance period. Monitoring nitrate concentrations in soils is carried out over several hundred reference fields, representative of the agricultural holding classification.

One of the means proposed for limiting exports of nitrates, both to underground water and to surface water via surface runoff, is the practice of intercropping. This involves planting a crop at the end of autumn in fields intended for spring crops, e.g., mustard, which grows by using the soil's residual nitrates and creates a vegetation cover enabling runoff effects to be limited. At the end of winter, the plant is buried in the soil at the time of preparing the spring crop.

The drinking water suppliers are financing one-half the cost of these advisory operations, the other half being accounted for by the farming profession; the funds are not therefore intended for the farmers themselves, but for the improvement of farming practices. In addition, the drinking water suppliers play an important part in directing and monitoring the operations and continue with their technical contributions in terms of validating the work carried out. Since nitrate transfers are the result of slow physical processes, a simulation of these transfers by modeling is necessary for assessing the impact of an improvement in practices upon water quality on the scale of the catchment basin.

References

EEC. 1980. Directive 80/778/EEC on the quality of drinking water for human consumption. *Commission of the European Communities,* Paris, France.

DG 11, Commission of the European Communities. 1993. Administrative structures for environmental management in the European Community, Paris, France.

Compagnie Générale des Eaux. 1993. Monitoring of water quality of the rivers Marne and Grand Morin—Report. Paris, France.

Fielding, M. 1992. Pesticides in ground and drinking water. Water pollution research report 27. *Commission of the European Communities* **12,** Paris, France.

Teniere-Buchot, P. F. 1995. European directives and their financial impact. Public water policies in the European Union. *Symposium of the French Water Syndicate,* Paris, France.

18

Issues in Developing and Implementing a Successful Multiparty Watershed Management Strategy

B. Fritts Golden, Ph.D., Vice President
E. A. Engineering, Science, and Technology, Inc.
Lafayette, California

Balancing Science and Values

Scientific inquiry and evaluation provide necessary information on how watersheds work and the effects wrought by changes within a watershed. However, scientific knowledge is not management. Actual resource management is driven by administrative or legislative mandates or directives. Both federal and state (including local) authorities have roles and responsibilities. Their authorities are found in resource agency organic law, specific legislative or executive directives, as well as regulations implementing the Clean Water Act, the Safe Drinking Water Act, the Endangered Species Act, the Forest Management Act, and other federal laws. Similar laws and authorities are in place at the state and local level. Regulatory authority is necessary for success, but is it sufficient? The value of regulatory mandates notwithstanding, the most successful programs are those driven by public values as well.

The facts discoverable by science and engineering about how a watershed operates must be balanced by social values and goals. However, both the science and the values underpinning our resource goals evolve as we learn more. Thus, the important resource management questions of our day cannot be addressed with a set of absolute and final answers. Rather, they must be addressed based on the best current knowledge and current values, with full knowledge that the policies and strategies adopted will be subject to constant revisions.

It is quixotic to believe that we will discover simple, elegant, and universal answers to resource management questions. There are two reasons for this. First, complexity, variability, and diversity are hallmarks of the natural environment. Second, at the same time, a wide spectrum of human interests extends across and intertwines within this far-from-simple environment. Therefore, management will be a perpetual and open-ended process of discovery and compromise, of substantial successes and backsliding. These are the realities of public resource management.

There are a large number of successful ecosystem and watershed management programs operating. They are the key steps needed to implement successful programs. This discussion relies most heavily on the New Jersey Pinelands management program and the Chesapeake Bay program. It also draws from the author's experiences, including a watershed management study of the Great Swamp in New Jersey, the Virginia Chesapeake Bay Program, the Northwest Watershed Restoration Partnership Conference, the Barnegat Bay Estuary Study, and Adirondack Park's Warren County water quality study.

Sustainability as a Vision

As managers, policymakers, and the public struggle to find effective strategies to maintain a stable and sustainable environment, the limits to a social philosophy rooted in subduing the earth and making economic use of all resources have become evident. As a result, the concept of sustainable development is gaining widespread acceptance as a vision of how we should function in our society. Disagreements exist on the sustainability of particular resource and land practices; people are loath to acknowledge that *they* engage in unsustainable behaviors. Nevertheless, a consensus is found on the central objective of sustainable development: economic activity that leaves an undiminished and unimpaired stock of environmental goods—such as topsoil, clean air, potable water, forests, and diverse plant and animal species—to future generations. It is clear that resource-depleting and environmentally destructive practices need to be replaced with strategies designed to restore resources, establish sustainable rates of use, and foster stewardship of the landscape.

Even the most cursory look at both human and ecological health in industrial regions where environmental protection has been minimal, such as the former eastern bloc countries, reveals that sustained wealth and "quality of life" are not possible without a healthful environment.

Managing ecosystems as landscapes, within which multiple species function as parts of integrated natural communities, may be the most effective way to ensure the survival of species at risk while also providing natural functions that benefit humans. Such natural functions include retarding floods, filtering and adsorbing nonpoint source pollution, and protecting water supplies.

Dampening the Rhetoric

In North America, resource management planning often takes place in a hothouse climate of conflicting views that are strongly held and passionately argued. These are based on a mixture of science, romanticism, personal philosophy, and economic need. The result can be paralysis in the decision-making process.

Amid uncertainty, one reaction is to be protective of all resources. This quickly extends to a blanket preservation of all areas. A lack of definitive knowledge and lack of a consensus on the actions needed tend to make us conservative in our approaches. As a result, we discount resource-based cultures and industries. These are no match for the thoughtful person who feels a compulsion to address the "crisis" besetting the natural world. Even the most obscure species, if not observed in significant numbers in enough places, is valued over employment or resource use.

Against a backdrop of hand wringing, contradictory claims, shrill arguments, perpetually "incomplete" studies, and conflicting objectives, is it any wonder that resource managers are stymied on how to go about their jobs? The basic physical law that for every action there is an equal and opposite reaction is nowhere more evident than in resource management. It seems that whatever management tactic is proposed, it is met with a howl of protest from some quarter. This is hardly inspiring to the dedicated manager trying to fulfill his or her commitment to manage resources for the common good. Fortunately, evolving practices of public involvement and of open, systematic exploration of issues and problem solutions are being effectively employed to achieve widely accepted and supported consensus on a basic program over an extended area.

Management Strategies

Concepts with similar aims

Integrated resource planning, ecosystem management, biodiversity, and watershed management are allied concepts. Arguably, these phrases describe concepts or programs that have extensive overlap in geographic areas of concern, philosophical approach, final objectives, and technical strategies. At their heart are similar goals of

- Minimizing adverse impacts on the landscape
- Understanding natural and physical relationships
- Establishing multiobjective goals that encompass ecological, community, and business needs
- Having a management strategy for achieving those goals

Whatever their resource focus, management programs must be flexible in responding to expanding knowledge and to the changing needs and values of society. They must be comprehensive in the scope of issues addressed and inclusive in determining who is involved in debating and setting goals and then developing strategies to achieve those goals.

Integrated resource planning (IRP) is typical of such programs. IRP's principal concepts, demonstrated in and borrowed from the electric utility industry, have been expanded and adapted to natural resource parameters and characteristics. The central features of IRP include

- Looking at multiple options for efficiently matching resource supply to consumer demand
- Opening up the decision-making process to new ideas and interests

- Recognizing that uncertainty and change are endemic to long-range planning
- Evaluating planning decisions from a regional and societal perspective as well as an agency perspective

The supply-demand aspect of IRP does not dictate that resources be consumed in the traditional sense of the word. Rather, supply is what is wanted by those making the demands. Wilderness, for instance, can be a "supplied" resource, and the consumer is the public at large.

Basic objectives of a program

A sound ecosystem management approach is one that can address ambiguity and uncertainty. Ambiguity comes from conflicting values and directives; uncertainty comes from the failure of a program to be definitive. Species protection offers an example. A challenge is to get away from single-species protection strategies and from "save it all until we figure it out" philosophies, and to develop approaches that address broad ecosystems and clearly define the goals and the strategies that will be employed.

Today, our forests, wetlands, headwater watersheds, and other sensitive ecosystems often appear as islands in a sea of disturbed land. Biologists and resource managers have been aware since Darwin's work in the early 19th century that a relationship exists between habitat size and species composition. An integrated system of connected large natural areas is necessary to protect biological diversity. Species-based strategies may only protect selected species over the short-term. Species-level approaches must be augmented by or even replaced by landscape-level strategies that recognize broader ecosystem patterns and processes.

The patterns and sizes of ecosystems must be managed to prevent extinction of plants and animals and to maintain and enhance ecosystem functions. The interplay of nodes of "wilderness" and connecting corridors for species flow between nodes is important when a landscape-scale management plan is designed. Adaptive use may allow areas to be managed in a dynamic pattern. The mirror image of this ecological model is the system of the built environment: the towns (nodes) and highways (corridors) of human culture that are vital to biological and economic success of our species. Nodes and corridors for sustainable ecosystems and water quality are as important as towns and roads are for human settlement and economic growth. The balance to be struck is between serving both the natural environment and the economic system needs. Difficult choices are required.

Education will play a major role. To effectively instill a sustainable development ethic will require extensive and prolonged education. Developing a meaningful stewardship ethic requires reexamining and reshaping how people in the United States engage the natural world. This is not work for a few policymakers, biologists, and volunteers. This will require significant debate and agreement at many levels in government, the community, and business.

Laws will have to be redrafted to acknowledge the interconnections among resource and social objectives. For example, most laws regarding the natural environment do not mandate protection of biological diversity; these laws are not directed to watersheds as a whole. Instead, they aim to protect particular resources, such as a species with a dangerously low population level or the quality of water supply sources used for drinking water. Likewise, resource-oriented laws are almost universally silent on economic and social concerns that must be incorporated into resource management.

Conservation or preservation or resource-use arguments pose either/or choices and are a legacy of adversarial approaches to policy development. This tradition is giving way to blended approaches stressing conservation and preservation and resource use. In the past, the management choice has been presented in terms of jobs *or* the environment. This is being shown to be a false dichotomy. It is emerging that it is possible, and even more desirable, to have jobs *and* the environment. People are tuning out single-objective arguments that polarized past debates. Now, it is broadly accepted that we need to achieve multiple objectives through programs that many persons have helped craft and actively support.

Establishing Successful Programs

By their nature, management initiatives addressing broad geographic areas must include a wide range of interests. They must encompass conflicting objectives and consider various approaches to solving problems. These initiatives must find ways to place values on many intangible qualities and conditions. The size of a region, the diversity and complexity of its environment, the breadth of interests, and the difficulty of having sufficient good data for supporting decisions are major contributors to program inertia.

Recent management-planning efforts surrounding the northern spotted owl and other timberland issues in the northwest are an example of what can happen when there is a complex brew of agencies, mandates, resources, and values. Some problems are institutional, such as agencies having changing, contradictory, and overlapping missions. These inter- and intraagency situations have fostered interagency coordination and cooperation. This process is not fully mature; it still struggles with the inherent difficulty of serving unreconciled objectives and multiple masters. Other problems are symptomatic of the philosophical debate over whether to permit limited use of a resource or to require absolute preservation. Even apparently clear-cut alternatives and policies have unintended consequences attached to them. For example, a debate is emerging over the condition of old-growth forests over time. Setting aside forest to achieve one set of objectives and suppressing fire have allowed fuel loads to increase and insect infestations to be largely unchecked. A fire in a heavily fueled forest causes greater damage to plants and the soil, and hence to fish and wildlife, than a fire sweeping through a managed area. As knowledge expands, policies deserve reexamination.

Ecosystem management programs have been implemented in many parts of the country. Two well-established programs are in the New Jersey Pinelands and the Chesapeake Bay. These are among the most informative models because of their relatively long histories. In the northwest, issues of forestry, fisheries, and species protection are energizing moves to develop integrated resource plans that achieve multiple objectives. The Pinelands and Chesapeake Bay experiences demonstrate the value of some guiding principles that any program would be well advised to adopt:

- Clearly expressed and understood needs
- Clearly expressed goals
- Good research
- An open climate for discussion of issues
- Genuine partnerships among stakeholders

Critical elements of a program

Those involved in establishing and nurturing a management program must address a number of subjects that, if left unattended, can slow a program or derail it altogether.

Trust. A major hurdle to moving a program forward to implementation is trust in leaders and trust that there is a process that will empower people to work together to define and achieve common goals that they understand and support.

Discovering leaders. Traditional elected and appointed leaders may have less well-defined roles and authority than in the past.

Defining roles. Responsibility and authority may be unclear when there are conflicting mandates and unforeseen gaps in authority or responsibility.

Changing roles. The identity and roles of various stakeholders are difficult to determine and may change as a program evolves.

Lack of clarity. It is difficult to establish a clear vision or image of what an ecosystem is and how it directly affects or benefits people.

Problem definition. Failure to define the correct problem due to a lack of a systematic process or wrong stakeholders can waste a lot of time and resources.

Persistence. Long-term planning horizons are difficult to reconcile with the short attention span of political bodies and the public. The political system tends to throw money and resources at a problem for a few years, then migrates to a new crisis, neglecting the previous crisis.

Abstraction and size of effort. Large-scale programs with multiple objectives and intricate interrelationships are too abstract to sustain attention. There tends to be a retreat to single-issue stances where people feel they have a grasp of basics.

Good data from good science. Good science is important to defining systems and their critical control factors, and to framing management options.

Separation of science and policy. Distinctions between science and policy are often confused or obscured in developing alternatives and making decisions.

Urgency. Management programs are difficult to implement without a crisis.

Measures of success. Success is exceedingly difficult to measure in the time frame within which most people expect to see results.

Leadership. Leaders of successful programs

1. Establish and reinforce credibility
2. Have a clear vision and a set of specific and comprehensive goals
3. Exhibit leadership that is willing to empower stakeholders and ensure an open climate for debate
4. Work to ensure that funding and resources are available

5. Identify the limits of a problem
6. Require a systematic decision process
7. Demonstrate their commitment to the process and to implementation of the solutions selected

The success of any program lies in the widespread trust of the program's leadership by the program's constituents. Trust is built through sustained credibility. In its simplest form, credibility is a measure of cumulative small successes forged from honest, fair, open, and even-handed pursuit of clearly articulated goals.

People expect leaders to be honest, to hold to an ethic of service, and to respect the intelligence and contributions of their constituents. People want leaders who will put principles ahead of politics and the interests of others before self-interests.

Agency leaders as well as politicians need to empower people to take independent initiative that fits within an agreed-upon framework. These actions should work toward a well-defined set of goals and should reinforce the actions of others.

In the Chesapeake Bay. e.g., a group of 12 people appointed by their respective governors to a panel of experts, to address population growth and its impacts, included public officials, elected politicians, academics, lawyers, and developers. All were leaders in their own right, but worked together to achieve a consensus and then returned to their leadership roles in their own communities. The group was able to take advantage of the leadership qualities of each person. Each person was able to contribute based on her or his leadership qualities, but did not seek to dominate.

In ecosystem management programs, leadership can come from any quarter. In both the Chesapeake and the Pinelands cases, leaders were fishers, businesspeople, developers, farmers, environmentalists and citizens, as well as public officials. Those who were empowered to implement specific activities showed leadership as well. They listened well. They ensured an open climate for transfer of insights and opinions and were not overly directive of the process. They were patient in the definition of problems and in the selection and use of data. They were open to the opinions of others. They were willing to deal with uncertainty and look at alternative outcomes.

The leadership pool can include agency personnel, retired professionals, dedicated self-trained citizens, and others. One often overlooked source of leadership is the business community. By tradition, we ascribe to business managers a streak of almost pure self-interest. In fact, this group is no less heterogeneous than the public at large when it comes to values, vision, and ideals. But they have talent and experience needed in consensus building, including negotiation skills, compromise identification, logical thinking, and a cost-benefit outlook.

From our experience, the most successful leaders are able to describe the need for specific projects or activities in ways that everyone involved can understand. During implementation, the description of need is continually reiterated so people always understand why the project is being undertaken.

Successful leaders also have the authority and ability to bring the necessary resources to bear when needed. Once momentum has developed in a program, leaders provide the resources needed to maintain it.

In order to implement a decision, a leader must

1. Map a clear decision pathway that includes the project team, the public, and the decision body
2. Participate at key points in the decision process
3. Help the project team break tasks into manageable short-term and long-term elements
4. Demonstrate and publicize success

In programs that cross jurisdictions and authorities, there are ill-defined lines of authority. As agencies increasingly work cooperatively and jointly, their roles become based more on a "natural" authority and are less defined by agency traditions. Setting goals and objectives that are broader than their mandate will be difficult for individual agencies; they will need to rely on the good offices and efforts of others to achieve their aims. Leaders within these agencies will need to find new ways to empower groups to work together. This will increase the pressure on leaders to have a coherent vision, explain it well, and motivate people to achieve it.

A proper balance of science and policy. Successful program managers

1. Understand the functioning and critical control points of their ecosystems
2. Maintain state-of-the-art research efforts
3. Make policy decisions based on the best science available at the moment
4. Are willing to monitor and reassess policy decisions
5. Demonstrate expertise

Ecosystem management must build on observational, analytical, and experimental science. Policy decisions have little chance of being sustained without a good foundation in science. Scientific study is needed to define system boundaries and describe a system's functions. These findings lead to the identification of critical elements for management and control. The process is seldom simple. Information is always incomplete and rarely points to a patently obvious management strategy. One risk is that the complexity of ecosystem analysis and understanding can become an end in itself, a sort of refuge from confronting tough management choices.

There is often a flow of personnel between the scientific and policy communities. Over time, those not formally trained as scientists gain substantial scientific knowledge. Conversely, a scientist is not unaware of the policy debates taking place. The difference between the role of science and the role of policy can get blurred and confused. Because science is so important in understanding a system and the consequences of different actions, scientific arguments often appear to take the place of public policy. Policy must be based upon the best scientific information, but must be clearly separated from scientific judgment. Policy weighs scientific information along with social objectives and the ethical and philosophical concepts abroad at the time.

Demonstration of scientific and management expertise is important. Both the Pinelands and the Chesapeake Bay programs have maintained extensive research and monitoring programs. The Pinelands also became a laboratory for land-use management techniques. Transfer of development rights, infrastructure boundaries, and a wide range of performance and incentive approaches have been used over the past 12 years.

Both programs were also willing to make the policy decisions needed to begin to meet their goals. More important is their willingness to reexamine earlier policy decisions in light of new information.

Public policy and public involvement. Successful programs

1. Have goals that are embraced by leaders at all levels of government, the community, and business
2. Ensure that diverse participants are brought into the process early, have a generous opportunity to participate, and are afforded a productive long-term role
3. Maintain continual public and institutional education programs
4. Create icons or symbols as part of the education process

Leaders and managers cannot develop, monitor, revise, and implement a program alone. No single group of landowners, organizations, or communities can implement a broad-based program. Everyone shares responsibility for getting extraordinary things done through ordinary means.

The Pinelands program, the Chesapeake Bay program, and similar programs everywhere undergo continual examination as a normal part of political processes. Some interests, generally those not extensively involved in building a program, are continually pressing for their particular objective or view. Because of their broad support and extensive public involvement, both the Chesapeake and the Pinelands programs have been through these challenges with good success. However, their battle is never-ending.

Implementation of ecosystem management programs requires strong support from political, community, and business leaders. Because of political pressures, diverse and broad-based support is critical to sustainable programs.

Unfortunately, it is difficult to energize a program without a crisis. A major challenge with ecosystem management will be to continually educate people about why certain practices and activities are necessary, especially those that affect people's livelihoods, their recreation, or their freedom to do what they want with land.

Successful programs create educational and public awareness campaigns that establish and promote symbols or icons identified with the program or the resource. In the Pinelands these include the pine tree, tree frog, and the traditional Piney culture. In the Chesapeake they are the skipjack sailing craft and the blue crab. Folk culture is also stressed. Each symbol has a purpose, to be a readily identified image or rallying point. These add a sense of romance and nostalgia that develops "connections" to the resource at a personal level. The programs strive to work at many levels to achieve management objectives, using a host of related, interconnected, and reinforcing programs and strategies.

Both the Pinelands and the Chesapeake Bay programs found ways to get watershed management accepted into the social and institutional fabric of their respective areas. While there is no set formula, successful programs appear to have several common characteristics. These include patience to allow the programs to mature and the constant involvement of people at all levels.

The public should be involved in the process of developing a program from the start, and its role should be maintained throughout the decision process. This is a key element in most successful programs. It is difficult to include all interests and a diversity of viewpoints, but this is necessary for a program to be credible and to become an accepted and adopted way of behaving and thinking about the managed resources.

Decision process. Successful programs

1. Work to frame the problem correctly
2. Look for integrated solutions
3. Develop balanced, comprehensive alternative approaches from which to choose
4. Use a full set of implementation and land-use tools
5. Involve the broader community and businesses
6. Demonstrate success

The quality and openness of a decision process are nearly as important as the quality of the results. Even with a host of decision tools and a stack of scientific reports, decision making remains fundamentally a human process subject to human error, bias, and plain folly.

Few people are trained in decision making. Most rely heavily on old habits or simplifications. These often lead to the adoption of advocacy or adversarial processes, rather than genuine problem-solving and decision processes.

Success is highly unlikely in the absence of a systematic decision process that frames problems well and identifies a comprehensive set of alternative actions. In both the Pinelands and the Chesapeake Bay programs, framing problems correctly was difficult. It required more time and effort than people anticipated.

A good decision process

1. Details goals, objectives, and values
2. Distinguishes the roles of science and policy making
3. Emphasizes which data are important to problem solving and decision making
4. Establishes a traceable and open record
5. Achieves credible results
6. Makes the best use of limited resources

Clear goals and a structured process for decision making provide all parties with a forum for exploring issues, conducting analyses, and integrating ideas and concerns. It is also a useful forum for linking policy to technical assessments, so that the analysis is directed by and informs policy.

Measuring success is difficult in natural resource management. Both the Pinelands and Chesapeake Bay programs have been working on research to show progress. This is difficult because most objectives are long term, with results off in the future. Rapid, dramatic results are infrequent. Whenever there is progress, it needs to be celebrated. For example, the rebound of the striped-bass population in the Chesapeake over a relatively short period is seen as real progress. Efforts to develop and publicize environmental "scorecards" are helpful in maintaining the public's attention and interest. It is important to have a mix of both short-term and long-term goals to allow for early demonstration of progress.

Implementation

Implementation needs money, and funding is always a problem. Support is required from all levels of government and the public. Most "programs" are not single programs, but clusters of strategies and independent programs and plans. These are worked out over time among a host of groups, within a generally accepted policy and goal framework. Funding for these individual efforts ebbs and flows. At the same time, the voluntary efforts of many individuals and groups play a significant role both in developing a public consciousness and in dealing directly with problems that are amenable to volunteer action.

Administrative barriers often cause natural resources and public lands to be administered differently by different agencies. Resource management is so complex and undergoes so much scrutiny that agencies are compelled to work together to address mutual problems. This is bringing down the barriers. Multiple objectives are replacing single or limited sets of objectives. For example, the Endangered Species Act (ESA) requires protection for listed species. Initially, implementation of this law focused on preservation of the species. This has evolved to a focus on protecting critical habitat for the organism, and not just the organism alone. In turn, there is now an understanding that multispecies systems or landscapes are vital for achieving a sustained population of a species and its related natural community. Thus, entire ecosystems need to be evaluated and have management strategies developed for them to successfully protect target species.

In a similar fashion, water is being looked at as part of a broader set of integrated resources. In nature, water shapes and is affected by environmental conditions. Its seasonal availability, volume, quality, and other characteristics are important to maintaining aquatic and terrestrial systems in a healthy state. Public water supply resources are now recognized as being integrated with other natural resources and resource values. There is integration within the water resource itself:

The need for an integrated strategy for supply development, storage, use, treatment, reclamation, reuse, and conservation is a common concern for water suppliers and managers.

Separate environmental laws designed to protect either natural conditions or human health are merging at their program implementation phases. Rather than wait for a species to achieve endangered status, we are looking at entire ecosystems as natural reserves of biological material to sustain a web of relationships. To reduce the need for filtration and chemical treatment to prepare water for human consumption, we are looking at water supply watersheds as assets to be managed, so as to reduce or eliminate the need for treatment, which itself can generate undesirable treatment by-products.

By combining the thrust of natural resource and public health laws with the requirements of the National Environmental Policy Act and its state-level counterparts, it is clear that integrated management of resources based on public involvement and analysis of alternatives is becoming the norm. Adversarial clashes in which champions compete to "win" a battle over how to manage a particular resource are anachronistic. Integrated planning will be a basic strategy to address a host of issues simultaneously. Public participation in decision making, calculating risks, monitoring of program results, and replanning will be common to all resource management programs.

The watershed restoration and forest management planning being undertaken in the northwest illustrate many of these points. Concern and energy to undertake programs reach to the grassroots in communities hard hit by the collapse of local resource-based economies. There is a need for partnerships among landowners and regulators to develop long-term strategies for land use and land protection. Incentives are needed, too, for private land managers to protect publicly valued goods in the land. These are being openly discussed and are beginning to emerge. Jobs increasingly include working on restoration projects and other management programs as well as traditional resource-based employment.

Conclusion

For the resource owner expecting to get a yield or product from his or her property, predictability in the administrative management of key resources is a critical factor. To avoid a continual "revolving door" in which landowners are never sure what the next issue or species will be that will prevent their economic use of the land, multispecies habitat protection plans are rapidly becoming the norm. Under these, land managers will have a predictable, sustainable yield of resources, while the publicly valued goods in the land remain protected.

This type of thinking goes beyond interagency coordination. Resource agencies are themselves members of a larger resource team and must begin to bring their special knowledge to bear as members of the team. For instance, the Environmental Protection Agency is now organizing to regulate impacts of concern to it on a watershed basis. Rather than segregate various pollutants by the medium they pollute, the EPA is looking at integrated land-water-air solutions. Neighboring landowners, environmental groups, county agencies, and state and federal resource agencies are seeking voluntary collaborations to create plans for watersheds. This is happening in rural and urban areas alike.

A clear example of bringing diverse parties together to review and explore options was the Northwest Watershed Restoration Conference held in Tacoma, Washington, in early 1994. Central messages coming from the conference are applicable elsewhere:

1. Watershed restoration activity will not be controlled by a central authority; it will be the result of a number of public and private programs.
2. Local political and opinion leaders must be involved in a coordination strategy.
3. Public agencies must abandon some of their traditional management focus and share resources, experience, and data.
4. Agencies have an opportunity to develop new project delivery systems to ensure that work is accomplished in a timely manner and that local skills are developed and used in practical management solutions.

Resource decisions are shaped by both technical considerations and individual or collective values. The concept of balancing technical and social inputs is central to resource management. For instance, knowledge of technical issues, such as development cost, risk, environmental impact, quality, and reliability, is essential to development of, say, a well-informed water supply plan. But this technical knowledge alone does not determine the plan's final shape. The most viable alternatives are those that have successfully balanced technical and economic criteria with local, regional, and national values.

Appendix A
Watershed Approach Framework

People working together
to protect public health and the environment
—community by community,
watershed by watershed.
CAROL M. BROWNER, ADMINISTRATOR
U.S. ENVIRONMENTAL PROTECTION AGENCY
JUNE 1996

*This appendix is reproduced in its entirety from "Watershed Approach Framework," Washington, D.C.: U.S. Environmental Protection Agency, June 1996. EPA840-S-96-001.

Introduction

Environmental protection programs in the United States have successfully improved water quality during the last quarter century, yet, many challenges remain. The most recent national water quality inventory shows that, as of 1994, nearly 40 percent of surveyed waters in the US remain too polluted for fishing, swimming and other uses. The leading causes of impairment found in the survey include silt, sewage, disease-causing bacteria, fertilizer, toxic metals, oil and grease.

Many public and private organizations are joining forces and creating multidisciplinary and multijurisdictional partnerships to focus on these problems, community by community and watershed by watershed. These *watershed approaches* are likely to result in significant restoration, maintenance and protection of water resources in the United States. Supporting them is a high priority for EPA's national water program.

This publication explains EPA's vision for watershed approaches and builds upon the Office of Water *Watershed Protection Approach Framework*, endorsed by senior EPA managers in 1991. It emphasizes the role EPA envisions for states and tribes. It also reflects the high priority that individual Office of Water programs have put on developing and supporting comprehensive state and tribal watershed approach strategies that actively involve public and private interests at all levels to achieve environmental protection.

What is a Watershed Approach?

The watershed approach is a coordinating framework for environmental management that focuses public and private sector efforts to address the highest priority problems within hydrologically-defined geographic areas, taking into consideration both ground and surface water flow.

Guiding Principles

EPA supports watershed approaches that aim to prevent pollution, achieve and sustain environmental improvements and meet other goals important to the community. Although watershed approaches may vary in terms of specific objectives, priorities, elements, timing, and resources, all should be based on the following guiding principles.

- *Partnerships*—Those people most affected by management decisions are involved throughout and shape key decisions.

This ensures that environmental objectives are well integrated with those for economic stability and other social and cultural goals. It also provides that the people who depend upon the natural resources within the watersheds are well informed of and participate in planning and implementation activities.

- *Geographic Focus*—Activities are directed within specific geographic areas, typically the areas that drain to surface water bodies or that recharge or overlay ground waters or a combination of both.

- ***Sound Management Techniques based on Strong Science and Data***—Collectively, watershed stakeholders employ sound scientific data, tools, and techniques in an iterative decision making process. This includes:
  - assessment and characterization of the natural resources and the communities that depend upon them;
  - goal setting and identification of environmental objectives based on the condition or vulnerability of resources and the needs of the aquatic ecosystem and the people within the community;
  - identification of priority problems;
  - development of specific management options and action plans;
  - implementation; and
  - evaluation of effectiveness and revision of plans, as needed.

  Because stakeholders work together, actions are based upon shared information and a common understanding of the roles, priorities, and responsibilities of all involved parties. Concerns about environmental justice are addressed and, when possible, pollution prevention techniques are adopted. The iterative nature of the watershed approach encourages partners to set goals and targets and to make maximum progress based on available information while continuing analysis and verification in areas where information is incomplete.

Need for Watershed Approaches

Over the past 20 years, substantial reductions have been achieved in the discharge of pollutants into the nation's air, lakes, rivers, wetlands, estuaries, coastal waters, and ground water. These successes have been achieved primarily by controlling point sources of pollution and, in the case of ground water, preventing contamination from hazardous waste sites. While such sources continue to be an environmental threat, it is clear that potential causes of impairment of a waterbody are as varied as human activity itself. For example, besides discharges from industrial or municipal sources, our waters may be threatened by urban, agricultural, or other forms of polluted runoff; landscape modification; depleted or contaminated ground water; changes in flow; overharvesting of fish and other organisms; introduction of exotic species; bioaccumulation of toxics; and deposition or recycling of pollutants between air, land and water.

The federal laws that address these problems have tended to focus on particular sources, pollutants, or water uses and have not resulted in an integrated environmental management approach. Consequently, significant gaps exist in our efforts to protect watersheds from the cumulative impacts of a multitude of activities. Existing air, waste and pesticide management, water pollution prevention and control programs and other related natural resource programs are, however, excellent foundations on which to build a watershed approach.

Benefits Derived from Taking a Watershed Approach

Operating and coordinating programs on a watershed basis makes good sense for environmental, financial, social, and administrative reasons. For example, by jointly reviewing the results of assessment efforts for drinking water protection, pollution control, fish and wildlife habitat protection and other aquatic resource protection programs, managers from all levels of government can better understand the cumulative impacts of various human activities and determine the most critical problems within each watershed. Using this information to set priorities for action allows public and private managers from all levels to allocate limited financial and human resources to address the most critical needs. Establishing environmental indicators helps guide activities toward solving those high priority problems and measuring success in making real world improvements rather than simply fulfilling programmatic requirements.

Besides driving results towards environmental benefits, the approach can result in cost savings by leveraging and building upon the financial resources and the willingness of the people with interests in the watershed to take action. Through improved communication and coordination the watershed approach can reduce costly duplication of efforts and conflicting actions. Regarding actions that require permits, specific actions taken within a watershed context (for example the establishment of pollutant trading schemes or wetlands mitigation banks and related streamlined permit review) enhances predictability that future actions will be permitted and reduces costs for the private sector. As a result, the watershed approach can help enhance local and regional economic viability in ways that are environmentally sound and consistent with watershed objectives.

Finally, the watershed approach strengthens teamwork between the public and private sectors at the federal, state, tribal and local levels to achieve the greatest environmental improvements with the resources available. This emphasis gives those people who depend on the aquatic resources for their health, livelihood or quality of life a meaningful role in the management of the resources. Through such active and broad involvement, the watershed approach can build a sense of community, reduce conflicts, increase commitment to the actions necessary to meet societal goals and, ultimately, improve the likelihood of sustaining long-term environmental improvements.

Implementing the Guiding Principles through State and Tribal Watershed Approaches

From EPA's perspective, states and tribes are in a pivotal position because they implement many existing water and natural resource protection programs and they are situated well to coordinate among other levels of government (e.g., local, regional and federal). For these reasons, EPA places special emphasis on supporting our state and tribal partners in developing and implementing comprehensive watershed approaches. This emphasis should not be construed as a lack of support for the involvement of other parties in watershed management, especially local stakeholders. As stated in the guiding principles, partnerships that promote the active participation of concerned parties from all levels of government and from across the public and private sectors is essential to the watershed approach.

EPA recognizes that each state or tribe may approach watershed management differently. The agency will not prescribe their actions; rather it supports watershed approaches that are tailored to the needs of the jurisdictions.

The agency has both a national interest in and responsibility for supporting watershed approaches. The interest stems from the belief that the diverse sources of aquatic ecosystem impacts will best be brought under control through a combination of cooperative and mandatory measures tailored to the needs in specific watersheds with wholehearted support from watershed stakeholders. EPA's responsibility includes definition and ensured compliance with basic water programs; development of national standards and tools; funding; and national assessment of status and progress.

For the long term, EPA envisions locally-driven, watershed-based activities embedded in comprehensive state and tribal watershed approaches all over the United States. Based on observation of the development of such comprehensive approaches in several jurisdictions, there are four key elements of state and tribal watershed approaches. These reflect and provide the operating structure for these guiding principles described earlier. They are:

Stakeholder Involvement
(providing structure for the *Partnership* principle)

Geographic Management Units
(providing structure for the *Geographic Focus* principle)

Coordinated Management Activities
(providing structure for the *Sound Management* principle)

A Management Schedule
(providing further structure for the *Sound Management* principle)

The following describes in more detail how the key elements implement the guiding principles.

Stakeholder Involvement

Broad involvement is critical. In many cases, the solutions to natural resource problems depend on voluntary actions on the part of the people who live, work and play in the watershed. Besides improving coordination among their own agencies, the watershed approach calls upon states and tribes to fully engage local government entities, sources of watershed impacts, users of watershed resources, environmental groups, and the public in the watershed management process to help them better understand the problems, identify and buy into goals, select priorities, and choose and implement solutions.

States and tribes work with other partners on watershed management issues in geographically-based watershed "teams." As appropriate, partnerships include representatives from local, regional, state, tribal, and federal agencies, conservation districts, public interest groups, industries, academic institutions, private landowners, concerned citizens, and others. There are a great many watershed partnerships already in effect across the country. Ideally, states and tribes will commission or build on these. Some examples of partnerships that have been formed under existing programs are:

- Local Wellhead Protection Programs or other source water protection efforts, including cooperative efforts to meet requirements to avoid filtration under the Surface Water Treatment Rule.
- National Estuary Program Management Conferences.
- Clean Lakes Program management teams.
- Tributary teams in the Chesapeake Bay.
- Watershed alliances formed through conservation districts and under various state and federal programs, for example the Watershed Protection and Flood Prevention Act (P.L.83-566) and comprehensive resource management teams working on forestry issues.

Geographic Management Units

The entire jurisdiction is divided into geographic management units. Ideally, these units are determined on the basis of hydrologic connections, as described under the geographic focus principle. Other factors such as political boundaries and existing partnership program areas are often factored into decisions about geographic management units, as well.

The size of the management unit is an important consideration because, depending on scale different parties may take different roles. For example, for large river basins or lakes, state and tribal agencies are likely to lead watershed planning efforts, while local government, conservation districts, and watershed councils may take the lead in developing and implementing solutions in smaller watersheds. "Nesting" smaller watersheds areas (such as those designated as drinking water source water protection areas or special management areas for wetlands protection) within larger watershed or river basins allows those involved at every level to scale their efforts up or down to address specific concerns and still maintain consistency with related efforts.

Coordinated Management Activities

State and tribal agencies have responsibility for many of the management activities described in the guiding principles. Ideally, the various agencies with responsibilities for wetlands protection, drinking water source protection, waste management, point and nonpoint source pollution control, air pollution, pesticide management and other programs such as water supply, agriculture, navigation, and transportation (in any given jurisdiction, these might be several different agencies) would jointly compare their lists of high priority areas, meet with each other and other stakeholders, and look for opportunities to leverage their limited resources to meet common goals. Watershed approaches should not be viewed as an additional layer of oversight; rather watershed approaches should constitute improvements in coordination of current programs, processes and procedures to increase efficiency and efficacy.

Working together cooperatively, state and tribal programs can support and facilitate many of the management activities likely to be taken by watershed teams. The activities described below suggest some of the ways that EPA-related water programs can support watershed approaches. It is important to keep in mind that many other activities and programs, both public and private, at all levels, may need to be included in watershed planning and management.

1. *Assessment and Characterization of Aquatic Resources, Problems, their Causes and Sources*

Ideally, monitoring parameters would be determined by water quality standards and other watershed goals and indicators, which are specified according to the needs and conditions of the area and reflect Clean Water Act and Safe Drinking Water Act goals and build on the environmental indicators that EPA and its public and private partners have adopted.

The state or tribal monitoring program should have a multiyear strategy to portray existing information on physical, chemical, biological, and habitat conditions and comprehensively monitor waters. Ideally, the strategy should recognize that responsibilities can be shared by many stakeholders and that monitoring must be done to fulfill distinct purposes: characterizing the watershed; identifying and locating specific problems; and determining if actions are effective and goals are met. A strong monitoring program should include:

- An inventory of key existing information on resources, including priority ground water, sources of drinking water, habitat, wetlands and riparian acreage, function and/or restoration sites.

- A monitoring design that confirms or updates existing information or fills gaps and can report trends.

- Reference conditions for biological monitoring programs to provide baseline data for water quality assessments and development of biological and nutrient criteria.

- Data collected using comparable methods to allow aggregation of data at various scales and stored so as to be readily accessible to others (e.g., in EPA's database STORET).

- Geographic references (using Reach File 3) so that monitored waters can be mapped using a Geographical Information System (GIS), allowing information to be aggregated on a watershed basis.

- Key information on condition of waters (e.g., impaired, in need of special protection, endangered species present, threatened sources of drinking water) and causes of impairment are reported in the national water quality inventory (305(b) report).

- Collaborative efforts on existing and planned monitoring activities with other public and private institutions to share information when goals are similar.

2. *Goal Setting*

In the process of identifying goals, water quality standards provide a legal baseline or starting point. These goals clearly identify the uses to be made of the waters, for example the protection and propagation of a warm water fishery. Water quality standards also include the appropriate chemical, physical and biological criteria to characterize and protect the uses and an antidegradation policy to preserve the uses and water improvements attained in the waters of their watersheds. As an outcome of watershed planning processes, a state or tribe may also adopt new or revised water quality standards for the waters within a watershed to reflect agreements made by the stakeholders to meet the watershed goals (this would likely take place as part of the triennial review process required by law). Actions by states and tribes that support watershed efforts include:

- Reviewing, and if appropriate, revising water quality standards within the watershed framework, consulting the other stakeholders involved in the watershed.

- Adopting precisely defined uses given the chemical, physical and biological characteristics of the waterbody.

- Expanding the suite of tools applicable to the development and implementation of their water quality standards and management programs. The expanded suite should include tools to address multiple stressors and their cumulative impacts, including criteria to protect human health, aquatic life, wildlife and sediment dwelling organisms; methodologies for sediment and whole effluent toxicity testing; and assessment methods for establishing Total Maximum Daily Loads (TMDLs) or waste load allocations, and evaluating ecological risk, nutrient enrichment and habitat.

3. *Problem Prioritization and Resource Targeting*

Staff in the various water-related programs in the state or tribe should work with other stakeholders to jointly set priorities for the particular suite of water resources concerns present in each identified management unit. Deliberations should consider:

- Drinking water source protection for both ground and surface water sources;

- Wetlands and riparian area protection and other ecological values;

- Nonpoint source pollution control;

- Point source pollution control;

- Living resource needs; and

- Other issues, such as waste and pesticide management, air pollution affects on water resources, and water supply, as appropriate.

The watershed approach should take into consideration the findings of and priorities established under preexisting initiatives, such as the Comprehensive State Ground Water Protection Program (CSGWPP), Wellhead Protection Program, State Wetlands Conservation Plans, NPDES watershed or basin strategy, National Estuary Program Comprehensive Conservation Management Plan, or Clean Lakes projects. In addition, states and tribes should take into consideration the goals and plans of relevant large-scale projects, such as the Chesapeake Bay, Great Lakes, and Gulf of Mexico programs and the Northwest Forest Plan and Everglades initiative. These projects may provide significant opportunities for "nesting" smaller projects within larger frameworks, yielding benefits to both.

The composition of watershed partnerships should reflect the agreed upon priorities for the watershed areas. Similarly, Clean Water Act funds, both grants and loans, should be applied to the development and implementation of watershed plans.

4. *Management Option Development and Watershed (or Basin) Plans*

Each watershed partnership should develop management options and set forth a watershed or basin management plan that should:

- Establish environmental objectives that are consistent with all applicable state, tribal, and federal statutes and regulations, including water quality standards and drinking water maximum contamination levels and health advisories. The environmental objectives should reflect the needs and concerns of the watershed stakeholders and thus may include objectives unrelated to EPA programs.

- Identify environmental indicators compatible or complementary to national indicators that can be used to monitor and report on attainment of the environmental objectives. (In June 1996 the agency issued *Environmental Indicators of Water Quality in the United States* EPA 841-R-96-002.)

- Identify specific implementation actions, including voluntary, mandatory, and educational efforts, that will attain and maintain the goals.

- Set forth milestones, assign responsibility, specify who will implement actions, and identify existing and potential sources of funding for implementation.

5. *Implementation*

Due to the participatory nature of watershed approaches, responsibility for implementation of watershed plans will fall to various parties relative to their particular interests, expertise and authorities. To the maximum extent possible, state and tribal water-related programs should support the implementation of watershed plans through their actions. They should consider the full range of tools available to them in programs as diverse as water quality protection, pesticide management, waste management, air pollution control, as well as natural resources protection, agriculture programs, water supply, transportation and other related programs. For example, under water quality and natural resource protection programs they may:

- Support watershed approaches to water quality permitting, nonpoint source pollution control, habitat protection and other water resource protection and restoration activities using Total Maximum Daily Load analyses.

- Issue NPDES permits in accordance with the state or tribal watershed management schedule.

- Tailor their Clean Water Act §319 nonpoint source management program to respond to watershed needs and ground water connections.

- Direct activities in the State Wetland Conservation Plan toward reducing wetland impacts from land and water-based activities.

- Integrate federal, state and/or local wetland permit programs with individual watershed plans that contain adequate wetland protection provisions.

- Promote the establishment of mitigation banks by providing funding for bank sponsors, identifying and prioritizing potential bank sites, and providing appropriate direction.

- Use their watershed approach to target overall source water protection areas and approved Wellhead Protection Program protection areas as high priority for various federal and state programs.

- Direct federal and state activities toward protection of high priority ground water (e.g., wellhead protection areas or other areas designated under endorsed Comprehensive State Ground Water Protection Program).

- Develop or use approved program under primacy for Phase I/II/V National Primacy Drinking Water Regulations for granting monitoring waivers under Public Water System Supervision program.

- As authorized, monitor, verify implementation, and, when necessary, enforce management actions.

6. *Monitoring and Evaluation*

To evaluate the effectiveness, the watershed management cycle should include monitoring to ascertain both the environmental and socioeconomic impacts of implemented watershed plans. Progress should be reported and results of monitoring help guide decisions about continued implementation. See Assessment and Characterization of Aquatic Resources, Problems, their Causes and Sources, above.

Management Schedule

A schedule for carrying out coordinated management activities within each of the management units helps organize the work states and tribes need to undertake. The schedule would lay out a long-term program for maintaining, restoring, and protecting water resources and provide other interested parties an opportunity to plan for their involvement.
To most effectively create an orderly system for focusing and coordinating watershed management activities on a continuous basis, the schedule should contain two features:

- A sequence for addressing watersheds that balances workloads from year to year; and

- A specified length of time planned for each major management activity (e.g. assessment, management option development, implementation).

The schedule should reflect the magnitude of activities to be carried out within any particular watershed or basin, which depends largely on the range and severity of problems found within that management unit. For example, some watersheds may require minimal actions to maintain high environmental quality, whereas others may require substantial effort to restore environmental quality.

Reorganizing workloads to take a watershed approach may take a considerable amount of time . During the early phases of reorientation (before the entire jurisdiction is covered by the watershed schedule), existing program activities to address high priority restoration, remediation and/or protection concerns, such as wellhead protection, may need to proceed in some places independently of the watershed schedule. Ideally, however, over time all relevant programs would be carried out within a jurisdiction-wide watershed approach.

EPA Support to Facilitate Watershed Approaches

EPA's National Water Program has examined its work in order to identify ways that the agency can better support watershed approaches. Besides the provision of basic national programs upon which watershed approaches are built, specific operational changes have been suggested. These include reduced water quality reporting requirements, priority consideration for Clean Water Act grants for watershed activities, use of funds under the Safe Drinking Water Act for source water protection , simplified wetlands permitting, allowances for NPDES permitting backlogs, longer cycles for reviewing and, if appropriate, revising water quality standards, reduced monitoring under the Safe Drinking Water Act, TMDL assistance, and facilitated development of wetlands mitigation banks and effluent trading. These programmatic changes are described in more detail in another EPA publication entitled, *Why Watersheds?* (EPA800-F-96-001)

The Office of Water offers assistance to help water quality managers and staff throughout the public and private sectors develop and implement watershed approaches. The four main areas covered include watershed management training, statewide watershed approach facilitation, watershed program scoping, and technical analysis assistance. Training and facilitation have been the most actively requested services of the watershed assistance program.

Watershed management training is available through the *Watershed Academy,* which offers a set of core courses and related reference materials about basic watershed management principles and techniques as well as contact information on more specialized and advanced courses. The core courses address watershed management fundamentals, watershed tools, the statewide approach to watershed management, and an executive overview course. During 1995, the two-day Statewide Watershed Management Course was offered in five locations to over 300 people. Completion of the other core courses is planned for late 1996. Although EPA itself offers only a few courses, dozens of watershed training opportunities exist. The Watershed Academy will continually update its *Catalogue of Watershed Training Opportunities* (available on internet) to spread information about watershed-oriented training courses offered by other local, state and federal agencies and private organizations. Participation in an interagency watershed training workgroup will be another source of joint planning, shared training materials and expertise.

Watershed approach facilitation is generally provided to states and tribes that intend to reorient their water resources management programs along watershed lines. Facilitation involves several onsite working meetings with water program managers and decision makers to help them develop a transition plan, schedule, and comprehensive organizational framework based on major river basins and their component watersheds. Twelve states contacted EPA for some form of facilitation assistance during 1995, and several have completed significant reorientations of their programs to implement a watershed approach.

In addition to training and facilitation, the Office of Water offers assistance in watershed program scoping and technical analysis to states and tribes. Scoping projects are preliminary to full-scale reorientation and involve one or two meetings with managers to determine what form a watershed approach might take, the effort involved, and the next steps needed. Technical analysis projects focus on scientific, economic or programmatic analysis as related to specific watershed management issues.

For information on the Watershed Academy, contact Doug Norton at 202-260-7017. For information on the statewide watershed management course, contact Greg Currey at 202-260-1718. For information on watershed facilitation, scoping, or technical analysis assistance, contact either Doug or Deborah Nagle at 202-260-2656.

Several EPA documents may be of particular interest.

Watershed Protection: A Statewide Approach (EPA841-R-95-004)

Watershed Protection: A Project Focus (EPA841-R-95-003)

The Watershed Protection Approach 1993-94 Activity Report (EPA840-S-94-001)

Watershed '93: A National Conference on Watershed Management (Proceedings) (EPA840-R-94-002)

Why Watersheds? (EPA800-F-96-001)

Catalogue of Watershed Training Opportunities (available on internet)

Watershed Tools Directory (EPA841-B-95-005)

These documents are or soon will be available on the internet at URL:http://www.epa.gov/OWOW. Printed copies of the final (non-draft) documents can be obtained by telephone at 513-489-8190, by fax at 513-489-8695, or by written request to NCEPI, 11029 Kenwood Road, Building 5, Cincinnati, OH 45242.

Frequently Asked Questions About the Watershed Approach

How can the watershed approach address both ground water and surface water protection?

When delineating geographic management units, boundaries should be constructed to accommodate hydrologic connections and processes and address the priority problems at hand. So, particular management areas may vary depending on the priority problems to be addressed. For example, when ground water contributes significantly to surface water flow, the management unit should include the ground water recharge area. When the vulnerability of drinking water to contamination is of primary concern, then the drinking water source (e.g., reservoir or well-head protection area) should be the area upon which attention is focused. When the protection of an aquifer is of primary concern, the management area should include the overlaying or recharging area and recognize impacts upon surface water. Interesting research is now underway in the State of Florida to delineate hydrogeological watersheds that accurately depict ground and surface water connections. Similarly, the US Army Corps of Engineers has developed new techniques for hydrogeomorphic analyses related to wetlands.

How does the watershed approach relate to other programs with similar characteristics, such as the National Estuary Program and Source Water Protection? And, how does the NPDES watershed strategy relate to the watershed approach?

States and tribes may want to build on the successes of geographically-focused programs and increasingly integrate assessments, sort out and establish joint priorities, and coordinate actions among programs while making a transition to the watershed approach. Whether a jurisdiction starts with a source water protection program like Wellhead Protection, a Wetlands Conservation Plan, a National Estuary Program , a NPDES watershed strategy or other water resource, place-based strategy, EPA will support them in moving to an even more comprehensive approach to protecting water resources. These more targeted programs can provide the community roots for broader watershed approaches. Ultimately, we hope to see comprehensive, jurisdiction-wide, and when appropriate cross-jurisdiction, watershed approaches that involve all appropriate agency staff working with local stakeholders while setting goals, establishing priorities, and implementing integrated and effective solutions.

What is the relationship between the watershed approach and community-based environmental protection?

Community-based environmental protection is an iterative approach in which diverse stakeholders strive to achieve environmental objectives. Typically it includes:

- Adoption of local environmental goals compatible with economic sustainability;
- Characterization of environmental problems and solutions; and
- Implementation of solutions that are coordinated and tailored to the goals and needs of the community.

The watershed approach is community-based environmental protection using watershed or hydrologic boundaries to define the problem area. In fact, the momentum and success of the watershed approach and its "predecessors," the National Estuary Program, Great Water Bodies programs, and the Clean Lakes Program, strongly influenced the development of EPA's community-based environmental protection approach.

How does the watershed approach relate to the National Environmental Performance Partnership System and Performance Partnerships Grants?

States that choose to adopt the National Environmental Performance Partnership System could choose to set water quality protection goals and priorities and organize their work on a watershed basis. Watershed plans could be incorporated or referenced in the required Environmental Performance Agreements.

Through Performance Partnership Grants (PPGs), states and tribes can combine funding from eligible grants to target high priority problems and address multimedia problems within their watersheds. States and tribes that combine categorical grants into PPGs must continue to address the core program requirements which those grants are meant to support. A final approved PPG will be the result of negotiations between the state or tribe and its EPA Regional office.

Index

About the Editor

Robert J. Reimold, Ph.D., is Vice President and National Technical Director for Risk-Based Environmental Decision Making for EA Engineering, Science, and Technology, a leading consulting/management firm solving complex environmental, energy, and public health problems for global clients. Dr. Reimold directs the company's practice in risk sciences and sustainable natural resources management. An internationally recognized expert in wetlands ecology, assessments, construction, and mitigation, Dr. Reimold has more than 30 years of international experience in managing ecological studies related to natural, social, and economic resources. He is a Certified Senior Ecologist of the Ecological Society of America and a Certified Senior Fisheries Scientist of the American Fisheries Society. He served as a Postdoctoral Fellow in Ecology at the University of Georgia under the direction of Professor Eugene P. Odum, the world-renowned ecologist.